D0874109

ASTRONOMY AND ASTROPHYSICS LIBRARY

Series Editors: M. O. Harwit, R. Kippenhahn, J.-P. Zahn

Advisory Board:

J. N. Bahcall
P. L. Biermann
S. Chandrasekhar
L. V. Kuhi
P. G. Mezger
P. A. Strittmatter

LIBRARY

ASTRONOMY AND ASTROPHYSICS LIBRARY

Series Editors: M. O. Harwit, R. Kippenhahn, J.-P. Zahn

Tools of Radio Astronomy
By K. Rohlfs

Preface

This is a book describing the tools that a radio astronomer needs to pursue his trade. These tools consist, on one hand, of the radio telescopes and the various kinds of receivers needed to analyse the cosmic radio signals, and on the other, of the physics of the radiation mechanisms responsible for both the continuous and line radiation. The book grew out of a one year graduate course that I gave repeatedly at the Ruhr University at Bochum. I hope, however, that this text will also be useful for all scientists that use data and results obtained by radio astronomical means and will help them to understand both their strong points and characteristic drawbacks. Finally, this book may occasionally save some scientists working in the field of radio astronomy from long searches in the literature when questions concerning their tools occur.

Although the students whom this course was addressed to had a rather thorough background knowledge of physics, a characteristic difficulty often turned up when the material tools (Chaps. 4–7) were discussed. Obviously there is a difference between how a subject is treated in a genuine physics book and the way it is dealt with in a text intended for engineers – one example is the way in which four-terminal networks are used. I have tried to explain everything using concepts that are familiar to both astrophysicists and general physicists.

Very little general astronomical background information is actually needed (the subject of the book is the tools and methods of radio astronomy, not the results) but it naturally helps if there is some general knowledge of astrophysics as, e.g., provided by the book of A. Unsöld: Der neue Kosmos, 3rd ed., Springer-Verlag (English translation: The New Cosmos, 3rd ed., Springer-Verlag), or by the excellent text of Frank H. Shu: The Physical Universe, 1982, University Science Books, Mill Valley.

Formulae and equations are numbered consecutively in each chapter with the chapter number as a prefix, e.g., the thermal emittivity of a cloud of ionized gas is referred to as (8.21) throughout the book.

As far as possible and convenient the equations are written so as to conform to the SI regulations, but since, especially in astronomy, a multitude of non-SI units are in permanent use, equations using mixed quantities could not be avoided altogether. In these cases the units to be used have been indicated, as for instance in (8.28).

For each chapter, a list of references is given. Usually this list has two parts: general references give a list of papers and books that cover the general aspects and which often give a more thorough treatment of the subjects covered, and special references document the sources for specific topics. However, these references do not give a complete or even nearly complete review of the relevant literature. The papers cited are those that I found to present the subject in a way convenient for the purpose intended here.

At the end of the book some useful vector relations and Fourier and Hankel transform pairs are given as well as a list of calibration radio sources. An index of notations giving the place in the book where a symbol first appears and an extensive subject index are provided in order to make this book useful as a reference source.

The book is organized into 13 Chapters. Chapter 1 presents some basic concepts used in radiation theory, while Chap. 2 gives some electromagnetic wave propagation fundamentals. Although this should be covered in any course of theoretical electrodynamics, my experience is that this part is often passed over in a rather cursory way in such courses. This is particularly true for wave polarization in Chap. 3.

Chapters 4–6 cover the main aspects of antenna theory, both that for filled aperture antennas (Chap. 5) and for interferometers and aperture synthesis (Chap. 6). Calibration problems as well as considerations that limit performance are discussed too.

In Chap. 7 a survey of the different constituents of a receiver is given. Because none of the books that are commonly used as references by the radio astronomy community contains a proof of the fundamental formula (7.48) of the limiting receiver sensitivity, this relation is proved here, even if this requires some effort. Other subjects treated in this chapter are the physical principles of parametric amplifiers, masers, autocorrelation spectrometers and acousto-optical spectrometers, without going into too much technical detail.

The remainder of the book, Chaps. 8 to 13, present some of the conceptual tools that a scientist needs in order to understand the significance conveyed by the received radiation. Both thermal emission and, in a more qualitative way, synchrotron emission are described. Chapter 9 applies these concepts to two different kinds of radio sources – the quiet sun and supernova remnants.

Line radiation fundamentals are covered in Chap. 10; Chap. 11 applies this to the line radiation of neutral hydrogen, while recombination line measurements and the problem of how to interpret them are discussed in Chap. 12. Here, in particular, questions about deviations from local thermodynamical equilibrium are treated. Problems concerned with interstellar molecules and their line radiation are finally described in Chap. 13. Beginning with the classification into rotational and vibrational transitions, and the complications introduced by the various level-splitting mechanisms, the arguments leading to cloud densities and temperatures are presented. Line formation by unsaturated and saturated masers is discussed, as well as problems in the interpretation of CO measurements.

This book could not have been written without the help of many people. I should mention particularly Dr. B. Sherwood who helped me very effectively to overcome the obstacle of writing a book in a language that is not my mother tongue. The text and its many corrections were typed quickly and neatly by Mrs. G. Trosbach, who never lost her patience. All formulae have been checked and recomputed independently by my student Dipl.-Phys. Th. Luks. He detected many errors and blunders; any that still remain are my responsibility.

Many colleagues helped me by providing diagrams and photographs, and by discussions. I should in particular mention Prof. Dr. R. Wielebinski, but many others could be referred to as well.

Bochum, February 1986 *Kristen Rohlfs*

Contents

1. Radio Astronomical Fundamentals

1.1 On the Role of Radio Astronomy in Astrophysics

Almost everything that we know about the universe, about stars and the space in between them, about stellar systems, their distribution, kinematics and dynamics has been obtained from information brought to the observer by electromagnetic radiation. Only a very small part of our knowledge stems from material information carriers, be they meteorites that hit the surface of the earth, cosmic ray particles or samples of material collected by manned or unmanned space probes.

For many thousands of years man was restricted to visible light when he made his observations, and only since the time of Herschel was this wavelength range slightly extended; in 1930, it reached from the near ultraviolet to the near infrared: $0.35\ \mu m \leqq \lambda \leqq 1\ \mu m$. Only light in this range could be investigated by the astronomers either because the terrestrial atmosphere blocks very effectively radiation of other wavelengths from reaching the telescopes, or because no detectors for this radiation were available. This situation was changed dramatically in 1931 when Jansky showed that radiation at a wavelength of 14.6 m received by him with a direction-sensitive antenna array must be emitted by some extraterrestrial radiation source. Jansky continued his observations over several years without achieving much scientific impact, and his observations were first taken up and improved in 1937 by another radio engineer, Grote Reber, who did his measurements at a shorter wavelength of $\lambda = 1.87$ m. These observations were published in an astronomical journal, and they were later, after the end of World War II, instrumental in opening up the new radio window in the blanket of the earth's atmosphere. Radiophysics had made great progress during the war years, mainly due to the efforts directed towards the development of sensitive and efficient radar equipment, and after the war some of the researchers turned their attention towards the radio-"noise" that they received from extraterrestrial sources.

We will not follow this historical development any further here; it should be sufficient to state that the radio window reaching from $\lambda \cong 10-15$ m to about $\lambda \cong 1$ mm or even less was the first new spectral range that became available to astronomy besides the slightly extended optical window. Soon exciting new and strange objects were found, and the new astronomical discipline of radio astronomy was changing our views on many subjects, requiring mechanisms for their explanation that differed considerably from those commonly used in "optical" astrophysics. While the objects of "optical" astrophysics usually radiate because they are hot and therefore "thermal" physics was needed for an explanation, in radio astronomy most often the radiation has a non-thermal origin and different physical mechanisms had to be considered.

In the years since 1945 technological advances permitted the opening up of additional "windows" transparent to electromagnetic radiation of widely different wavelength ranges. Balloons or high-flying aircraft permit observations in the far infrared, or extend the accessible part of the ultraviolet wavelength range, while telescopes and detectors installed in satellites open up the full spectral range from γ-rays to infrared or even to very long wavelengths greater than 10^4 m.

Each spectral window requires its own technology and the tradition and the art of doing measurements differ for each. Therefore astronomers have developed the habit of viewing at these different techniques as forming different sections of astronomy: they talk of them as radio astronomy, X-ray astronomy, infrared astronomy and so on.

But it is really only the wavelength range and to some extend the technology that differs in these different "astronomies". The objects that emit the radiation are the same for all branches. It is true, some kinds of objects are "visible" only in certain spectral windows and not in others: neutral hydrogen of low excitation is visible only because it emits the forbidden hyperfine structure line at $\lambda = 21$ cm, and it cannot be detected by any other means. Therefore the aspect of the universe differs for each spectral window; but it is one of the fundamental axioms of science that there is one single world behind all the different aspects. If an astrophysicist is investigating a specific object, he must collect information obtained by all the various means, irrespective of whether they have been obtained with the help of optical astronomy, radio astronomy or by other techniques. It is with regard to this situation that one can state: there is no such thing as a separate scientific discipline of radio astronomy.

When new experimental techniques open up new ways to attack the old problems, and even more so, when new kinds of exciting objects are detected by these means, methods and results are often collected into a new discipline like radio astronomy. But when the experimental methods have become mature and both the strong points and limitations of the methods are becoming clearer, it seems to be appropriate to reintegrate the specialized field into main-stream astrophysics.

In my opinion radio astronomy is in such a situation now. The first, vigorous years when the pioneers did everything themselves are over now. Today a radio astronomer very seldom builds his telesope and receiver all by himself, and this has quite naturally profound effects on the way research is done. Quite often a researcher now starts with the problem and then searches for the observational means to attack it, while in the pioneer days it usually started with the instrument collecting strange observations that required an explanation. But even then there were noted exceptions to this.

Therefore radio astronomy is today the science of the instruments with which the data are gathered, their properties, strong points and limitations, but not a collection of the results. And since these instruments are usually no longer built by the astronomer himself but by an engineering staff, it is the main job of the astronomer to supervise their functioning in a critical way, while he need not be able to improve the instruments himself. But he has to have a very clear idea how they work and what their limitations are. Radiotelescope and receivers are his tools.

But a radio astronomer uses not only material tools: in interpreting his measurements theoretical concepts have to be applied to his data. These concepts belong to a wide variety of physical fields, from plasma physics to molecular physics. All these concepts are his tools, and therefore it might be useful to collect them in a unified way into a "toolbox" of the radio astronomer. To help towards this is the aim of this book.

1.2 The Radio Window

The atmosphere surrounding the earth is transparent to radio waves as long as none of its constituents is able to absorb this radiation to a noticeable extent. This radio window extends roughly from a lower frequency limit of 15 MHz ($\lambda \cong 20$ m) to a high frequency cut-off at about 300 GHz ($\lambda \cong 1$ mm), but these limits are not sharp (Fig. 1.1); they can vary both with the geographical position and with time.

The high frequency cut-off occurs because the resonant absorption of the lowest rotation bands of molecules in the troposphere fall into this frequency range. There are mainly two molecules responsible for this: water vapour H_2O

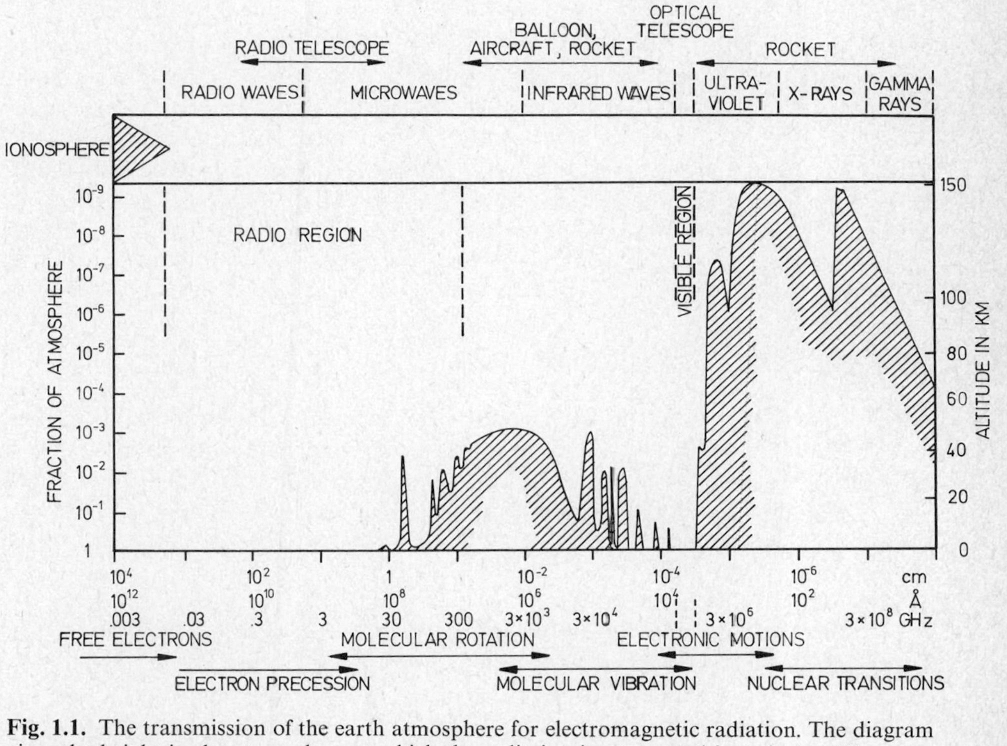

Fig. 1.1. The transmission of the earth atmosphere for electromagnetic radiation. The diagram gives the height in the atmosphere at which the radiation is attenuated by a factor 1/2

with strong bands at 22.2 GHz ($\lambda = 13.5$ mm) and 184 GHz (1.63 mm) and O_2 with an exceedingly strong band at 60 GHz (5 mm) consisting of closely-spaced rotational levels of the ground electronic state resulting in two interleaved series of absorption lines near 60 GHz ($\lambda = 5$ mm) and a single line near 119 GHz ($\lambda = 2.52$ mm). The absorption of other abundant molecules in the atmosphere, namely N_2 and CO_2, occurs at higher frequencies where the atmosphere is not transparent anyhow and therefore does not effect this high frequency cut-off of the radio window.

As we will see later it is of great interest for the radio astronomer to push the high-frequency limits of his observations to as high a value as possible, since most lines emitted or absorbed by interstellar molecules lie in this range. The circumstance that water vapour is one of the determining factors for this cut-off makes it possible to extend the accessible frequency range somewhat by doing the measurements from places where the air has a low total water content. With respect to the absorption caused by oxygen the observing site is of much less influence, but high-altitude observatories with a dry climate are the best one can do.

At the low frequency end, the terrestrial atmosphere ceases to be transparent because the free electrons in the ionosphere absorb electromagnetic radiation in a massive way if the frequency of the radiation is below the plasma frequency ν_p. As we will show later this frequency is given by

$$\frac{\nu_\mathrm{p}}{\mathrm{Hz}} = 9\sqrt{\frac{N_\mathrm{e}}{\mathrm{m}^{-3}}}$$

where N_e is the electron density of the plasma in m^{-3} and ν_p is given in Hz. Thus the low frequency limit of the radio window will be near 4.5 MHz at night when the F_2 layer of the ionosphere has an average maximum density of $N_\mathrm{e} \cong 2.5 \cdot 10^{11}\ \mathrm{m}^{-3}$, and near 11 MHz at daytime, because then $N_\mathrm{e} \cong 1.5 \cdot 10^{12}\ \mathrm{m}^{-3}$. But the electron densities in the ionosphere depend on the solar activity, and therefore this low-frequency limit is variable. Only if the observing frequency is well above this limit are ionospheric properties without noticeable effect.

1.3 Some Basic Definitions

Electromagnetic radiation in the radio window is a wave phenomenon, but when the scale of the system involved is much larger than the wavelength, we can consider the radiation to travel in straight lines called rays. The infinitesimal power dW intercepted by an infinitesimal surface $d\sigma$ (Fig. 1.2) then is

$$dW = B_\nu \cos\theta \, d\Omega \, d\sigma \, d\nu \quad \text{where} \tag{1.1}$$

dW = infinitesimal power, watts
$\mathrm{d}\Omega$ = infinitesimal solid angle from which radiation is coming, sr
$d\sigma$ = infinitesimal area of surface, m^2
$d\nu$ = infinitesimal bandwidth, Hz
θ = angle between normal to $d\sigma$ and $d\Omega$
B_ν = brightness or specific intensity, $\mathrm{W\,m^{-2}\,Hz^{-1}\,sr^{-1}}$.

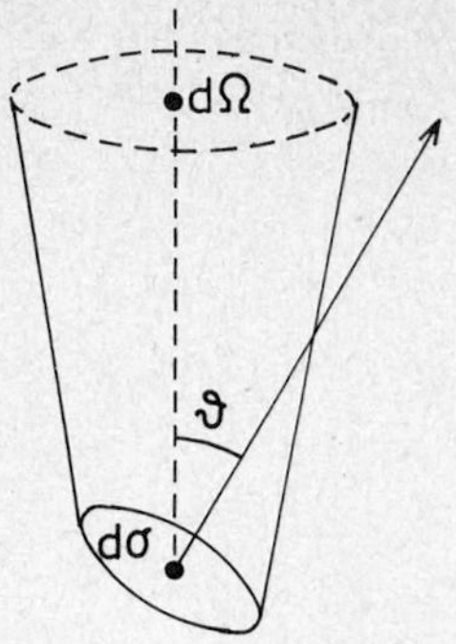

Fig. 1.2 The definition of brightness

Equation (1.1) should be considered to be the definition of the brightness B_ν. Quite often the term *intensity* or *specific intensity* I_ν is used instead of the term "brightness". We will use all three designations interchangeably.

The total flux of a source is obtained by integrating (1.1) over the total solid angle Ω_s subtended by the source

$$S_\nu = \int_{\Omega_s} B_\nu(\theta, \varphi) \cos\theta \, d\Omega \ , \tag{1.2}$$

and this flux is measured in units of W m^{-2} Hz^{-1}. Since the flux of radio sources is usually very small, a special radioastronomical flux unit, the Jansky (abbreviated Jy) has been introduced

$$1 \text{ Jy} = 10^{-26} \text{ W m}^{-2} \text{ Hz}^{-1} \ . \tag{1.3}$$

Very few sources are as bright as 1 Jy, but even such a source would produce a signal of only 10^{-15} W in the 100 m Telescope ($A \cong 5 \cdot 10^3$ m^2, $\Delta\nu = 2 \cdot 10^7$ Hz).

The brightness of an extended source is a quantity very much akin to the surface brightness in optical astronomy: it is independent of the distance to the source, as long as diffraction and extinction processes can be neglected.

Consider a bundle of rays emitted by a source (Fig. 1.3). They have the power dW, and as long as the surface element $d\sigma$ stays inside the ray-tube the power remains constant:

$$dW_1 = dW_2 \ . \tag{1.4}$$

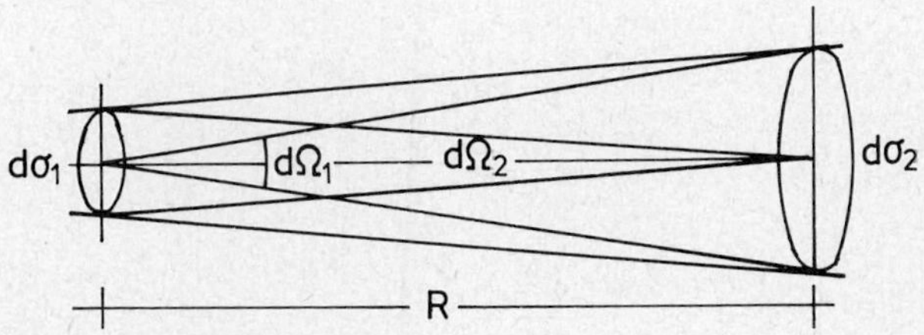

Fig. 1.3. The brightness is independent of the distance along a ray

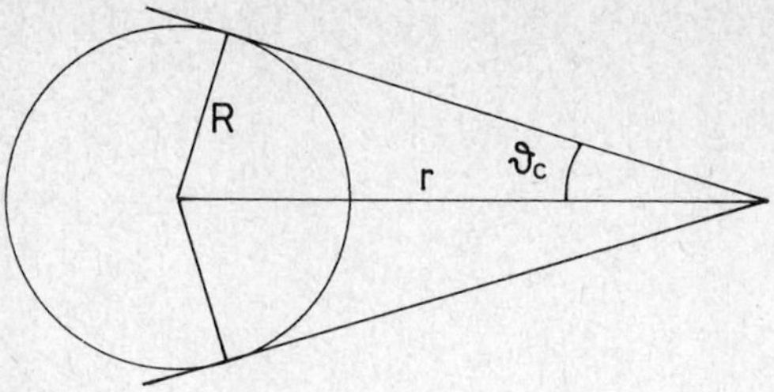

Fig. 1.4. Total flux received from an uniformly bright sphere

For each of these we have

$$dW_1 = B_{\nu_1}\, d\sigma_1\, d\Omega_1\, d\nu \quad \text{and}$$

$$dW_2 = B_{\nu_2}\, d\sigma_2\, d\Omega_2\, d\nu \ .$$

If the distance between $d\sigma_1$ and $d\sigma_2$ is R, then the solid angles $d\Omega_2 = d\sigma_1/R^2$, $d\Omega_1 = d\sigma_2/R^2$ and thus

$$dW_1 = B_{\nu_1}\, d\sigma_1 \frac{d\sigma_2}{R^2}\, d\nu \quad \text{and} \quad dW_2 = B_{\nu_2}\, d\sigma_2 \frac{d\sigma_1}{R^2}\, d\nu \ .$$

Using (1.4) we thus obtain

$$B_{\nu_1} = B_{\nu_2} \tag{1.5}$$

so that the brightness is independent of the distance. The total flux S_ν, on the other hand, shows the expected dependence with $1/r^2$.

Consider a sphere with uniform brightness B_ν with a radius R (Fig. 1.4). The total flux received by an observer at the distance r then is, according to (1.2),

$$S_\nu = \int\limits_{\Omega_s} B_\nu \cos\theta\, d\Omega = B_\nu \int\limits_0^{2\pi} \left(\int\limits_0^{\theta_c} \sin\theta \cos\theta\, d\theta \right) d\varphi$$

where

$$\sin\theta_c = \frac{R}{r}$$

defines the angle θ_c that the radius of the sphere subtends at r. We obtain

$$S_\nu = \pi B_\nu \sin^2\theta_c \quad \text{or} \quad S_\nu = B_\nu \frac{\pi R^2}{r^2} \ . \tag{1.6}$$

Another useful quantity related to the brightness is the radiation density u_ν. Since radiation energy propagates with the velocity of light c, we obviously have for the *energy density per solid angle*

$$u_\nu(\Omega) = \frac{1}{c} B_\nu \tag{1.7}$$

while integrating over the whole sphere, (1.7) results in the total *energy density*

$$u_\nu = \int_{(4\pi)} u_\nu(\Omega)\, d\Omega = \frac{1}{c} \int_{(4\pi)} B_\nu\, d\Omega \ . \tag{1.8}$$

1.4 Radiative Transfer

Equation (1.5) shows that for radiation in free space the specific intensity I_ν remains independent of the distance along a ray. If I_ν changes, it can do this only because radiation is absorbed or emitted, and this change of I_ν is described by the equation of transfer.

The theory to be outlined here is phenomenological one: for a change in I_ν, certain expressions are adopted which contain free parameters. Only experience will then show whether these expressions are appropriate, or whether different ones should be preferred.

For a change in I_ν along the line of sight, a loss term $dI_{\nu -}$ and a gain term $dI_{\nu +}$ are introduced, and we adopt the form

$$dI_{\nu -} = -\kappa_\nu I_\nu\, ds$$

$$dI_{\nu +} = \varepsilon_\nu\, ds$$

so that the change of intensity in a slab of material of the thickness ds will be

$$[I_\nu(s + ds) - I_\nu(s)]\, d\sigma\, d\Omega\, d\nu = [-\kappa_\nu I_\nu + \varepsilon_\nu]\, d\sigma\, d\Omega\, d\nu$$

resulting in the *equation of transfer*

$$\boxed{\frac{dI_\nu}{ds} = -\kappa_\nu I_\nu + \varepsilon_\nu} \ . \tag{1.9}$$

From general experience, the opacity κ_ν is independent of the intensity I_ν leading to the adoption of the above form for $dI_{\nu -}$; similar arguments hold for the emissivity ε_ν.

Certainly situations are conceivable that will result in an ε_ν that depends strongly on I_ν, such as an environment in which radiation is strongly scattered, there are other important situations, where ε_ν is independent of I_ν.

There are several limiting cases, where the solution of the differential equation (1.9) is especially simple.

1) Emission only: $\kappa_\nu = 0$

$$\frac{dI_\nu}{ds} = \varepsilon_\nu \ , \qquad I_\nu(s) = I_\nu(s_0) + \int_{s_0}^{s} \varepsilon_\nu(s)\, ds \ . \tag{1.10}$$

2) Absorption only: $\varepsilon_\nu = 0$

$$\frac{dI_\nu}{ds} = -\kappa_\nu I_\nu$$

$$I_\nu(s) = I_\nu(s_0)\exp\left\{-\int_{s_0}^{s} \kappa_\nu(s)\,ds\right\} . \tag{1.11}$$

3) Thermodynamic equilibrium (TE): If there is complete equilibrium of the radiation with its surrounding the brightness distribution is described by the Planck function of the thermodynamic temperature of the surroundings (black-body radiation)

$$\frac{dI_\nu}{ds} = 0 , \quad I_\nu = B_\nu(T) = \varepsilon_\nu/\kappa_\nu \tag{1.12}$$

$$\boxed{B_\nu(T) = \frac{2h\nu^3}{c^2}\frac{1}{e^{h\nu/kT}-1}} . \tag{1.13}$$

4) Local thermodynamic equilibrium (LTE): Full thermodynamic equilibrium will be realized only in very special circumstances such as in a black enclosure or, say, in stellar interiors. But quite often "Kirchhoff's law" which states that

$$\boxed{\frac{\varepsilon_\nu}{\kappa_\nu} = B_\nu(T)} \tag{1.14}$$

will be applicable independent of the material just like in full thermodynamic equilibrium, although I_ν in general will differ from $B_\nu(T)$.

Defining now the *optical depth* $d\tau_\nu$ (Fig. 1.5) by

$$d\tau_\nu = -\kappa_\nu\,ds \tag{1.15}$$

then the equation of transfer (1.9) can be written as

$$\boxed{-\frac{1}{\kappa_\nu}\frac{dI_\nu}{ds} = \frac{dI_\nu}{d\tau_\nu} = I_\nu - B_\nu(T)} . \tag{1.16}$$

The solution of (1.16) is obtained by first multiplying (1.16) by $\exp(-\tau_\nu)$ and then partially integrating over τ_ν:

$$\int_0^{\tau_\nu(0)} e^{-\tau}\frac{dI_\nu}{d\tau}d\tau = I_\nu e^{-\tau}\Big|_0^{\tau_\nu(0)} + \int_0^{\tau_\nu(0)} I_\nu e^{-\tau}d\tau = \int_0^{\tau_\nu(0)} (I_\nu - B_\nu)e^{-\tau}d\tau$$

$$I_\nu(\tau_\nu(0))\,e^{-\tau_\nu(0)} - I_\nu(\tau_\nu(s_0))\,e^{-0} = -\int_0^{\tau_\nu(0)} B_\nu(T(\tau))\,e^{-\tau}d\tau$$

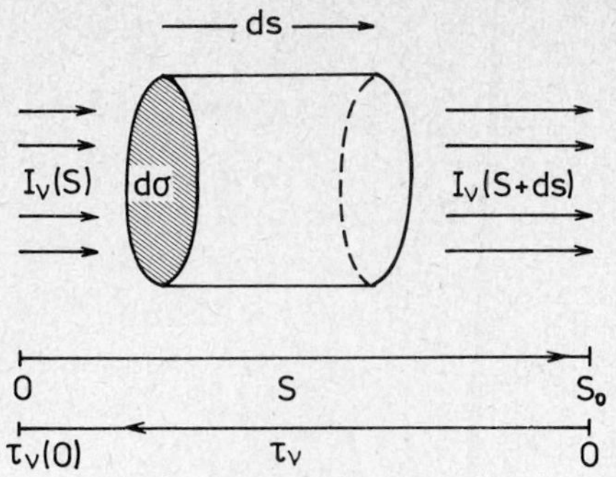

Fig. 1.5. The equation of transfer

or finally

$$I_\nu(s) = I_\nu(0)\,\mathrm{e}^{-\tau_\nu(0)} + \int_0^{\tau_\nu(0)} B_\nu(T(\tau))\,\mathrm{e}^{-\tau}\,d\tau \quad . \tag{1.17}$$

Due to the definition (1.15), s and τ increase in opposite directions as indicated in Fig. 1.5.

If the medium is isothermal, that is, if

$$T(\tau) = T(s) = T = \text{const}$$

the integral in (1.17) can be computed explicitly resulting in

$$I_\nu(s) = I_\nu(0)\,\mathrm{e}^{-\tau_\nu(0)} + B_\nu(T)\,(1 - \mathrm{e}^{-\tau_\nu(0)}) \quad . \tag{1.18}$$

For a large optical depth, that is for $\tau_\nu(0) \to \infty$, (1.18) approaches the limit

$$I_\nu = B_\nu(T) \ . \tag{1.19}$$

The observed brightness I_ν for the optically thick case is equal to the Planck black-body brightness distribution independent of the material.

1.5 Black-Body Radiation and the Brightness Temperature

The spectral distribution of the radiation of a black body in thermodynamic equilibrium is given by the Planck law

$$B_\nu(T) = \frac{2h\nu^3}{c^2}\frac{1}{\mathrm{e}^{h\nu/kT} - 1} \quad . \tag{1.13}$$

It gives the power per unit frequency interval. Converting this into the wavelength scale we obtain $B_\lambda(T)$. Because $B_\nu(T)\,d\nu = -\,B_\lambda(T)\,d\lambda$ and $d\nu = (-\,c/\lambda^2)\,d\lambda$

this is

$$\boxed{B_\lambda(T) = \frac{2hc^2}{\lambda^5} \frac{1}{e^{hc/\lambda kT} - 1}} \quad . \tag{1.20}$$

Integrating either (1.13) over ν or (1.20) over λ, the total brightness of a black body is obtained

$$B(T) = \frac{2h}{c^2} \int_0^\infty \frac{\nu^3}{\exp\{h\nu/kT\} - 1} d\nu \ .$$

Putting

$$x = \frac{h\nu}{kT} \ , \tag{1.21}$$

we get

$$B(T) = \frac{2h}{c^2} \left(\frac{kT}{h}\right)^4 \int_0^\infty \frac{x^3}{e^x - 1} dx \ .$$

The integral has the value $\pi^4/15$ as shown in many books on mathematical physics [an explicit demonstration of this is given in Reif (1965) Sect. A.11]. Thus

$$\boxed{\begin{aligned} &B(T) = \sigma T^4 \\ &\sigma = \frac{2\pi^5 k^4}{15 c^2 h^3} = 5.6697 \cdot 10^{-8}\ \mathrm{W\,m^{-2}\,K^{-4}} \end{aligned}} \quad . \tag{1.22}$$

This is the Stefan-Boltzmann radiation law which was found experimentally in 1879 by J. Stefan and derived theoretically in 1884 by L. Boltzmann before Planck's radiation law was known. Both (1.13) and (1.20) have maxima (Fig. 1.6) which are found by solving $\partial B_\nu/\partial\nu = 0$ and $\partial B_\lambda/\partial\lambda = 0$ respectively. Using (1.21), these correspond to solving $3(1 - e^{-x}) - x = 0$ and $5(1 - e^{-x}) - x = 0$ respectively with the solutions

$$x_\mathrm{m} = 2.82143937 \quad \text{and}$$

$$\hat{x}_\mathrm{m} = 4.96511423 \ .$$

Thus (1.13) attains its maximum at

$$\boxed{\frac{\nu_\mathrm{max}}{\mathrm{Hz}} = 5.8789 \cdot 10^{10} \left(\frac{T}{\mathrm{K}}\right)} \tag{1.23}$$

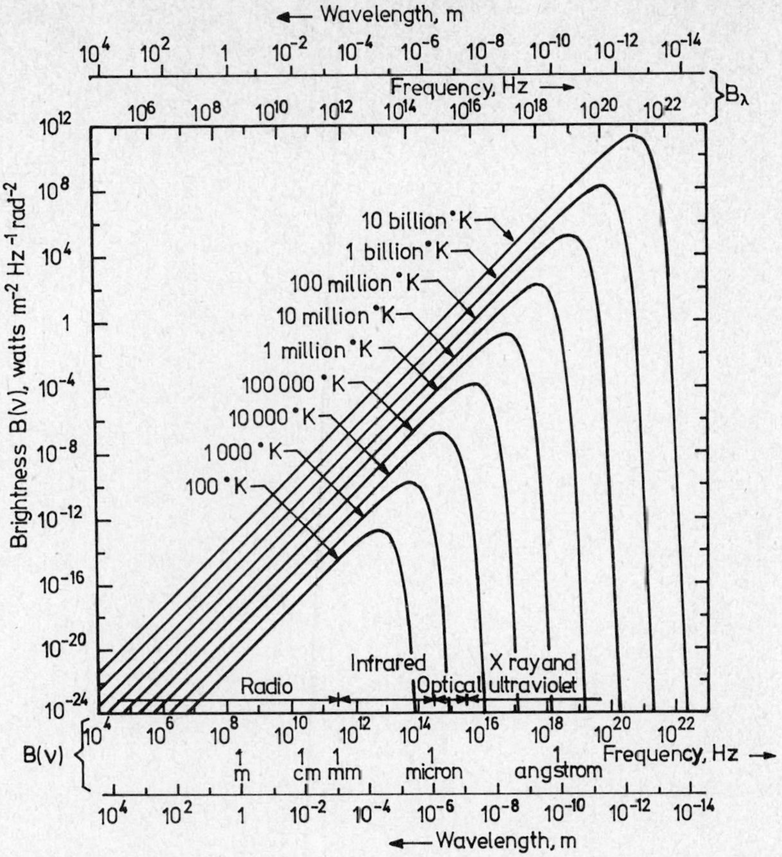

Fig. 1.6. The Planck spectrum for black bodies of different temperatures [adapted from Kraus (1966)]

while (1.20) peaks at

$$\boxed{\left(\frac{\lambda_{\max}}{\mathrm{m}}\right)\left(\frac{T}{\mathrm{K}}\right) = 2.8978 \cdot 10^{-3}} \tag{1.24}$$

(1.23) and (1.24) are both known as *Wien's displacement law.*

If $x = h\nu/kT$ is well off the value appropriate for the maximum, (1.13) can be approximated by simpler expressions (Fig. 1.7)

1) $h\nu \ll kT$: *Rayleigh-Jeans Law.* An expansion of the exponential

$$e^{h\nu/kT} \cong 1 + \frac{h\nu}{kT} + \cdots$$

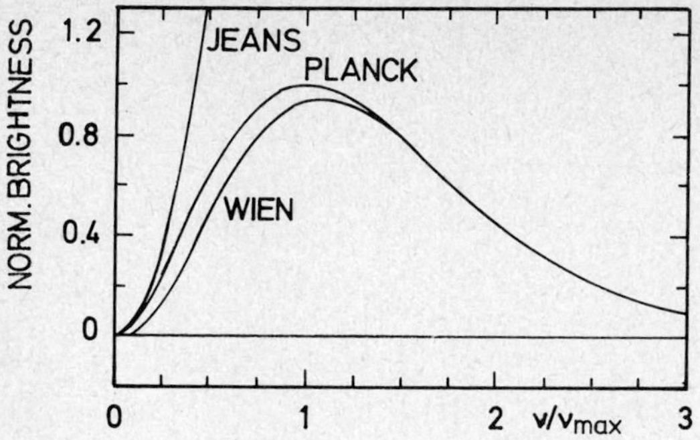

Fig. 1.7. Normalized Planck curve and the Jeans and Wien approximation

results in

$$\boxed{B_J(\nu, T) = \frac{2\nu^2}{c^2}\, k\, T} \quad . \tag{1.25}$$

This is the classical limit of the Planck law since it does not contain Planck's constant.

Inserting numerical values for k and h, we see that Jeans' law is applicable for frequencies

$$\frac{\nu}{\mathrm{GHz}} \ll 20.84 \left(\frac{T}{\mathrm{K}}\right) \quad . \tag{1.26}$$

It can thus be used for all thermal radio sources except perhaps for low-temperature thermal sources in the millimeter wavelength range.

2) $h\nu \gg k\,T$: *Wien's Law*. In this case $e^x \gg 1$, so that

$$\boxed{B_W(\nu, T) = \frac{2h\nu^3}{c^2}\, e^{-h\nu/kT}} \quad . \tag{1.27}$$

While this limit is quite useful for stellar measurements in the visual and ultraviolet range, it plays no great role in radio astronomy.

One of the important features of Jeans' law is that it states that the brightness and the thermodynamic temperature of the black body that emits this radiation are strictly proportional (Fig. 1.8). This feature is so useful that it has become the custom in radio astronomy to measure the brightness of an extended source by its *brightness temperature* T_b. This is the temperature which would result in the given brightness if inserted into Jeans' law

$$T_\mathrm{b} = \frac{c^2}{2k}\frac{1}{\nu^2} B_\nu \quad . \tag{1.28}$$

If B_ν is emitted by a black body and $h\nu \gg k\,T$ then (1.28) indeed gives the thermodynamic temperature of the source, a value that is independent of ν. But

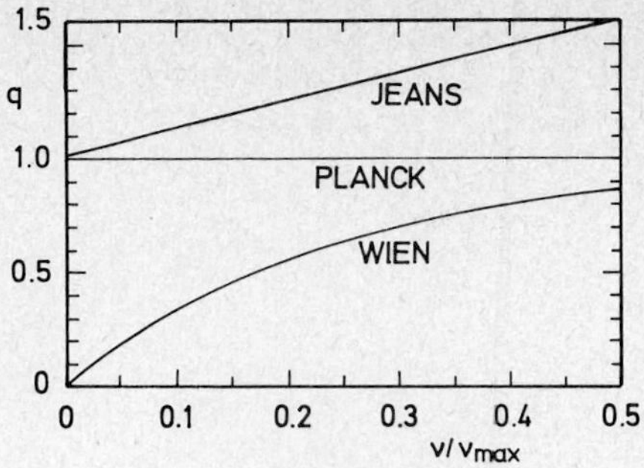

Fig. 1.8. Ratio q of the true thermodynamic temperature of a black body to the values inferred from the Rayleigh-Jeans and Wien approximation

$$T_{\text{thermodyn}} = q\, T_{\text{approx}}\ .$$

$\nu_{\max}$ is given by (1.23)

if other processes are responsible for the emission of the radiation, T_b will depend on the frequency; it is, however, still a useful quantity.

This is true also, if the frequency is so high that the condition (1.26) is invalidated. (1.28) can still be applied, but it should be understood that, in this case, T_b is different from the thermodynamic temperature of a black body. It is, however, rather simple to give correction factors applicable for this case.

It is also convenient to introduce the concept of brightness temperature into the radiative transfer equation (1.16). With (1.28) this becomes

$$\boxed{\frac{dT_b(s)}{d\tau_\nu} = T_b(s) - T(s)}\ , \tag{1.29}$$

where $T(s)$ is the thermodynamic temperature of the medium at the position s. The general solution (1.17) is then

$$\boxed{T_b(s) = T_b(0)\,\mathrm{e}^{-\tau_\nu(0)} + \int\limits_0^{\tau_\nu(0)} T(s)\,\mathrm{e}^{-\tau}\,d\tau} \tag{1.30}$$

and, if the medium is isothermal, this becomes

$$\boxed{T_b(s) = T_b(0)\,\mathrm{e}^{-\tau_\nu(0)} + T(1 - \mathrm{e}^{-\tau_\nu(0)})}\ . \tag{1.31}$$

For the sake of simplicity, let us assume that $T_b(0) = 0$. Then two limiting cases that are often applicable are (Fig. 1.9):

1) for optically thin $\tau \ll 1$,

$$T_b = \tau_\nu\, T \quad \text{and}\ , \tag{1.32}$$

2) for optically thick $\tau \gg 1$,

$$T_b = T\,. \tag{1.33}$$

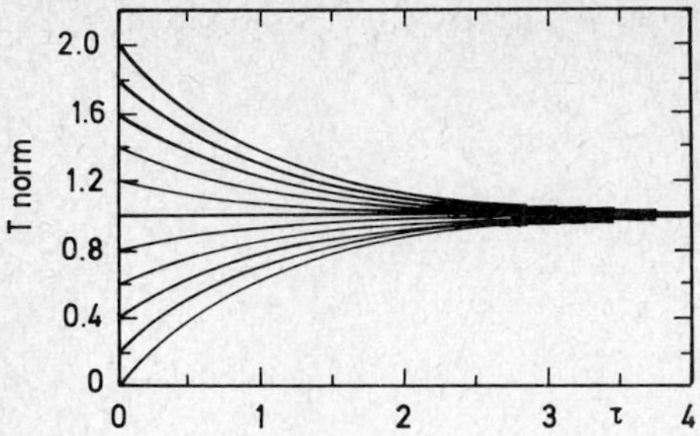

Fig. 1.9. Variation of the brightness temperature in an isothermal medium if radiation with T_{b0} is entering the medium at $s = 0$

$T_{norm} = T_{b0}/T$,

T temperature of the medium

1.6 The Nyquist Theorem and the Noise Temperature

Suppose an electrical resistor R is connected across the input terminals of a linear amplifier that accepts frequencies between ν_1 and ν_2. The thermal motion of the electrons in the resistor will produce a current i which forms a random input to the amplifier. The mean value of this current will be zero, but general thermodynamic considerations show that the rms value of this current is different from zero. Since $\langle i^2 \rangle \neq 0$ represents a power, the first law of thermodynamics requires that this power must be supplanted by other, thermal energy. The available, or to use the appropriate technical term, the exchangeable power of the resistor is then

$$W\,d\nu = \frac{h\,\nu}{e^{h\nu/kT} - 1}\,d\nu \ , \tag{1.34}$$

independent of the value of the resistor. Here T is the thermodynamic temperature of the resistor.

If we again use the series expansion $\exp(h\,\nu/k\,T) \cong 1 + h\,\nu/k\,T$, retaining only the linear term, (1.34) results in

$$W d\nu = k\,T\,d\nu \ . \tag{1.35}$$

The available noise power of a resistor is thus proportional to its temperature, the *noise temperature* T_N.

From (1.35) we see that in the radio range the spectral density of the noise power is flat; that is, the noise spectrum is *white*.

2. Electromagnetic Wave Propagation Fundamentals

2.1 Maxwell's Equation

Maxwell's theory of electrodynamics describes the effects and forces of an electromagnetic field and its variations in terms of electromagnetic field components, their space distributions and time-variations. In most treatises on electrodynamics, this theory is derived in a more-or-less inductive way starting from static situations, although there are notable exceptions to this. So in Sommerfeld's well-known books on theoretical physics, electrodynamics starts right away with the full set of Maxwell's equation.

Here no systematic and complete exposition of electrodynamic theory is intended. Only those features of the theory that are needed to understand the formation, emission and propagation of electromagnetic waves will be presented. A uniform set of quantities will be used for this. This may sound trivial, but since part of this subject is usually covered by theoretical physics proper while the rest is usually found only in books intended for the engineer, different systems of units and designations are often used in the different disciplines.

The interrelations of the five vector and one scalar fields which are required to (properly) describe the electromagnetic phenomena are given by Maxwell's equations. These can conveniently be divided into three groups. Some of the field components are related by the properties of the medium in which they exist. These are the so-called material equations

$$\boldsymbol{J} = \sigma \boldsymbol{E} \tag{2.1}$$

$$\boldsymbol{D} = \varepsilon \boldsymbol{E} \tag{2.2}$$

$$\boldsymbol{B} = \mu \boldsymbol{H} \quad . \tag{2.3}$$

σ, ε and μ are scalar functions that are almost constant in most materials. Their values for a vacuum are marked by the sub-index 0

$$\sigma_0 = 0 \tag{2.4}$$

$$\varepsilon_0 = 8.854196 \cdot 10^{-12}\ \mathrm{A\,s/V\,m} \tag{2.5}$$

$$\mu_0 = 1.256637 \cdot 10^{-6}\ \mathrm{V\,s/A\,m} \quad . \tag{2.6}$$

For other materials ε and μ are usually given as relative quantities

$$\varepsilon = \varepsilon_r \, \varepsilon_0 \tag{2.7}$$

$$\mu = \mu_r \, \mu_0 \ . \tag{2.8}$$

ε_r is the "relative permittivity constant" which is also called the "dielectric constant"; μ_r is the relative permeability, while (2.1) is the differential form of Ohm's law, and σ is the specific conductivity.

Maxwell's equations proper can now be further divided into two groups – one that involves only the spatial structure of the fields

$$\boxed{\begin{aligned} \nabla \cdot \boldsymbol{D} &= \varrho \qquad (2.9) \\ \nabla \cdot \boldsymbol{B} &= 0 \qquad (2.10) \end{aligned}} \ ,$$

and one that includes time derivaties

$$\boxed{\begin{aligned} \nabla \times \boldsymbol{E} &= -\dot{\boldsymbol{B}} \qquad (2.11) \\ \nabla \times \boldsymbol{H} &= \boldsymbol{J} + \dot{\boldsymbol{D}} \qquad (2.12) \end{aligned}} \ .$$

Taking the divergence, (2.12), is found to be equal to zero if (2.10) is considered. Using (2.9) therefore

$$\boxed{\nabla \cdot \boldsymbol{J} + \dot{\varrho} = 0} \ ; \tag{2.13}$$

that is, charge density and current obey a continuity equation.

2.2 The Energy Law and the Poynting Vector

By considering the forces that a static electric or magnetic field imposes on a test charge it can be shown that the energy-density of an electromagnetic field is given by

$$\boxed{u = \tfrac{1}{2}\left(\boldsymbol{E} \cdot \boldsymbol{D} + \boldsymbol{B} \cdot \boldsymbol{H}\right) = \tfrac{1}{2}\left(\varepsilon \boldsymbol{E}^2 + \mu \boldsymbol{H}^2\right)} \ . \tag{2.14}$$

If both ε and μ are time-independent, the time derivative of u is given by

$$\dot{u} = \varepsilon \boldsymbol{E} \cdot \dot{\boldsymbol{E}} + \mu \boldsymbol{H} \cdot \dot{\boldsymbol{H}} = \boldsymbol{E} \cdot \dot{\boldsymbol{D}} + \boldsymbol{H} \cdot \dot{\boldsymbol{B}} \ . \tag{2.15}$$

Substituting both $\dot{\boldsymbol{D}}$ and $\dot{\boldsymbol{B}}$ from Maxwell's equations (2.11) and (2.12), this becomes

$$\dot{u} = \boldsymbol{E} \cdot (\nabla \times \boldsymbol{H}) - \boldsymbol{H} \cdot (\nabla \times \boldsymbol{E}) - \boldsymbol{E} \cdot \boldsymbol{J}$$

$$\dot{u} = -\nabla \cdot (\boldsymbol{E} \times \boldsymbol{H}) - \boldsymbol{E} \cdot \boldsymbol{J} \tag{2.16}$$

if the vector identity (A 9) given in Appendix A is applied. By introducing the Poynting vector $\boldsymbol{S}$ (Poynting 1884)

$$\boxed{\boldsymbol{S} = \boldsymbol{E} \times \boldsymbol{H}} \quad , \tag{2.17}$$

(2.16) can be written as an equation of continuity for $\boldsymbol{S}$:

$$\boxed{\frac{\partial u}{\partial t} + \nabla \cdot \boldsymbol{S} = -\boldsymbol{E} \cdot \boldsymbol{J}} \quad . \tag{2.18}$$

The time variation of the energy density u thus consists of two parts: a spatial change of the Poynting vector or energy flux $\boldsymbol{S}$ and a conversion of electromagnetic energy into thermal energy (Joule's energy).

The significance of (2.18) may become more evident if it is applied to a simple example. Consider the conditions in a straight wire of circular cross section carrying a steady current I (Fig. 2.1). If all conditions are constant, the total electromagnetic energy density u must be constant too, so that $\dot{u} = 0$. But if a constant current I is flowing in the wire there is a constant transformation of electric energy into thermal energy. Per unit length l of the wire, this thermal energy is formed at a rate

$$\frac{dW}{dl} = r\, I^2 \tag{2.19}$$

where r is the specific resistance of the wire. Obviously

$$r = \frac{1}{\sigma \pi R^2} \quad \text{so that}$$

$$\frac{dW}{dl} = \frac{I^2}{\sigma \pi R^2} \quad .$$

But

$$|\boldsymbol{J}| = \frac{I}{\pi R^2} \, ,$$

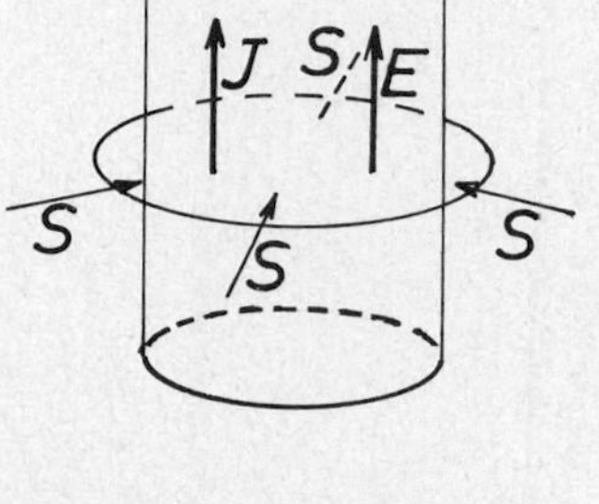

Fig. 2.1. Energy conservation and Poynting vector for a straight wire with circular cross-section carrying a steady current I

and according to (2.1)

$$\boldsymbol{E} = \frac{1}{\sigma}\boldsymbol{J} \ ,$$

so that the thermal loss rate is

$$\frac{dW}{dl} = |\boldsymbol{E}|\,I \ . \tag{2.20}$$

But according to the definition of the magnetic field intensity in Appendix D

$$|\boldsymbol{H}| = \frac{I}{2\pi R}$$

with a direction perpendicular to I. Therefore

$$|\boldsymbol{S}| = |\boldsymbol{E}\times\boldsymbol{H}| = \frac{|\boldsymbol{E}|\,I}{2\pi R} \ ,$$

where $\boldsymbol{S}$ is orientated such that $\boldsymbol{E}$, $\boldsymbol{H}$ and $\boldsymbol{S}$ form a right handed system. Therefore

$$|\boldsymbol{E}|\,I = 2\pi R\,|\boldsymbol{S}| \ . \tag{2.21}$$

This is obviously the total flux of $\boldsymbol{S}$ at the surface of the wire and, from the direction of $\boldsymbol{J}$ and $\boldsymbol{H}$, we see that $\boldsymbol{S}$ flows into it (Fig. 2.1). But according to (2.20) this is just the conversion rate of electrical energy into thermal energy. Therefore the Poynting flux just compensates for this loss as it should in a steady state.

2.3 Complex Field Vectors

In situations, where electromagnetic wave phenomena are considered, the field vectors usually show a harmonic time dependence described by sine or cosine functions. But since these functions are related to the exponential function by the Euler relation

$$\cos x + \mathrm{i}\sin x = \mathrm{e}^{\mathrm{i}x} \ ,$$

the inconvenience of having to apply the rather complicated trigonometric addition theorem can be avoided, if complex field vectors are introduced by

$$\boldsymbol{E} = (\boldsymbol{E}_1 + \mathrm{i}\,\boldsymbol{E}_2)\,\mathrm{e}^{-\mathrm{i}\omega t} \ ; \qquad \boldsymbol{E}_1,\ \boldsymbol{E}_2 \text{ real vector fields} \ , \tag{2.22}$$

and

$$\boldsymbol{H} = (\boldsymbol{H}_1 + \mathrm{i}\,\boldsymbol{H}_2)\,\mathrm{e}^{-\mathrm{i}\omega t} \ ; \qquad \boldsymbol{H}_1,\ \boldsymbol{H}_2 \text{ real vector fields} \ . \tag{2.23}$$

In any application the electric or magnetic field considered is then identified with the real part of $\boldsymbol{E}$ and $\boldsymbol{H}$ or the imaginary part, whichever is more convenient. All mathematical operations can then be performed on $\boldsymbol{E}$ or $\boldsymbol{H}$ directly, as long as they are restricted to linear operations. Only if non-linear operations are involved one must return to real quantities. But even here convenient short-cuts sometimes exist. Such is the case for the Poynting vector.

For $\boldsymbol{S}$ obviously the expression

$$\boldsymbol{S} = \mathrm{Re}\{\boldsymbol{E}\} \times \mathrm{Re}\{\boldsymbol{H}\} + \mathrm{Im}\{\boldsymbol{E}\} \times \mathrm{Im}\{\mathbf{H}\} \tag{2.24}$$

should be used.

But since

$$\mathrm{Re}\{\boldsymbol{E}\} = \boldsymbol{E}_1 \cos\omega t + \boldsymbol{E}_2 \sin\omega t \quad \text{and}$$

$$\mathrm{Re}\{\boldsymbol{H}\} = \boldsymbol{H}_1 \cos\omega t + \boldsymbol{H}_2 \sin\omega t$$

this is

$$\mathrm{Re}\{\boldsymbol{E}\} \times \mathrm{Re}\{\boldsymbol{H}\} = (\boldsymbol{E}_1 \times \boldsymbol{H}_1)\cos^2\omega t + (\boldsymbol{E}_2 \times \boldsymbol{H}_2)\sin^2\omega t$$
$$+ (\boldsymbol{E}_1 \times \boldsymbol{H}_2 + \boldsymbol{E}_2 \times \boldsymbol{H}_1)\cos\omega t \sin\omega t \; .$$

If we now do not consider the instantaneous value of $\boldsymbol{S}$, but the mean value over a full oscillation, and if such mean values are designated by $\langle\;\rangle$, then one obtains, since

$$\langle \sin^2\omega t\rangle = \langle \cos^2\omega t\rangle = \tfrac{1}{2} \quad \text{and}$$

$$\langle \sin\omega t \cos\omega t\rangle = 0 \; ,$$

$$\langle \mathrm{Re}\{\boldsymbol{E}\} \times \mathrm{Re}\{\boldsymbol{H}\}\rangle = \tfrac{1}{2}(\boldsymbol{E}_1 \times \boldsymbol{H}_1 + \boldsymbol{E}_2 \times \boldsymbol{H}_2) \; . \tag{2.25}$$

On the other hand

$$\boldsymbol{E} \times \boldsymbol{H}^* = (\boldsymbol{E}_1 + \mathrm{i}\boldsymbol{E}_2)\mathrm{e}^{-\mathrm{i}\omega t} \times (\boldsymbol{H}_1 - \mathrm{i}\boldsymbol{H}_2)\mathrm{e}^{i\omega t}$$
$$= (\boldsymbol{E}_1 + \mathrm{i}\boldsymbol{E}_2) \times (\boldsymbol{H}_1 - \mathrm{i}\boldsymbol{H}_2)$$

so that

$$\mathrm{Re}\{\boldsymbol{E} \times \boldsymbol{H}^*\} = \boldsymbol{E}_1 \times \boldsymbol{H}_1 + \boldsymbol{E}_2 \times \boldsymbol{H}_2$$

where $\boldsymbol{H}^*$ denotes the complex conjugate of $\boldsymbol{H}$. Inserting this in (2.25) the average value of $\boldsymbol{S}$ is

$$\boxed{\langle \boldsymbol{S}\rangle = \mathrm{Re}\{\boldsymbol{E} \times \boldsymbol{H}^*\}} \; . \tag{2.26}$$

It should, however, be remembered that this formula applies only to complex electromagnetic fields that have a harmonic time variation according to the definitions (2.22) and (2.23).

2.4 The Wave Equation

Maxwell's equations (2.9)–(2.12) give a connection between the spatial and the time variation of the electromagnetic field. But the situation is complicated by the fact that the equations form conditions for different fields: e.g. curl $\boldsymbol{E}$ is related to $\dot{\boldsymbol{B}}$ (2.11), and the other equations show a similar behaviour.

A better insight into the behaviour of the fields can be expected, if the equations are manipulated so that only one vector field appears in each equation. This is achieved by the wave equations.

To simplify the derivation, the conductivity σ, the permittivity ε and the permeability μ will be assumed to be constants both in time and in space. Taking the curl of (2.12)

$$\nabla\times(\nabla\times\boldsymbol{H}) = \nabla\times\boldsymbol{J} + \frac{\partial}{\partial t}\nabla\times\boldsymbol{D} = \nabla\times(\sigma\boldsymbol{E}) + \frac{\partial}{\partial t}\nabla(\varepsilon\boldsymbol{E}) = \left(\sigma + \varepsilon\frac{\partial}{\partial t}\right)\nabla\times\boldsymbol{E}\ ,$$

where the order of ∇ and time derivation have been interchanged, and $\boldsymbol{J}$ and $\boldsymbol{D}$ have been replaced by $\sigma\boldsymbol{E}$ and $\varepsilon\boldsymbol{E}$ respectively by application of (2.1) and (2.2). Using (2.11) and (2.3), this can be further modified to

$$\nabla\times(\nabla\times\boldsymbol{H}) = -\mu\left(\sigma + \varepsilon\frac{\partial}{\partial t}\right)\frac{\partial}{\partial t}\boldsymbol{H} = -\mu(\sigma\dot{\boldsymbol{H}} + \varepsilon\ddot{\boldsymbol{H}})\ . \tag{2.27}$$

By a similar procedure from (2.11)

$$\nabla\times(\nabla\times\boldsymbol{E}) = -\frac{\partial}{\partial t}(\nabla\times\boldsymbol{B}) = -\mu\frac{\partial}{\partial t}(\nabla\times\boldsymbol{H})\ .$$

Using (2.12) this becomes

$$\begin{aligned}\nabla\times(\nabla\times\boldsymbol{E}) &= -\mu\frac{\partial}{\partial t}(\boldsymbol{J} + \dot{\boldsymbol{D}}) = -\mu\frac{\partial}{\partial t}(\sigma\boldsymbol{E} + \varepsilon\dot{\boldsymbol{E}})\\ &= -\mu(\sigma\dot{\boldsymbol{E}} + \varepsilon\ddot{\boldsymbol{E}})\ .\end{aligned} \tag{2.28}$$

The left-hand side of (2.27) and (2.28) can be reduced to a more easily recognizable form by using the vector identity [see Appendix (A 13)].

$$\nabla\times(\nabla\times\boldsymbol{P}) = \nabla(\nabla\cdot\boldsymbol{P}) - \nabla^2\boldsymbol{P}$$

so that because of (2.10)

$$\nabla\times(\nabla\times\boldsymbol{H}) = \nabla(\nabla\cdot\boldsymbol{H}) - \nabla^2\boldsymbol{H} = -\nabla^2\boldsymbol{H}$$

and, if it can be assumed that there are no free charges in the medium, that is, if

$\nabla\cdot\boldsymbol{D} = 0\ ,$ similarly

$$\nabla\times(\nabla\times\boldsymbol{E}) = \nabla(\nabla\cdot\boldsymbol{E}) - \nabla^2\boldsymbol{E} = -\nabla^2\boldsymbol{E}\ ,$$

so that finally

$$\boxed{\begin{aligned} \nabla^2 \boldsymbol{H} &= \varepsilon\mu\ddot{\boldsymbol{H}} + \sigma\mu\dot{\boldsymbol{H}} \\ \nabla^2 \boldsymbol{E} &= \varepsilon\mu\ddot{\boldsymbol{E}} + \sigma\mu\dot{\boldsymbol{E}} \end{aligned}} \quad . \tag{2.29, 2.30}$$

Both $\boldsymbol{E}$ and $\boldsymbol{H}$ therefore obey the same inhomogeneous wave equation, a linear second order partial differential equation. These equations are derived from Maxwell's equations, and therefore every solution of these will also be a solution of the wave equation. The reverse conclusion is true only if some restrictions are made.

In (2.29) and (2.30) the $\boldsymbol{E}$ and the $\boldsymbol{H}$ fields are decoupled, and therefore any arbitrary solution for $\boldsymbol{E}$ can be coupled to any solution for $\boldsymbol{H}$ provided that they obey the initial conditions. In Maxwell's equations this is not true; here $\boldsymbol{E}$ and $\boldsymbol{H}$ are interdependent. For simple cases it is rather easy to specify which $\boldsymbol{H}$ solution belongs to a given $\boldsymbol{E}$ solution of Maxwell's equations; for more complicated situations other methods for the solution will have to be used. Some of these will be outlined in Chap. 4; here a direct solution of the wave equation should suffice to show the principle.

2.5 Plane Waves in Isolating Media

In a homogeneous, isolating medium ($\sigma = 0$) that is free of currents and charges, each rectangular component u of $\boldsymbol{E}$ and $\boldsymbol{H}$ obeys the homogeneous wave equation

$$\boxed{\nabla^2 u - \frac{1}{v^2}\ddot{u} = 0} \quad \text{where} \tag{2.31}$$

$$\boxed{v = \frac{1}{\sqrt{\varepsilon\mu}}} \tag{2.32}$$

is a constant with the dimension of a velocity. For the vacuum this becomes

$$\boxed{c = \frac{1}{\sqrt{\varepsilon_0\mu_0}}} \quad . \tag{2.33}$$

When Kohlrausch and Weber in 1856 obtained this result experimentally, it became one of the basic facts used by Maxwell when he developed his electromagnetic theory predicting the existence of electromagnetic waves. This prediction eventually was confirmed by the experiments of Hertz (1888).

Equation (2.31) is a homogeneous linear partial differential equation of second order and its complete family of solutions is rather wide and complicated. No attempt will be made here to discuss this general solution, it will suffice to outline the properties of the harmonic wave.

$$u = u_0 \, e^{i(kx \pm \omega t)} \tag{2.34}$$

is a solution of (2.31) if the wavenumber k obeys the relation

$$\boxed{k^2 = \varepsilon \mu \omega^2} \tag{2.35}$$

as can be easily checked by substituting (2.34) into (2.31).

If

$$\varphi = kx \pm \omega t \tag{2.36}$$

where φ is the phase of the wave, we see that points of constant phase move with the phase velocity

$$\boxed{v = \frac{\omega}{k} = \frac{1}{\sqrt{\varepsilon \mu}}} \ , \tag{2.37}$$

thus giving a physical meaning to the constant v appearing in (2.31). Introducing the index of refraction n as the ratio of c to v this becomes

$$\boxed{n = \frac{c}{v} = \sqrt{\frac{\varepsilon}{\varepsilon_0} \frac{\mu}{\mu_0}} = \frac{c}{\omega} k} \ . \tag{2.38}$$

In a plane electromagnetic wave each component of $\boldsymbol{E}$ and $\boldsymbol{H}$ will consist of a solution such as (2.34) though with an amplitude u_0 that generally is complex. But even then, a form like (2.34) permits the introduction of some important simplifications, since for a travelling plane wave

$$\boldsymbol{A}(\boldsymbol{x}, t) = \boldsymbol{A}_0 \, e^{i(\boldsymbol{k} \cdot \boldsymbol{x} - \omega t)} \ , \qquad \boldsymbol{A}_0, \ \boldsymbol{k}, \ \omega = \text{const} \tag{2.39}$$

$$\dot{\boldsymbol{A}} = - i \omega \boldsymbol{A} \tag{2.40}$$

$$\ddot{\boldsymbol{A}} = - \omega^2 \boldsymbol{A} \tag{2.41}$$

$$\nabla \cdot \boldsymbol{A} = i \boldsymbol{k} \boldsymbol{A} \quad \text{and} \tag{2.42}$$

$$\nabla^2 \boldsymbol{A} = - \boldsymbol{k}^2 \boldsymbol{A} \ . \tag{2.43}$$

The $\boldsymbol{E}$ and $\boldsymbol{H}$ fields of an electromagnetic wave not only have to be solutions of the wave equation (2.31), they have in addition to obey Maxwell's equations. Because of the decoupling of the two fields in the wave equation this produces some additional constraints.

In order to investigate the properties of plane waves as simply as possible, let us orientate the rectangular coordinate system such that the wave propagates into the positive z-direction.

A wave will be considered to be plane, if the surfaces of constant phase form planes $z = \text{const}$. Thus all components of the $\boldsymbol{E}$ and the $\boldsymbol{H}$ field will be independent of x and y for fixed z; that is,

$$\begin{aligned} &\frac{\partial E_x}{\partial x} = 0, \quad \frac{\partial E_y}{\partial x} = 0, \quad \frac{\partial E_z}{\partial x} = 0 \\ &\frac{\partial E_x}{\partial y} = 0, \quad \frac{\partial E_y}{\partial y} = 0, \quad \frac{\partial E_z}{\partial y} = 0 \end{aligned} \tag{2.44}$$

and a similar set of equations for $\boldsymbol{H}$. But according to Maxwell's equations (2.9) and (2.10) with $\varrho = 0$ and $\varepsilon = \text{const}$

$$\frac{\partial E_x}{\partial x} + \frac{\partial E_y}{\partial y} + \frac{\partial E_z}{\partial z} = 0 \quad \text{and} \quad \frac{\partial H_x}{\partial x} + \frac{\partial H_y}{\partial y} + \frac{\partial H_z}{\partial z} = 0\ .$$

Because of (2.44) this results to

$$\boxed{\frac{\partial E_z}{\partial z} = 0 \quad \text{and} \quad \frac{\partial H_z}{\partial z} = 0}\ . \tag{2.45}$$

From the remaining Maxwell's equations (2.11) and (2.12) we similarly obtain

$$\boxed{\frac{\partial E_z}{\partial t} = 0 \quad \text{and} \quad \frac{\partial H_z}{\partial t} = 0}\ . \tag{2.46}$$

Therefore both the longitudinal components E_z and H_z must be constant both in space and time. Since such a constant field is of no significance here, we may conclude that

$$\boxed{E_z \equiv 0\ , \quad H_z \equiv 0} \tag{2.47}$$

that is, the plane electromagnetic wave in a nonconducting medium is *transversal* (Fig. 2.2).

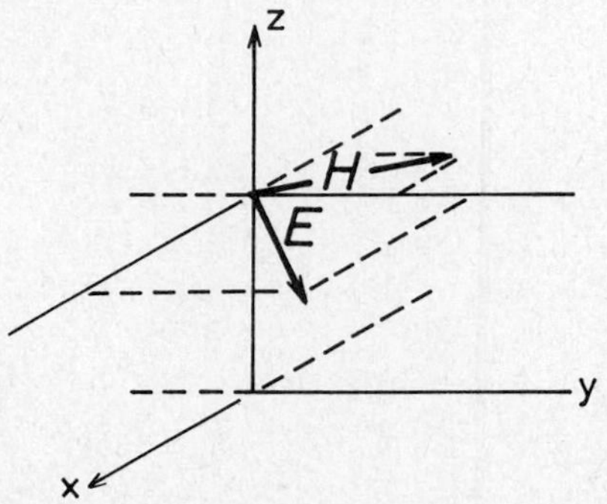

Fig. 2.2. The field vectors in a plane electromagnetic wave propagating in the z-direction

The remaining components have the form of travelling harmonic waves as given by (2.34), and the only components of Eqs. (2.11) and (2.12) that are different from zero are

$$\begin{aligned} \frac{\partial E_x}{\partial z} &= -\mu \frac{\partial H_y}{\partial t} & & & \frac{\partial H_x}{\partial z} &= \varepsilon \frac{\partial E_y}{\partial t} \\ & & \text{and} & & & \\ \frac{\partial E_y}{\partial z} &= \mu \frac{\partial H_x}{\partial t} & & & \frac{\partial H_y}{\partial z} &= -\varepsilon \frac{\partial E_x}{\partial t} \end{aligned} \quad . \tag{2.48}$$

Applying now the relations (2.40) and (2.42) for plane harmonic waves, we find

$$\begin{aligned} \frac{\partial E_x}{\partial z} &= \mathrm{i} k E_x = -\mu \dot{H}_y = \mathrm{i} \omega \mu H_y \\ \frac{\partial E_y}{\partial z} &= \mathrm{i} k E_y = \mu \dot{H}_x = -\mathrm{i} \omega \mu H_x \end{aligned} \tag{2.49}$$

resulting in

$$\boldsymbol{E} \cdot \boldsymbol{H} = E_x H_x + E_y H_y = -\frac{k}{\omega \mu} E_x E_y + \frac{k}{\omega \mu} E_y E_x = 0$$

$$\boxed{\boldsymbol{E} \cdot \boldsymbol{H} = 0} \quad . \tag{2.50}$$

E and ***H*** are thus always perpendicular to each other, and together with the wave vector ***k*** form an orthogonal system. For the ratio of their absolute values, (2.49) and (2.35) result in

$$\frac{|\boldsymbol{E}|}{|\boldsymbol{H}|} = \sqrt{\frac{\mu}{\varepsilon}} \quad . \tag{2.51}$$

The unit of this *intrinsic impedance* of the medium in which the wave propagates is the Ohm (Ω). In a vacuum it has the value

$$Z_0 = \sqrt{\frac{\mu_0}{\varepsilon_0}} = 376.73\ \Omega \quad . \tag{2.52}$$

Finally, the energy flux of the Poynting vector of this wave is of interest. As given by (2.26) we find

$$|\boldsymbol{S}| = \sqrt{\frac{\varepsilon}{\mu}} \boldsymbol{E}^2 \quad , \tag{2.53}$$

and ***S*** points in the direction of the propagation vector ***k***. The (time averaged) energy density of the wave given by (2.14) is then [1]

$$u = \tfrac{1}{2} (\varepsilon \boldsymbol{E} \cdot \boldsymbol{E}^* + \mu \boldsymbol{H} \cdot \boldsymbol{H}^*) \quad . \tag{2.54}$$

[1] This energy density should not be confused with the cartesian component u of ***E*** or ***H*** in (2.31) and succ.

The argument used in this is quite similar to that used in deriving (2.26). Using (2.51), (2.54) becomes

$$u = \varepsilon \boldsymbol{E}^2 \tag{2.55}$$

so that the energy flows in a harmonic wave with the phase velocity (2.37).

The time averaged Poynting vector is often used as a measure of the intensity of the wave; its direction represents the direction of the wave propagation.

2.6 Wave Packets and the Group Velocity

A monochromatic plane wave [2]

$$u(x, t) = A\, \mathrm{e}^{\mathrm{i}(kx - \omega t)} \tag{2.56}$$

propagates with the phase velocity

$$v = \frac{\omega}{k} \,. \tag{2.57}$$

If this velocity is the same for a whole range of frequencies, then a wave packet formed by the superposition of a whole range of these waves will propagate with the same velocity. In general, however, the propagation velocity v will depend on the wavenumber k, and then such wave packets have some new and interesting properties.

Let a wave of a more general shape be formed by superposing simple harmonic waves

$$u(x, t) = \frac{1}{\sqrt{2\pi}} \int_{-\infty}^{\infty} A(k)\, \mathrm{e}^{\mathrm{i}(kx - \omega t)}\, dk \,, \tag{2.58}$$

where $A(k)$ is the amplitude of the wave with the wave number k. The angular frequency of these waves will be different for different k; this function

$$\omega = \omega(k) \tag{2.59}$$

will be called the *dispersion equation* of the waves.

Now if $A(k)$ is a fairly sharply peaked function around some k_0, only waves with wavenumbers not too different from k_0 will contribute to (2.58), and quite often a linear approximation for (2.59)

$$\omega(k) = \omega_0 + \left.\frac{d\omega}{dk}\right|_0 (k - k_0) \tag{2.60}$$

will be sufficient. Substituting this into (2.58) we can extract all those factors that do not depend on k from the integral and obtain

$$u(x, t) = \frac{1}{\sqrt{2\pi}} \exp\left[\mathrm{i}\left(\left.\frac{d\omega}{dk}\right|_0 k_0 - \omega_0\right) t\right] \int_{-\infty}^{\infty} A(k) \exp\left[\mathrm{i}k\left(x - \left.\frac{d\omega}{dk}\right|_0 t\right)\right] dk \,. \tag{2.61}$$

[2] See footnote at p. 24.

According to (2.58), at the time $t = 0$ the wave packet has the shape

$$u(x,0) = \frac{1}{\sqrt{2\pi}} \int_{-\infty}^{\infty} A(k)\, e^{ikx}\, dk \; .$$

Therefore the integral in (2.61) can be seen to designate

$$u(x',0) \quad \text{of} \quad x' = x - \left.\frac{d\omega}{dk}\right|_0 t$$

and the whole expression as

$$u(x,t) = u\left(x - \left.\frac{d\omega}{dk}\right|_0 t, 0\right) \exp\left[i\left(k_0 \left.\frac{d\omega}{dk}\right|_0 - \omega_0\right)t\right] . \tag{2.62}$$

The exponential in (2.62) has a purely imaginary argument and therefore is only a phase factor. Therefore, the wave packet travels undistorted in shape except for an overall phase factor with the group velocity

$$\boxed{v_g = \frac{d\omega}{dk}} \; . \tag{2.63}$$

This is strictly true if the angular frequency is a linear function of k. If $\omega(k)$ is of a more general character, the group velocity will itself depend on the wave number, and the form of the wave packet made up of waves with a finite range of wavenumbers will be distorted in time. The pulse will disperse.

Whether in a medium the phase velocity (2.57) or the group velocity (2.63) is the greater of the two depends on the properties of the medium in which the wave propagates. Writing (2.57) as

$$\omega = k v \; ,$$

one finds

$$\frac{d\omega}{dk} = v_g = v + k\frac{dv}{dk} \; . \tag{2.64}$$

Remembering the definition of the index of refraction (2.38)

$$n = \frac{c}{v}$$

and realizing that the wavelength is given by

$$\lambda = \frac{2\pi}{k} \; , \tag{2.65}$$

we see that normal dispersion $dn/d\lambda < 0$ in the medium corresponds to $dv/dk < 0$. In a medium with normal dispersion therefore $v_g < v$. Only for anomalous dispersion will we have $v_g > v$.

Energy and information are usually propagated with the group velocity. The situation is, however, fairly complicated if propagations in dispersive media are considered. These problems have been investigated by Sommerfeld (1914) and Brillouin (1914). Details can be found in Sommerfeld (1959).

2.7 Plane Waves in Dissipative Media

In Sect. 2.5 the propagation properties of plane harmonic waves in an *isolating* ($\sigma = 0$) medium have been investigated. Now this assumption will be dropped so that $\sigma \neq 0$, but we still restrict the investigation to strictly harmonic waves propagating in the direction of increasing x

$$\boldsymbol{E}(x,t) = \boldsymbol{E}_0\, \mathrm{e}^{\mathrm{i}(kx - \omega t)} \; . \tag{2.66}$$

Both $\boldsymbol{E}_0$ and k are complex constants. Making use of (2.40) to (2.43), the wave equations (2.29) and (2.30) become

$$\boxed{[k^2 - (\varepsilon\,\mu\,\omega^2 + \mathrm{i}\,\sigma\,\mu\,\omega)]\begin{Bmatrix}\boldsymbol{E}\\ \boldsymbol{H}\end{Bmatrix} = 0} \; . \tag{2.67}$$

If these equations are to be valid for arbitrary $\boldsymbol{E}$ or $\boldsymbol{H}$ [of the form (2.66)] the square bracket must be zero, so that the *dispersion equation* becomes

$$\boxed{k^2 = \mu\,\varepsilon\,\omega^2\left(1 + \mathrm{i}\,\frac{\sigma}{\omega\,\varepsilon}\right)} \; . \tag{2.68}$$

The wave number k thus is indeed a complex number. Writing

$$k = \alpha + \mathrm{i}\,\beta \; , \tag{2.69}$$

we find

$$\alpha = \omega\sqrt{\varepsilon\,\mu}\,\sqrt{\frac{1}{2}\left(\sqrt{1 + \left(\frac{\sigma}{\varepsilon\,\omega}\right)^2} + 1\right)} \tag{2.70}$$

$$\beta = \omega\sqrt{\varepsilon\,\mu}\,\sqrt{\frac{1}{2}\left(\sqrt{1 + \left(\frac{\sigma}{\varepsilon\,\omega}\right)^2} - 1\right)} \tag{2.71}$$

and the field therefore can be written

$$\boldsymbol{E}(x,t) = \boldsymbol{E}_0\, \mathrm{e}^{-\beta x}\, \mathrm{e}^{\mathrm{i}(\alpha x - \omega t)} \; . \tag{2.72}$$

Thus the conductivity gives rise to an exponential damping of the wave. If Eq. (2.72) is written using the index of refraction n and the absorption coefficient κ,

$$\boxed{\boldsymbol{E}(x,t) = \boldsymbol{E}_0 \exp\left(-\frac{\omega}{c}\, n\kappa x\right) \exp\left[\mathrm{i}\,\omega\left(\frac{n}{c}x - t\right)\right]} \quad , \tag{2.73}$$

we obtain

$$\boxed{n\kappa = c\sqrt{\varepsilon\mu}\,\sqrt{\frac{1}{2}\left(\sqrt{1+\left(\frac{\sigma}{\varepsilon\,\omega}\right)^2} - 1\right)}} \tag{2.74}$$

$$\boxed{n = c\sqrt{\varepsilon\mu}\,\sqrt{\frac{1}{2}\left(\sqrt{1-\left(\frac{\sigma}{\varepsilon\,\omega}\right)^2} + 1\right)}} \quad . \tag{2.75}$$

2.8 The Dispersion Measure of a Tenuous Plasma

The simplest model for a dissipative medium is that of a tenuous plasma where free electrons and ions are evenly mixed so that the total space charge density is zero. This model was first given by Drude (1900) to explain the propagation of ultraviolet light in a transparent medium, but has later found application in the propagation of transversal electromagnetic radio waves in a tenuous plasma.

The free electrons are accelerated by the electric field intensity; their equation of motion is

$$m_\mathrm{e}\,\dot{\boldsymbol{v}} = m_\mathrm{e}\,\ddot{\boldsymbol{r}} = -\,e\,\boldsymbol{E}_0\,\mathrm{e}^{-\mathrm{i}\omega t} \tag{2.76}$$

with the solution

$$\boldsymbol{v} = \frac{e}{\mathrm{i}\,m_\mathrm{e}\,\omega}\,\boldsymbol{E}_0\,\mathrm{e}^{-\mathrm{i}\omega t} = -\,\mathrm{i}\,\frac{e}{m_\mathrm{e}\,\omega}\,\boldsymbol{E} \ . \tag{2.77}$$

Equation (2.77) describes the motion of the electrons. Moving electrons, however, carry a current, whose current density is

$$\boldsymbol{J} = -\sum_\alpha e\,\boldsymbol{v}_\alpha = -\,N\,e\,\boldsymbol{v} = \mathrm{i}\,\frac{N\,e^2}{m_\mathrm{e}\,\omega}\,\boldsymbol{E} = \sigma\,\boldsymbol{E} \ . \tag{2.78}$$

This expression explains why the ions can be neglected in this investigation. Due to their large mass ($m_\mathrm{i} \gtrsim 2 \cdot 10^3\, m_\mathrm{e}$), the induced ion velocity (2.77) is smaller than that of the electrons by the same factor, and since the charge of the ions is the same as that of the electrons, the ion current (2.78) will be smaller than the electron-current by the same factor.

According to (2.78) the conductivity of the plasma is purely imaginary:

$$\sigma = \mathrm{i}\,\frac{N\,e^2}{m_\mathrm{e}\,\omega}\ . \tag{2.79}$$

Inserting this into (2.68) we obtain, for a thin medium with $\varepsilon \approx \varepsilon_0$ and $\mu \approx \mu_0$

$$\boxed{k^2 = \frac{\omega^2}{c^2}\left(1 - \frac{\omega_\mathrm{p}^2}{\omega^2}\right)}\ , \quad \text{where} \tag{2.80}$$

$$\boxed{\omega_\mathrm{p}^2 = \frac{N\,e^2}{\varepsilon_0\,m_\mathrm{e}}} \tag{2.81}$$

is the *plasma frequency*. It gives a measure of the mobility of the electron gas. Inserting numerical values we obtain

$$\frac{\nu_\mathrm{p}}{\mathrm{Hz}} = 8.97\sqrt{\frac{N}{\mathrm{m}^{-3}}} \tag{2.82}$$

if we convert (2.81) to frequencies by $\nu = \omega/2\,\pi$. In plasma physics the CGS system of units is usually used, and the expression for the plasma frequency then becomes

$$\omega_\mathrm{p}^2 = \frac{4\,\pi\,N\,e^2}{m_\mathrm{e}} \quad \text{[CGS units]}\ . \tag{2.81 a}$$

For $\omega > \omega_\mathrm{p}$, k is real, and we obtain from (2.57)

$$\boxed{v = \frac{c}{\sqrt{1 - \dfrac{\omega_\mathrm{p}^2}{\omega^2}}}} \tag{2.83}$$

for the *phase velocity* v and so $v > c$ for $\omega > \omega_\mathrm{p}$. *For the group velocity* follows from (2.63)

$$v_\mathrm{g} = \frac{d\omega}{dk} = \frac{1}{dk/d\omega} \quad \text{so that}$$

$$\boxed{v_\mathrm{g} = c\sqrt{1 - \frac{\omega_\mathrm{p}^2}{\omega^2}}} \tag{2.84}$$

and $v_\mathrm{g} < c$ for $\omega > \omega_\mathrm{p}$. Both v and v_g thus depend on the frequency ω. For $\omega = \omega_\mathrm{p}$, $v_\mathrm{g} = 0$, and for waves with a frequency lower than ω_p no wave propagation in the plasma is possible. The frequency-dependence of v and v_g are in the opposite sense, and taking (2.83) and (2.84) together the relation

$$\boxed{v\, v_{\mathrm{g}} = c^2} \tag{2.85}$$

is obtained.

For some applications the *index of refraction* is a useful quantity. According to (2.38) and (2.80) it is

$$\boxed{n = \sqrt{1 - \frac{\omega_{\mathrm{p}}^2}{\omega^2}}}\;, \tag{2.86}$$

thus giving a direct physical meaning to this expression.

The fact that electromagnetic pulses propagate with the group-velocity and that this velocity varies with the frequency so that there is a dispersion in the pulse propagation in a plasma became of fundamental importance when the radio-pulses from pulsars were detected in 1967. The arrival time of these pulses depends on the frequency in which they were measured in the sense that, the lower the observing frequency the later the pulse arrives. This behaviour can easily be explained as being due to wave propagation in a tenuous plasma.

The plasma frequency of the interstellar medium is much lower than the observing frequency. With N typically in the range $10-10^3\ \mathrm{m}^{-3}$, ν_{p} is in the range 28–285 Hz while the observing frequency is $\nu > 10$ MHz in order to be able to propagate through the ionosphere of the earth. For v_{g} therefore the series expansion of (2.84)

$$\frac{1}{v_{\mathrm{g}}} = \frac{1}{c}\left(1 + \frac{1}{2}\frac{\nu_{\mathrm{p}}^2}{\nu^2}\right) \tag{2.87}$$

is valid with high precision. A pulse emitted by a pulsar at the distance L therefore will be received after a delay

$$\tau_{\mathrm{D}} = \int_0^L \frac{dl}{v_{\mathrm{g}}} \cong \frac{1}{c}\int_0^L \left(1 + \frac{1}{2}\left(\frac{\nu_{\mathrm{p}}}{\nu}\right)^2\right) dl = \frac{1}{c}\int_0^L \left(1 + \frac{e^2}{8\pi^2 \varepsilon_0 m_{\mathrm{e}}}\frac{1}{\nu^2} N(l)\right) dl$$

$$= \frac{L}{c} + \frac{e^2}{8\pi^2 c \varepsilon_0 \mathrm{m_e}}\frac{1}{\nu^2}\int_0^L N(l)\, dl\ .$$

The difference of the pulse arrival times measured at two frequencies ν_1 and ν_2 therefore is given by

$$\boxed{\Delta\tau_{\mathrm{D}} = \frac{e^2}{8\pi^2 c \varepsilon_0 \mathrm{m_e}}\left[\frac{1}{\nu_1^2} - \frac{1}{\nu_2^2}\right]\int_0^L N(l)\, dl}\ . \tag{2.88}$$

The quantity $\int_0^L N(l)\, dl$ is the column-density of the electrons in the intervening space between pulsar and observer. Its dimension in SI units is obviously m^{-2}.

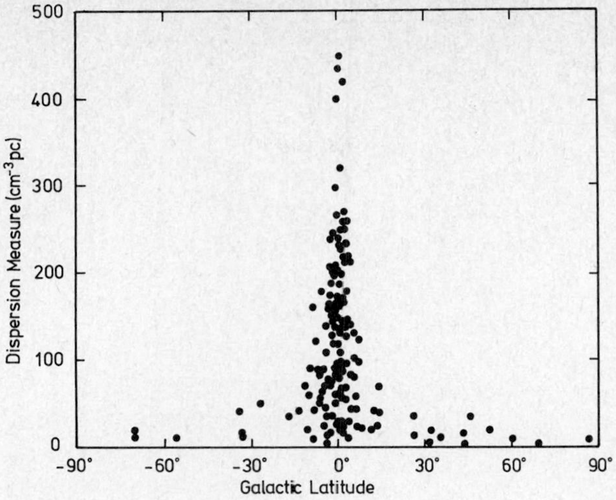

Fig. 2.3. Dispersion measure for pulsars at different galactic latitudes [adapted from Manchester and Taylor (1977)]

But since distances in astronomy are measured in parsecs (1 pc $= 3.085677 \cdot 10^{16}$ m), it has become customary to measure $N(l)$ in cm^{-3} but dl in pc. The integral then is called the *dispersion measure* (Fig. 2.3)

$$\mathrm{DM} = \int_0^L \left(\frac{N}{\mathrm{cm}^{-3}}\right) d\left(\frac{l}{\mathrm{pc}}\right) = 3.2408 \cdot 10^{-23} \int_0^L N\, dl \;[\mathrm{SI}] \tag{2.89}$$

and

$$\boxed{\frac{\mathrm{DM}}{\mathrm{cm}^{-3}\,\mathrm{pc}} = 2.410 \cdot 10^{-4} \left(\frac{\Delta\tau_{\mathrm{D}}}{\mathrm{s}}\right) \left[\frac{1}{\left(\frac{\nu_1}{\mathrm{MHz}}\right)^2} - \frac{1}{\left(\frac{\nu_2}{\mathrm{MHz}}\right)^2}\right]^{-1}} \;. \tag{2.90}$$

Both the time delay $\Delta\tau_{\mathrm{D}}$ and the observing frequencies ν_1 and ν_2 can be measured with high precision resulting in a highly accurate value of DM for a given pulsar. Provided the distance L to the pulsar is known this then results in a measure for the average electron density between observer and pulsar. But since L is usually known only very approximately, only rough guesses for N are obtained in this way. Quite often the procedure is even turned around. From reasonable guesses of N, a measured DM provides information on the unknown distance L of the pulsar.

3. Wave Polarization

3.1 Vector Waves

In the preceding chapter we have shown that plane electromagnetic waves in a dielectric medium are transversal and that the x and the y component of both $\boldsymbol{E}$ and $\boldsymbol{H}$ for a wave propagating in the z-direction obey the same wave equation. For the sake of simplicity, we have investigated the propagation of only one component of these fields.

In general both the x- and the y-component have to be specified but, in a strictly monochromatic wave, they are not independent, since both share the same harmonic dependence, albeit with a different phase term:

$$\begin{aligned} E_x &= E_1 \cos(k z - \omega t + \delta_1) \\ E_y &= E_2 \cos(k z - \omega t + \delta_2) \\ E_z &= 0 \ . \end{aligned} \tag{3.1}$$

Regarding (E_x, E_y, z) as the coordinates of a point in a rectangular coordinate system we realize that (3.1) describes a helical path on the surface of a cylinder. The cross section of this cylinder can be determined by eliminating the phase of this wave, abbreviated by

$$\tau = k z - \omega t \ . \tag{3.2}$$

Rewriting the first two equations of (3.1) as

$$\begin{aligned} \frac{E_x}{E_1} &= \cos\tau \cos\delta_1 - \sin\tau \sin\delta_1 \\ \frac{E_y}{E_2} &= \cos\tau \cos\delta_2 - \sin\tau \sin\delta_2 \end{aligned} \tag{3.3}$$

gives

$$\frac{E_x}{E_1} \sin\delta_2 - \frac{E_y}{E_2} \sin\delta_1 = \cos\tau \sin(\delta_2 - \delta_1)$$

$$\frac{E_x}{E_1} \cos\delta_2 - \frac{E_y}{E_2} \cos\delta_1 = \sin\tau \sin(\delta_2 - \delta_1) \ .$$

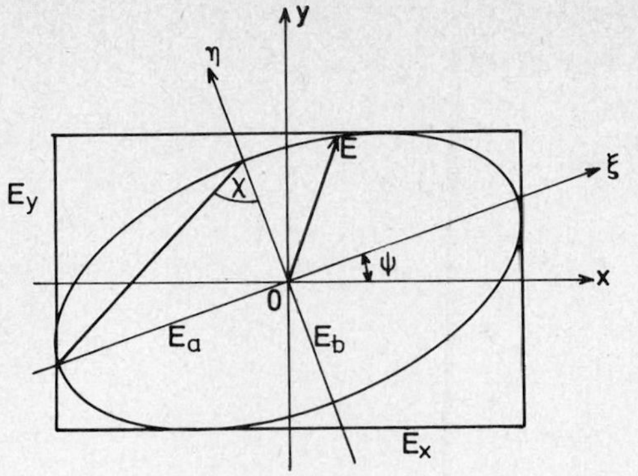

Fig. 3.1. The polarization ellipse for the electric vector of an elliptically polarized wave

By squaring and adding we obtain

$$\left(\frac{E_x}{E_1}\right)^2 + \left(\frac{E_y}{E_2}\right)^2 - 2\frac{E_x}{E_1}\frac{E_y}{E_2}\cos\delta = \sin^2\delta \tag{3.4}$$

$$\delta = \delta_1 - \delta_2 \quad . \tag{3.5}$$

This is the equation of an ellipse, since the discriminant is not negative

$$\begin{vmatrix} \dfrac{1}{E_1^2} & -\dfrac{\cos\delta}{E_1 E_2} \\ -\dfrac{\cos\delta}{E_1 E_2} & \dfrac{1}{E_2^2} \end{vmatrix} = \frac{1-\cos^2\delta}{E_1^2 E_2^2} = \frac{\sin^2\delta}{E_1^2 E_2^2} \geqq 0 \; . \tag{3.6}$$

The wave therefore is said to be elliptically polarized, and this applies to both the electric and the magnetic field of the wave; $\sin\delta$ determines the sense in which the electric vector rotates.

The ellipse (3.4) usually is not oriented in any preferential way with respect to the coordinate system. Its geometric properties are seen best by selecting a coordinate system oriented along the major and minor axes (Fig. 3.1). In this system the ellipse equation is

$$\begin{aligned} E_\xi &= E_a \cos(\tau + \delta) \\ E_\eta &= E_b \sin(\tau + \delta) \; , \end{aligned} \tag{3.7}$$

and the relation between the coordinate systems (x, y) and (ξ, η) is given by the linear transformation

$$\begin{aligned} E_\xi &= E_x \cos\psi + E_y \sin\psi \\ E_\eta &= -E_x \sin\psi + E_y \cos\psi \; . \end{aligned} \tag{3.8}$$

The intrinsic parameters of the polarization ellipse E_a and E_b, as well as the angle ψ by which the major axis is tilted with respect to the x-axis, can then be determined by requiring that (3.4) transformed by (3.8) should lead to (3.7). Substituting (3.3) and (3.7) into (3.8) while simultaneously expanding the $\cos(\tau + \delta)$ term leads to

$$E_a(\cos\tau\cos\delta - \sin\tau\sin\delta) = E_1(\cos\tau\cos\delta_1 - \sin\tau\sin\delta_1)\cos\psi + E_2(\cos\tau\cos\delta_2 - \sin\tau\sin\delta_2)\sin\psi \tag{3.9}$$

and

$$E_b(\sin\tau\cos\delta + \cos\tau\sin\delta) = -E_1(\cos\tau\cos\delta_1 - \sin\tau\sin\delta_1)\sin\psi + E_2(\cos\tau\cos\delta_2 - \sin\tau\sin\delta_2)\cos\psi\ . \tag{3.10}$$

These equation are valid for all τ, i.e. also for $\tau = 0$ and $\tau = \frac{\pi}{2}$, resulting in

$$E_a\cos\delta = E_1\cos\delta_1\cos\psi + E_2\cos\delta_2\sin\psi \tag{3.11}$$

$$-E_a\sin\delta = -E_1\sin\delta_1\cos\psi - E_2\sin\delta_2\sin\psi \tag{3.12}$$

$$E_b\cos\delta = E_1\sin\delta_1\sin\psi - E_2\sin\delta_2\cos\psi \tag{3.13}$$

$$E_b\sin\delta = -E_1\cos\delta_1\sin\psi + E_2\cos\delta_2\cos\psi\ . \tag{3.14}$$

Squaring these equations and adding we obtain

$$\boxed{S_0 \equiv E_a^2 + E_b^2 = E_1^2 + E_2^2}\ . \tag{3.15}$$

This can be interpreted remembering (2.55) that the total Poynting flux of the polarized wave is equal to the sum of the fluxes of two orthogonal, but otherwise arbitrary directions.

Multiplying (3.11) by (3.13) and (3.12) by (3.14) and subtracting the results, we obtain

$$E_a E_b = E_1 E_2 \sin\delta \tag{3.16}$$

while division and addition of the same pairs of equations result in

$$-(E_1^2 - E_2^2)\sin\psi\cos\psi = E_1 E_2\cos\delta(\sin^2\psi - \cos^2\psi)$$

$$(E_1^2 - E_2^2)\sin 2\psi = 2E_1 E_2\cos\delta\cos 2\psi\ . \tag{3.17}$$

If we now define α by

$$\boxed{\frac{E_1}{E_2} = \tan\alpha}\ , \tag{3.18}$$

(3.17) can be rewritten

$$\tan 2\psi = \frac{2E_1E_2}{E_1^2 - E_2^2}\cos\delta = -\frac{2\tan\alpha}{1-\tan^2\alpha}\cos\delta$$

or

$$\boxed{\tan 2\psi = -\tan 2\alpha \cos\delta} \quad . \tag{3.19}$$

Dividing (3.16) by (3.15) results in

$$\frac{2E_a E_b}{E_a^2 + E_b^2} = \frac{2E_1 E_2}{E_1^2 + E_2^2} \sin\delta \ .$$

Defining

$$\boxed{\frac{E_a}{E_b} = \tan\chi} \quad , \tag{3.20}$$

(3.19) is equivalent to

$$\boxed{\sin 2\chi = \sin 2\alpha \sin\delta} \quad . \tag{3.21}$$

Equations (3.15, 3.18, 3.19, 3.20 and 3.21) now permit the computation of all intrinsic polarization properties of the elliptically polarized wave from its specifications in an arbitrary coordinate system. Given E_1, E_2 and δ (3.15) gives S_0, (3.19) with (3.18) allows the determination of the angle ψ, while χ is determined from (3.21). E_a and E_b are then computed from (3.20) and (3.15).

The phase-difference δ is important in several respects. Its sign determines the sense in which the wave-vector is rotating. If $\sin\delta > 0$ or equivalently $\tan\chi > 0$, the polarization is called *right-handed*; conversely $\sin\delta < 0$ or $\tan\chi < 0$ describes *left-handed* elliptical polarization. For right-handed polarization, the rotation of the $\boldsymbol{E}$-vector and the direction of propagation form a right-handed screw.

This is the convention adopted generally in microwave physics and modern physical optics. It is convenient too that, according to this definition, right-handed helical-beam antennas radiate or receive right-circular polarization.

Traditional optics used a different definition resulting in just the opposite sense of rotation based on the apparent behaviour of $\boldsymbol{E}$ when "viewed" face-on by the observer. Here we will conform to the modern definition. Therefore care should be taken when comparing some of the results to older texts.

If the phase difference is

$$\delta = \delta_1 - \delta_2 = m\pi \ , \qquad m = 0, \pm 1, \pm 2 \ldots \tag{3.22}$$

the polarization ellipse degenerates into a straight line and $\boldsymbol{E}$ is *linearly polarized*. As we have seen, an elliptically polarized wave can be regarded as the superposition of two orthogonal linearly polarized waves.

Another important special case is that of a *circularly polarized* wave. For this

$$E_1 = E_2 = E \quad \text{and} \tag{3.23}$$

$$\delta = \frac{m}{2}\pi \ , \qquad m = \pm 1, \pm 2 \ldots \tag{3.24}$$

so that (3.4) reduces to the equation of a circle

$$E_x^2 + E_y^2 = E^2 \tag{3.25}$$

with the orthogonal linear components

$$\begin{aligned} E_x &= E\cos\tau \\ E_y &= \pm\, E\cos\left(\tau - \tfrac{\pi}{2}\right) . \end{aligned} \tag{3.26}$$

From this we see that an arbitrary elliptically polarized wave can be decomposed into the sum of two circularly polarized ones, because (3.7) can be written as

$$\begin{aligned} E_\xi &= E_a\cos(\tau+\delta) = (E_r + E_l)\cos(\tau+\delta) \\ E_\eta &= E_b\sin(\tau+\delta) = (E_r - E_l)\cos\left(\tau+\delta-\tfrac{\pi}{2}\right) . \end{aligned}$$

Solving for E_r and E_l we find that

$$\begin{aligned} E_r &= \tfrac{1}{2}(E_a + E_b) \\ E_l &= \tfrac{1}{2}(E_a - E_b) \end{aligned} \tag{3.27}$$

and, for the total Poynting flux of the wave, we obtain

$$\boxed{S_0 = E_a^2 + E_b^2 = E_r^2 + E_l^2} \;. \tag{3.28}$$

3.2 The Poincaré Sphere and the Stokes Parameters

The results of the preceding paragraph show that three independent parameters are needed to describe the state of the polarization of a monochromatic vector wave. Various sets for these parameters have been introduced:

i) the amplitudes E_1, E_2 and the relative phase δ of two orthogonal, linearly polarized waves;
ii) the amplitudes E_r and E_l and the relative phase δ of one right- and one left-hand circularly polarized wave;
iii) the major and minor axis E_a, E_b and the position angle ψ of the polarization ellipse.

Poincaré (1892) introduced another representation that permits an easy visualization of all the different states of polarization of a vector wave. If we interpret the angles 2ψ of (3.19) and 2χ of (3.21) as longitude and latitude of a sphere with the radius S_0 of (3.15) there is a one-to-one relation between polarization states and points on the sphere (Fig. 3.2). The equator represents linear polarization; the

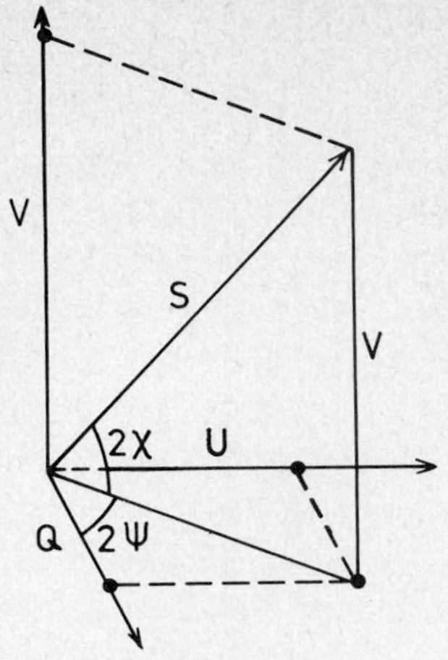

Fig. 3.2
Definition of the Stokes parameters

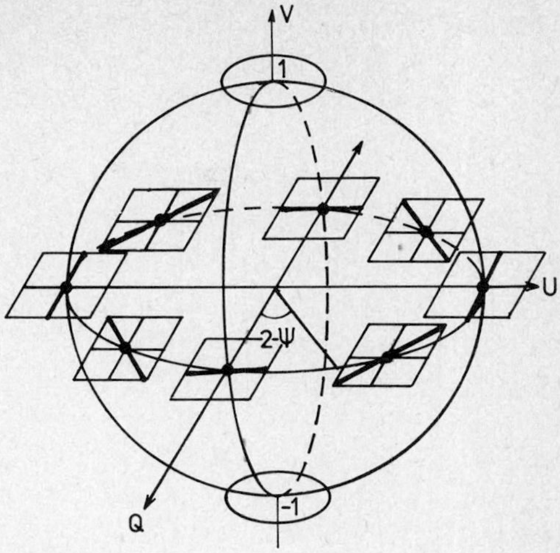

Fig. 3.3
Polarization and the Poincaré sphere

north pole corresponds to right-circular and the south pole to left-circular polarization (Fig. 3.3).

From the Poincaré sphere, it follows naturally to the Stokes parameters (1852). They are simply the cartesian coordinates of the points on the sphere with the obvious definitions:

$$\boxed{\begin{aligned} S_0 &= I = E_a^2 + E_b^2 \\ S_1 &= Q = S_0 \cos 2\chi \cos 2\psi \\ S_2 &= U = S_0 \cos 2\chi \sin 2\psi \\ S_3 &= V = S_0 \sin 2\chi \end{aligned}} \quad . \tag{3.29}$$

Only three of these parameters are independent, since according to the construction of the Poincaré sphere

$$\boxed{\begin{aligned} S_0^2 &= S_1^2 + S_2^2 + S_3^2 \\ I^2 &= Q^2 + U^2 + V^2 \end{aligned}} \quad . \tag{3.30}$$

The Stokes parameters can also be directly expressed by the parameters of the polarization ellipse (3.4). To do this we derive from (3.18)

$$\tan 2\alpha = \frac{2\tan\alpha}{1 - \tan^2\alpha} = -\frac{2E_1E_2}{E_1^2 - E_2^2} \tag{3.31}$$

$$\cos 2\alpha = \frac{1}{\sqrt{1+\tan^2 2\alpha}} = -\frac{E_1^2 - E_2^2}{E_1^2 + E_2^2} \tag{3.32}$$

$$\sin 2\alpha = \frac{2E_1E_2}{E_1^2 + E_2^2} \ . \tag{3.33}$$

Then from (3.21), using (3.33) and (3.15),

$$\sin 2\chi = \frac{2E_1E_2}{E_1^2 + E_2^2}\sin\delta = \frac{2E_1E_2}{I}\sin\delta \tag{3.34}$$

$$\cos 2\chi = \frac{1}{I}\sqrt{I^2 - (2E_1E_2)^2\sin^2\delta} \ . \tag{3.35}$$

Then, from (3.19) with (3.31),

$$\tan 2\psi = \frac{2E_1E_2}{E_1^2 - E_2^2}\cos\delta \quad \text{and} \tag{3.36}$$

$$\cos 2\psi = \frac{1}{\sqrt{1+\tan^2 2\psi}} = \frac{E_1^2 - E_2^2}{\sqrt{I^2 - (2E_1E_2)^2\sin^2\delta}} \tag{3.37}$$

$$\sin 2\psi = \frac{2E_1E_2\cos\delta}{\sqrt{I^2 - (2E_1E_2)^2\sin^2\delta}} \ . \tag{3.38}$$

Substituting (3.34), (3.35) and (3.37), (3.38) into (3.29) we then obtain the desired result

$$\boxed{\begin{aligned} S_0 &= I = E_1^2 + E_2^2 \\ S_1 &= Q = E_1^2 - E_2^2 \\ S_2 &= U = 2E_1E_2\cos\delta \\ S_3 &= V = 2E_1E_2\sin\delta \end{aligned}} \ . \tag{3.39}$$

These equations permit the expression of the Stokes parameters directly in terms of observable quantities.

A few special cases will illustrate the use.

1) For a right-handed circularly polarized wave we have $E_1 = E_2$ and $\delta = \frac{\pi}{2}$, so that

$$\begin{aligned} S_0 &= I = S \\ S_1 &= Q = 0 \\ S_2 &= U = 0 \\ S_3 &= V = S \ . \end{aligned}$$

2) For a left-handed circularly polarized wave we have

$$S_0 = I = S$$
$$S_1 = Q = 0$$
$$S_2 = U = 0$$
$$S_3 = V = -S \ .$$

3) For a linearly polarized wave we have $E_b = E$ and $E_a = 0$, so that $\chi = 0$ and

$$S_0 = I = E = S$$
$$S_1 = Q = I \cos 2\psi$$
$$S_2 = U = I \sin 2\psi$$
$$S_3 = V = 0 \ .$$

Finally, a fact that has so far been implied in the developments of this chapter but not stated explicitely should be noted: A strictly monochromatic wave is always polarized; there is no such thing as an unpolarized monochromatic wave. This becomes immediately evident if we remember that, in a monochromatic plane harmonic wave, E_1, E_2, δ_1 and δ_2 are strictly constants.

This situation is different when we turn to quasi-monochromatic radiation, in which ω is restricted to some small but finite bandwidth. Radiation of this kind can be unpolarized or partially polarized. But to realize this, a convenient way to describe such radiation has first to be developed. This will be done in the next section.

3.3 Quasi-Monochromatic Plane Waves

The description of the polarization properties of electromagnetic waves as given until now applies only to strictly monochromatic waves. The problem now is, how to modify these to be valid for signals with a finite bandwidth.

Both the electric and the magnetic field intensity of the wave at a given fixed position can then be expressed by an integral of the form

$$V^{(r)}(t) = \int_0^\infty a(\nu) \cos[\phi(\nu) - 2\pi\nu t] d\nu \ . \tag{3.40}$$

This integral has precisely the form of a Fourier integral, and therefore it is convenient to associate with $V^{(r)}$ the complex function

$$\boxed{V(t) = \int_0^\infty a(\nu) e^{i[\phi(\nu) - 2\pi\nu t]} d\nu} \tag{3.41}$$

where

$$V(t) = V^{(r)}(t) + \mathrm{i}\, V^{(i)}(t) \tag{3.42}$$

$$V^{(i)}(t) = \int_0^\infty a(\nu)\sin[\phi(\nu) - 2\pi\nu t]\,d\nu \ . \tag{3.43}$$

$V^{(i)}$ does not contain any information that is not already contained in $V^{(r)}$ and V is called the *analytic signal* belonging to $V^{(r)}$. The integral in (3.41) formally extends over an infinite range in frequency, but frequently $a(\nu)$ has such a shape that this range is effectively limited to an interval $\Delta\nu$ which is small compared with its mean frequency $\bar{\nu}$; i.e.,

$$\Delta\nu/\bar{\nu} \ll 1 \ . \tag{3.44}$$

If this condition is fulfilled, the signal is said to be *quasi-monochromatic*. If we express V in the form

$$V(t) = A(t)\,\mathrm{e}^{\mathrm{i}[\Phi(t) - 2\pi\bar{\nu}t]} \ , \qquad \text{then} \tag{3.45}$$

$$A(t)\,\mathrm{e}^{\mathrm{i}\Phi(t)} = \int_0^\infty a(\nu)\,\mathrm{e}^{\mathrm{i}[\phi(\nu) - 2\pi(\nu-\bar{\nu})t]}\,d\nu$$

$$= \int_{-\bar{\nu}}^\infty a(\bar{\nu}+\mu)\,\mathrm{e}^{\mathrm{i}[\phi(\bar{\nu}+\mu) - 2\pi\mu t]}\,d\mu \tag{3.46}$$

where

$$\mu = \nu - \bar{\nu} \ .$$

$a(\bar{\nu}+\mu)$ will differ significantly from zero only for small μ. But then (3.46) is the superposition of harmonic components of low frequencies and both $A(t)$ and $\Phi(t)$ will vary slowly compared with $\exp(\mathrm{i}\,2\pi\nu t)$.

In terms of the analytic signal, the envelope $A(t)$ and the phase factor $\Phi(t)$ are given by

$$A(t) = \sqrt{V^{(r)2} + V^{(i)2}} = \sqrt{VV^*} = |V| \tag{3.47}$$

and

$$\Phi(t) = 2\pi\bar{\nu}t + \arctan\frac{V^{(i)}}{V^{(r)}} = 2\pi\bar{\nu}t + \arctan\left(\mathrm{i}\,\frac{V^* - V}{V^* + V}\right) \ . \tag{3.48}$$

A is independent of the mean frequency $\bar{\nu}$ and Φ depends on it only through the additive term $2\pi\bar{\nu}t$.

If the bandwidth $\Delta\nu$ of the signal is small, $A(t)$ will only vary slowly with t. But even this variation is often too fast to be directly measured; all that is really needed, is some kind of time average. Such an average will be denoted by sharp brackets

$$\langle F(t)\rangle = \lim_{T\to\infty}\frac{1}{2T}\int_{-T}^{T} F(t)\,dt \tag{3.49}$$

so that

$$\langle A^2(t)\rangle = \langle VV^*\rangle = \lim_{T\to\infty}\frac{1}{2T}\int_{-T}^{T} V(t)\,V^*(t)\,dt\ . \tag{3.50}$$

If we require that $\langle A^2\rangle$ should have a finite value, then obviously $\int_{-\infty}^{\infty} VV^*\,dt$ diverges. But according to Wiener (1949), the techniques of Fourier analysis can be extended to such cases of generalized harmonic analysis; therefore we will assume that time averaged values for A can be computed from (3.46) (see also Sect. 7.1).

3.4 The Coherency Matrix

The observable intensity of a wave is given by its time averaged Poynting flux which is, apart from a constant that is of no importance in this connection given by

$$I(P) = \langle V(P,t)\,V^*(P,t)\rangle\ . \tag{3.51}$$

Let us now consider a quasi-monochromatic wave of frequency $\bar{\nu}$ propagating in the z-direction:

$$E_x(t) = a_1(t)\,\mathrm{e}^{\mathrm{i}(\phi_1(t)-2\pi\bar{\nu}t)}\ , \qquad E_y(t) = a_2(t)\,\mathrm{e}^{\mathrm{i}(\phi_2(t)-2\pi\bar{\nu}t)} \tag{3.52}$$

where E_x and E_y are the "analytic signals" associated with the components $E_x^{(r)}(t) = a_1(t)\cos[\phi_1(t) - 2\pi\bar{\nu}t]$ and $E_y^{(r)}(t) = a_2(t)\cos[\phi_2(t) - 2\pi\bar{\nu}t]$. Now if the y-component is retarded in phase by ε relative to the x-component, then the electric vector in the θ-direction is

$$E(t;\theta,\varepsilon) = E_x\cos\theta + E_y\,\mathrm{e}^{\mathrm{i}\varepsilon}\sin\theta \tag{3.53}$$

and the intensity in this polarization angle is

$$\begin{aligned} I(\theta,\varepsilon) &= \langle E(t;\theta,\varepsilon)\,E^*(t;\theta,\varepsilon)\rangle \\ &= J_{xx}\cos^2\theta + J_{yy}\sin^2\theta + J_{xy}\,\mathrm{e}^{-\mathrm{i}\varepsilon}\cos\theta\sin\theta + J_{yx}\,\mathrm{e}^{\mathrm{i}\varepsilon}\sin\theta\cos\theta \end{aligned} \tag{3.54}$$

where the $J_{xx}\ldots$ are matrix elements. This matrix is different from the current density $\boldsymbol{J}$ of Chap. 2. This lack of uniqueness for the designations is unfortunate, but cannot be completely avoided if we want to stick to the commonly used designations.

$$\boldsymbol{J} = \begin{bmatrix} \langle E_x E_x^*\rangle & \langle E_x E_y^*\rangle \\ \langle E_y E_x^*\rangle & \langle E_y E_y^*\rangle \end{bmatrix} = \begin{bmatrix} \langle a_1^2\rangle & \langle a_1 a_2\,\mathrm{e}^{\mathrm{i}(\varphi_1-\varphi_2)}\rangle \\ \langle a_1 a_2\,\mathrm{e}^{-\mathrm{i}(\varphi_1-\varphi_2)}\rangle & \langle a_2^2\rangle \end{bmatrix}\ . \tag{3.55}$$

This matrix is called the *coherency matrix* of the quasi-monochromatic wave. The elements on the the main diagonal are real, those off the diagonal are usually complex but conjugates, so that the matrix is *Hermitian*.

Using Schwarz's inequality we can derive

$$|J_{xy}|^2 \leqq |J_{xx}||J_{yy}| \quad \text{so that}$$

$$|\boldsymbol{J}| = J_{xx} J_{yy} - J_{xy} J_{yx} \geqq 0 \ . \tag{3.56}$$

For a given quasi-monochromatic wave the coherency matrix can be expressed by a set of intensities at selected angles θ and retardation ε. From (3.54) we derive

$$\begin{aligned} J_{xx} &= I(0°, 0) \\ J_{yy} &= I(90°, 0) \\ J_{xy} &= \tfrac{1}{2}\{I(45°,0) - I(135°,0\} + \tfrac{1}{2}\mathrm{i}\{I(45,\tfrac{\pi}{2}) - I(135°,\tfrac{\pi}{2})\} \\ J_{yx} &= \tfrac{1}{2}\{I(45°,0) - I(135°,0\} - \tfrac{1}{2}\mathrm{i}\{I(45°,\tfrac{\pi}{2}) - I(135°,\tfrac{\pi}{2})\} \ . \end{aligned} \tag{3.57}$$

To gain some feeling for the significance of the different terms in the coherency matrix, some wavefields of particular interest will now be discussed.

1. Unpolarized Light

The intensity is independent of both the polarization angle and a retardation ε. From (3.54) we then see that

$$\boldsymbol{J} = \tfrac{1}{2} I \begin{bmatrix} 1 & 0 \\ 0 & 1 \end{bmatrix} . \tag{3.58}$$

2. Fully Polarized Monochromatic Waves

In this case, a_1 and a_2 in (3.52) are independent of t, and thus

$$\boldsymbol{J} = \begin{bmatrix} a_1^2 & a_1 a_2 \mathrm{e}^{\mathrm{i}\delta} \\ a_1 a_2 \mathrm{e}^{-\mathrm{i}\delta} & a_2^2 \end{bmatrix} \quad \text{where} \tag{3.59}$$

$$\delta = \varphi_1 - \varphi_2 \tag{3.60}$$

so that

$$|\boldsymbol{J}| = 0 \ . \tag{3.61}$$

For linear polarization, $\delta = m\pi$ ($m = 0, \pm 1 \ldots$) so that in this case

$$\boldsymbol{J} = \begin{bmatrix} a_1^2 & (-1)^m a_1 a_2 \\ (-1)^m a_1 a_2 & a_2^2 \end{bmatrix} . \tag{3.62}$$

$$\boldsymbol{J} = I \begin{bmatrix} 1 & 0 \\ 0 & 0 \end{bmatrix} \quad \text{and} \quad \boldsymbol{J} = I \begin{bmatrix} 0 & 0 \\ 0 & 1 \end{bmatrix}$$

denote linear polarization in the x and y direction respectively, while

$$\tfrac{1}{2} I \begin{bmatrix} 1 & 1 \\ 1 & 1 \end{bmatrix} \quad \text{and} \quad \tfrac{1}{2} I \begin{bmatrix} 1 & -1 \\ -1 & 1 \end{bmatrix}$$

represent linear polarization at 45° and 135° respectively. Circularly polarized waves have $a_1 = a_2$ and $\varphi_1 - \varphi_2 = \pm\frac{\pi}{2}$ so that the coherency matrix is

$$\frac{1}{2} I \begin{bmatrix} 1 & \pm \mathrm{i} \\ \mp \mathrm{i} & 1 \end{bmatrix} . \tag{3.63}$$

If several independent vector waves which propagate in the same direction are superposed, the coherency matrix of the resulting wave is equal to the sum of the matrices of the individual waves. Let $E_x^{(n)}$ and $E_y^{(n)}$ be the x and y components of the wave field n. Then the resulting wave is

$$E_x = \sum_{n=1}^{N} E_x^{(n)} , \qquad E_y = \sum_{n=1}^{N} E_y^{(n)} ,$$

and the elements of the coherency matrix are

$$\begin{aligned} J_{kl} = \langle E_k E_l^* \rangle &= \sum_{n=1}^{N} \sum_{m=1}^{N} \langle E_k^{(n)} E_l^{(m)*} \rangle \\ &= \sum_{n=1}^{N} \langle E_k^{(n)} E_l^{(n)*} \rangle + \sum_{n \neq m} \langle E_k^{(n)} E_l^{(m)*} \rangle . \end{aligned}$$

Now, the individual waves were assumed to be independent; that is $\langle E_k^{(n)} E_l^{(m)} \rangle = 0$ for $n \neq m$ and therefore

$$J_{kl} = \sum_n J_{kl}^{(n)} \quad \text{with} \quad J_{kl}^{(n)} = \langle E_k^{(n)} E_l^{(n)*} \rangle . \tag{3.64}$$

Conversely, any wave can be regarded as the sum of independent waves. These "components" can obviously be chosen in many different ways. So partially polarized quasi-monochromatic waves can be regarded as the superposition of completely polarized and completely unpolarized waves. This decomposition is unique. Let

$$\boldsymbol{J} = \boldsymbol{J}^{(u)} + \boldsymbol{J}^{(p)}$$

where $\boldsymbol{J}^{(u)}$ is the coherence matrix of the unpolarized part and $\boldsymbol{J}^{(p)}$ that of the fully polarized part. Then

$$\boxed{\begin{bmatrix} J_{xx} & J_{xy} \\ J_{yx} & J_{yy} \end{bmatrix} = \begin{bmatrix} u & 0 \\ 0 & u \end{bmatrix} + \begin{bmatrix} p_{xx} & J_{xy} \\ J_{yx} & p_{yy} \end{bmatrix}} . \tag{3.65}$$

According to (3.61)

$$\begin{aligned} &p_{xx} p_{yy} - J_{xy} J_{yx} = 0 \quad \text{and} \\ &J_{xx} = u + p_{xx} , \qquad J_{yy} = u + p_{yy} \quad \text{so that} \\ &(J_{xx} - u)(J_{yy} - u) - J_{xy} J_{yx} = 0 \\ &u = \tfrac{1}{2} [(J_{xx} + J_{yy}) \pm \sqrt{(J_{xx} + J_{yy})^2 - 4 |\boldsymbol{J}|}] . \end{aligned} \tag{3.66}$$

It can be shown that u_+ leads to $p_{xx} < 0$ and $p_{yy} < 0$; the decomposition therefore is unique.

Such a decomposition could be used, for example, to define a degree of polarization. The necessary definition can be found in Born and Wolf (1959).

3.5 The Stokes Parameters for Quasi-Monochromatic Waves

The Stokes parameters of a quasi-monochromatic wave are straightforward generalizations of the expressions in (3.39). For the wavefield (3.52), they are

$$\boxed{\begin{aligned} S_0 &= I = \langle a_1^2 \rangle + \langle a_2^2 \rangle \\ S_1 &= Q = \langle a_1^2 \rangle - \langle a_2^2 \rangle \\ S_2 &= U = 2\,\langle a_1\, a_2 \cos\delta \rangle \\ S_3 &= V = 2\,\langle a_1\, a_2 \sin\delta \rangle \end{aligned}} \quad . \tag{3.67}$$

Comparing this with (3.55), we see that the Stokes parameters and the coherency matrix are closely related and are connected by the formulae

$$\begin{aligned} S_0 &= I = J_{xx} + J_{yy} \\ S_1 &= Q = J_{xx} - J_{yy} \\ S_2 &= U = J_{xy} + J_{yx} \\ S_3 &= V = \mathrm{i}\,(J_{yx} - J_{xy}) \end{aligned} \tag{3.68}$$

and

$$\begin{aligned} J_{xx} &= \tfrac{1}{2}(S_0 + S_1) = \tfrac{1}{2}(I + Q) \\ J_{yy} &= \tfrac{1}{2}(S_0 - S_1) = \tfrac{1}{2}(I - Q) \\ J_{xy} &= \tfrac{1}{2}(S_2 + \mathrm{i}\, S_3) = \tfrac{1}{2}(U + \mathrm{i}\, V) \\ J_{yx} &= \tfrac{1}{2}(S_2 - \mathrm{i}\, S_3) = \tfrac{1}{2}(U - \mathrm{i}\, V) \ . \end{aligned} \tag{3.69}$$

Just like the elements of the coherency matrix the Stokes parameters of a plane wave can be calculated from 6 intensity measurements. Using (3.57) we find

$$\boxed{\begin{aligned} S_0 &= I = I\,(0°, 0) + I\,(90°, 0) \\ S_1 &= Q = I\,(0°, 0) - I\,(90°, 0) \\ S_2 &= U = I\,(45°, 0) - I\,(135°, 0) \\ S_3 &= V = I\,(45°, \tfrac{\pi}{2}) - I\,(135°, \tfrac{\pi}{2}) \end{aligned}} \quad . \tag{3.70}$$

These are the relationships used in radio polarimeters. We will return to this later in Chap. 6. The Stokes parameters are usually preferred to the coherency matrix because most expressions are simpler in terms of the S_i than in terms of J_{kl}.

For partially polarized light we find, if we rewrite (3.56) in terms of the S_i,

$$\begin{aligned} S_0^2 &\geqq S_1^2 + S_2^2 + S_3^2 \\ I^2 &\geqq Q^2 + U^2 + V^2 \end{aligned} \tag{3.71}$$

instead of (3.30), which is valid for strictly monochromatic waves. It is then easy to express the *degree of polarization*

$$p = \frac{\sqrt{S_1^2 + S_2^2 + S_3^2}}{S_0} \quad . \tag{3.72}$$

Since the Stokes parameters can be expressed according to (3.68) as the sum or difference of the components of the coherency matrix, the Stokes parametes of the superposition of several independent vector waves will be the sum of the Stokes parameters of the individual waves.

Similarly, a partially polarized wave can be regarded as the sum of a completely polarized wave and a completely unpolarized wave

$$\begin{aligned} S_0 &= S_0(1-p) + S_0 p \\ S_1 &= 0 + S_1 \\ S_2 &= 0 + S_2 \\ S_3 &= 0 + S_3 \end{aligned} \tag{3.73}$$

where p is given according to (3.72).

3.6 Faraday Rotation

In 1845, Faraday detected that the polarization angle of dielectric materials will rotate if a magnetic field is applied to the material in the direction of the light propagation. This indicated to him that light must be an electromagnetic phenomenon. In radio astronomy this Faraday rotation has become an important tool to investigate the interstellar magnetic field (see, e.g., Fig. 3.4).

Interstellar gas must be treated as a tenuous plasma as in Sect. 2.9. Wave propagation in such a medium in the presence of an external magnetic field is a rather complicated subject with many different wave modes, cut-offs, etc. It is treated rather extensively in most textbooks on plasma physics and it may suffice to refer to a few of these in the reference list of this chapter.

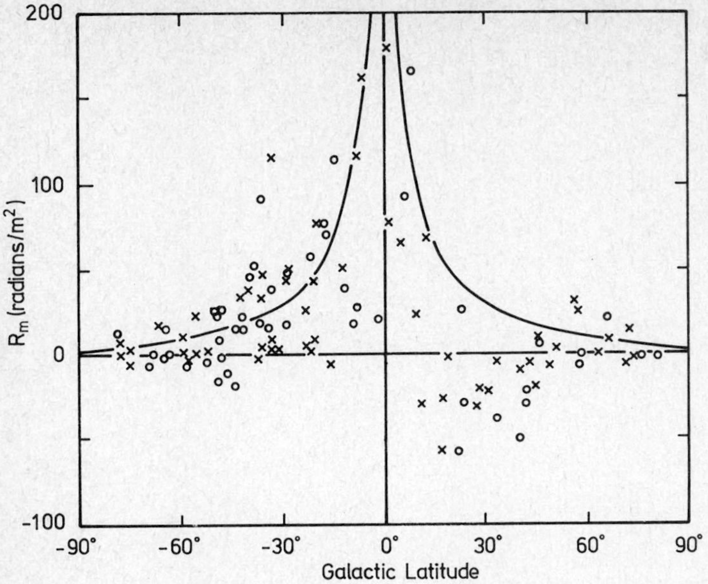

Fig. 3.4. Rotation measure of extragalactic radio sources [after Spitzer (1968)]

Here we will disregard all these complications and treat only the one remaining mode in the high frequency limit where the frequency of the electromagnetic wave is well above all the resonances, though still low enough that the interaction of the free electrons in the plasma with the external magnetic field cannot be neglected altogether. Since the effects of wave propagation in the direction of the magnetic field are so much larger than those of propagation perpendicular to the field, only this case will be considered.

Wave propagation in a dissipative medium and the resulting dispersion equation linking the wave number k and the circular frequency ω have been derived in Sect. 2.7. In Sect. 2.8, wave propagation in a tenous plasma was investigated by examining the effects on the conductivity σ of an electromagnetic wave in a medium with free electrons. Here we will do the same but will include an external magnetic field.

Strictly speaking, the material constants ε, μ and σ should be treated as tensors with 9 components each, but choosing a small angle between the direction of the magnetic field and the propagation direction and a high enough frequency permits us to work with scalars for ε, μ and σ.

So let us assume that the interstellar gas is a tenuous plasma with free electrons and ions. As in Sect. 2.8, only the electrons need to be considered, since the motion of the ions, due to their much larger mass, is at least three orders of magnitude less than that of the electrons. The equation of motion for an electron in the presence of a magnetic field $\boldsymbol{B}$ is

$$m\dot{\boldsymbol{v}} = m\ddot{\boldsymbol{r}} = -e(\boldsymbol{E} + \dot{\boldsymbol{r}} \times \boldsymbol{B}) \ . \tag{3.74}$$

If the magnetic field $\boldsymbol{B}$ is oriented in the z-direction, Eq. (3.74) becomes

$$\begin{aligned} \ddot{r}_x + \frac{e}{m} B \dot{r}_y &= -\frac{e}{m} E_x \qquad \Big| \quad 1 \\ \ddot{r}_y - \frac{e}{m} B \dot{r}_x &= -\frac{e}{m} E_y \qquad \Big| \quad \pm \mathrm{i}\ . \end{aligned} \tag{3.75}$$

Multiplying (3.75) by the factors given to the right of the equation and adding, this becomes

$$\boxed{\begin{aligned} \ddot{r} \mp \mathrm{i} \frac{e}{m} B \dot{r} &= -\frac{e}{m} E \\ r &= r_x \pm \mathrm{i}\, r_y \\ E_{\pm} &= E_x \pm \mathrm{i}\, E_y \end{aligned}}\ . \tag{3.76}$$

(3.76) is a differential equation for the complex quantities r and E. Depending on the sign of $\mathrm{i}(e/m) B \dot{r}$, we distinguish the solutions E_+ and E_-. These can be regarded as circularly polarized waves because the rectangular coordinates are given by

$$E_x = \frac{1}{2}(E_+ + E_-)\ , \quad E_y = \frac{1}{2\mathrm{i}}(E_+ - E_-)\ . \tag{3.77}$$

We are looking for solutions of (3.76) in the form of a harmonic wave and therefore put

$$E_{\pm} = A\, \mathrm{e}^{\mathrm{i}(k_{\pm} z - \omega t)} \tag{3.78}$$

where A is assumed to be real. Inserting this into (3.76), we see that a solution for r of the form

$$r_{\pm} = r_0\, \mathrm{e}^{\mathrm{i}(k_{\pm} z - \omega t)} \tag{3.79}$$

with r_0 being in general a complex quantity, is possible provided that

$$r_{\pm}\left(-\omega^2 \mp \frac{e}{m} B \omega\right) = -\frac{e}{m} E_{\pm} \quad \text{or}$$

$$\begin{aligned} r_{\pm} &= \frac{-\dfrac{e}{m}}{-\omega^2 \mp \dfrac{e}{m} B \omega} E_{\pm} \quad \text{and} \\ \dot{r}_{\pm} &= \frac{\mathrm{i}\dfrac{e}{m}}{-\omega^2 \mp \dfrac{e}{m} B \omega} \omega E_{\pm}\ . \end{aligned} \tag{3.80}$$

Thus, we find for the current density

$$|\boldsymbol{J}| = -\,N e \dot{r} = \mathrm{i}\frac{N e^2}{m\left(\omega \pm \frac{e}{m} B\right)} E_\pm = \sigma_\pm E_\pm \quad \text{with}$$

$$\boxed{\sigma_\pm = \mathrm{i}\frac{N e^2}{m\left(\omega \pm \frac{e}{m} B\right)}} \; . \tag{3.81}$$

The conductivity therefore is purely imaginary. For $\omega = \omega_\mathrm{c}$, where

$$\boxed{\begin{aligned} \omega_\mathrm{c} &= \frac{e}{m} B \\ \nu_\mathrm{c} &= \frac{e}{2\pi m} B \end{aligned}} \tag{3.82}$$

is the *cyclotron frequency*, the frequency of the wave is in resonance with the gyration frequency of the electrons in the magnetic field. Then $|\sigma| \to \infty$, and the wave cannot propagate. This is seen most easily if (3.81) is substituted in the dispersion equation (2.68), again assuming $\varepsilon \cong \varepsilon_0$ and $\mu \cong \mu_0$. Then

$$\boxed{k_\pm^2 = \frac{\omega^2}{c^2}\left(1 - \frac{\omega_\mathrm{p}^2}{\omega(\omega \pm \omega_\mathrm{c})}\right)} \tag{3.83}$$

where we have introduced the plasma frequency (2.81). The index of refraction thus becomes according to (2.38)

$$\boxed{n_\pm^2 = 1 - \frac{\omega_\mathrm{p}^2}{\omega(\omega \pm \omega_\mathrm{c})}} \; , \tag{3.84}$$

and consequently, the two modes E_+ and E_- have slightly different phase propagation velocities $v_\pm = c/n_\pm$. The two circularly polarized waves E_+ and E_- will thus achieve a relative phase difference $2\,\Delta\psi$ after a propagation length Δz given by

$$2\,\Delta\psi = (k_+ - k_-)\,\Delta z \; . \tag{3.85}$$

The two circularly polarized waves can be superposed to form an elliptically polarized wave. If this is done once for the original wave, and a second time after it has left the slab Δz, the polarization angle will have changed by $\Delta\psi$.

Truncating the series expansions of (3.83) after the second term, which is permissible for $\omega \gg \omega_p$ and $\omega \gg \omega_c$, we obtain

$$\Delta\psi = \frac{\omega_p^2\,\omega_c}{2\,c\,\omega^2}\,\Delta z = \frac{N\,e^3\,B}{2\,\varepsilon_0\,m^2\,c\,\omega^2}\,\Delta z \; . \tag{3.86}$$

For a finite slab with variable density $N(z)$ and magnetic flux density $B(z)$, we thus obtain the total rotation of the polarization direction

$$\boxed{\Delta\psi = \frac{e^3}{8\,\pi^2\,m^2\,c\,\varepsilon_0}\,\frac{1}{\nu^2}\int_0^L B_\parallel(z)\,N(z)\,dz} \; . \tag{3.87}$$

For SI units the coefficient becomes

$$\frac{e^3}{8\,\pi^2\,m^2\,c\,\varepsilon_0} = 23.80 \cdot 10^3 \quad [\mathrm{SI}] \; . \tag{3.88}$$

In astronomy, however, a system of mixed units is usually employed resulting in

$$\boxed{\frac{\Delta\psi}{\mathrm{rad}} = 8.1 \cdot 10^5 \left(\frac{\lambda}{\mathrm{m}}\right)^2 \int_0^{L/\mathrm{pc}} \left(\frac{B_\parallel}{\mathrm{Gauss}}\right)\left(\frac{N\,e}{\mathrm{cm}^{-3}}\right) d\left(\frac{z}{\mathrm{pc}}\right)} \; . \tag{3.89}$$

The dependence of $\Delta\psi$ on ν^{-2} can be used to determine the value of $\int BN\,dz$ from the measurement of the polarization direction at two frequencies:

$$\boxed{\int_0^{L/\mathrm{pc}} \left(\frac{B_\parallel}{\mathrm{Gauss}}\right)\left(\frac{Ne}{\mathrm{cm}^{-3}}\right) d\left(\frac{z}{\mathrm{pc}}\right) = 1.23 \cdot 10^{-6}\, \frac{\left(\frac{\Delta\psi_1}{\mathrm{rad}}\right) - \left(\frac{\Delta\psi_2}{\mathrm{rad}}\right)}{\left(\frac{\lambda_1}{\mathrm{m}}\right)^2 - \left(\frac{\lambda_2}{\mathrm{m}}\right)^2}} \; . \tag{3.90}$$

In this the unknown intrinsic polarization angle of the source cancels. Conversely, (3.90) then can be used to determine this intrinsic polarization angle from (3.89) and thus to correct the measured polarization.

4. Fundamentals of Antenna Theory

4.1 Electromagnetic Potentials

The direct solution of Maxwell's equations (2.9) to (2.12) is very difficult if not impossible for all but the simplest configurations. One way out of this dilemma is to decouple the different vector fields of $\boldsymbol{E}$ and $\boldsymbol{H}$ and thus to arrive at the wave equations (2.29) and (2.30). In these equations the $\boldsymbol{E}$ and $\boldsymbol{H}$ fields are independent. Therefore if we have found an appropriate solution for the $\boldsymbol{E}$ field of a given problem, there still remains the problem of how to determine the appropriate $\boldsymbol{H}$ field solution so that both jointly fulfil Maxwell's equations.

In simple cases like a plane harmonic wave this is not difficult, but if the configuration is more complicated, this is not at all easy. Our aim is therefore to introduce new functions which can be determined from the imposed current density $\boldsymbol{J}$ and charge density ϱ, and from which both $\boldsymbol{E}$ and $\boldsymbol{H}$ can then be determined in a straightforward way. These functions are the electrodynamic potentials Φ and $\boldsymbol{A}$.

In electromagnetic theory, potential functions were first used by Green 1828, but no notice was taken of this by the scientific community. This only changed in 1846, when Lord Kelvin directed attention to this paper. But one year before this Franz Neumann in Königsberg had already used this method successfully.

According to Maxwell's equation (2.10), we always have $\nabla \cdot \boldsymbol{B} = 0$. Due to Stokes' theorem (A 22) of vector analysis, we can then always write

$$\boxed{\boldsymbol{B} = \nabla \times \boldsymbol{A}} \quad , \tag{4.1}$$

so that (2.11) becomes

$$\nabla \times (\boldsymbol{E} + \dot{\boldsymbol{A}}) = 0$$

where the order of time and spatial differentiation have been interchanged. But Gauss' theorem states that a vector whose rotation vanishes can always be expressed as the gradient of a scalar, so that

$$\boldsymbol{E} + \dot{\boldsymbol{A}} = -\nabla \Phi \quad \text{or}$$

$$\boxed{\boldsymbol{E} = -\nabla \Phi - \dot{\boldsymbol{A}}} \quad . \tag{4.2}$$

Both $\boldsymbol{B}$ and $\boldsymbol{E}$ therefore can be expressed in terms of $\boldsymbol{A}$ and Φ, but if these expressions are to be of any use, we obviously must require that the resulting fields $\boldsymbol{B}$ and $\boldsymbol{E}$ should obey Maxwell's equations. In order to see which additional restrictions this imposes on $\boldsymbol{A}$ and Φ, let us introduce (4.1) and (4.2) into Maxwell's equations. For the sake of simplicity, let us assume constant ε, μ and σ.

From (2.12) we then obtain

$$\nabla \times (\nabla \times \boldsymbol{A}) + \varepsilon\mu \frac{\partial}{\partial t}[\nabla \Phi + \dot{\boldsymbol{A}}] = \mu \boldsymbol{J}$$

$$\nabla (\nabla \cdot \boldsymbol{A}) - \Delta \boldsymbol{A} + \varepsilon\mu \frac{\partial}{\partial t}[\nabla \Phi + \dot{\boldsymbol{A}}] = \mu \boldsymbol{J} \tag{4.3}$$

$$\nabla^2 \boldsymbol{A} - \varepsilon\mu \ddot{\boldsymbol{A}} - \nabla(\nabla \cdot \boldsymbol{A} + \varepsilon\mu\dot{\Phi}) = -\mu \boldsymbol{J} \, .$$

Using (4.2) the remaining equation (2.9) gives

$$\nabla \cdot \nabla \Phi + \nabla \cdot \dot{\boldsymbol{A}} = -\frac{1}{\varepsilon}\varrho \quad \text{or}$$

$$\nabla^2 \Phi - \varepsilon\mu\ddot{\Phi} + \frac{\partial}{\partial t}[\nabla \cdot \boldsymbol{A} + \varepsilon\mu\dot{\Phi}] = -\frac{1}{\varepsilon}\varrho \, . \tag{4.4}$$

Neither $\boldsymbol{A}$ nor Φ are completely determined by the defining equations (4.1) and (4.2). Obviously an arbitrary vector can be added to $\boldsymbol{A}$ without changing the resulting $\boldsymbol{B}$ provided this additive term has a vanishing rotation. This is certainly the case if this vector is the gradient of a scalar function

$$\boxed{\hat{\boldsymbol{A}} = \boldsymbol{A} + \nabla \Lambda} \tag{4.5}$$

for then $\boldsymbol{B}$ will not be changed at all. But according to (4.2), $\boldsymbol{E}$ will be affected, unless Φ is also changed into

$$\boxed{\hat{\Phi} = \Phi - \dot{\Lambda}} \, . \tag{4.6}$$

In (4.5) and (4.6) we are then completely free in choosing Λ and we can use this freedom in Λ to simplify equations (4.3) and (4.4).

An obvious choice is

$$\boxed{\nabla \cdot \boldsymbol{A} + \varepsilon\mu\dot{\Phi} = 0} \, . \tag{4.7}$$

In obeying this, $\boldsymbol{A}$ and Φ are said to be in the Lorentz-gauge. This can always be fulfilled by choosing a gauge-function Λ that obeys (4.5), (4.6) and (4.7);

$$\nabla \cdot \boldsymbol{A} + \nabla \cdot \nabla \Lambda + \varepsilon\mu\dot{\Phi} - \varepsilon\mu\ddot{\Lambda} = 0$$

$$\boxed{\nabla^2 \Lambda - \varepsilon\mu\ddot{\Lambda} = 0} \, . \tag{4.8}$$

Electrodynamic potentials in the Lorentz-gauge then obey the equations

$$\nabla^2 \boldsymbol{A} - \varepsilon\mu\ddot{\boldsymbol{A}} = -\mu\boldsymbol{J} \tag{4.9}$$

$$\nabla^2 \Phi - \varepsilon\mu\ddot{\Phi} = -\frac{1}{\varepsilon}\varrho \quad . \tag{4.10}$$

These Eqs. (4.9) and (4.10) with (4.7), are completely equivalent to Maxwell's equations (2.9) to (2.12) together with the constitutive equations (2.1) to (2.3), and they have the advantages that they are decoupled and that all four equations have the same form: that of an inhomogeneous wave equation.

4.2 Green's Function for the Wave Equation

The wave equations (4.9) and (4.10) have the form

$$\nabla^2 \psi - \frac{1}{v^2}\ddot{\psi} = -f(\boldsymbol{x}, t) \tag{4.11}$$

where $f(\boldsymbol{x}, t)$ is the given source distribution and v is the propagation velocity

$$v = \frac{1}{\sqrt{\varepsilon\mu}} \tag{4.12}$$

as derived in (2.37). Since the time dependence of (4.11) complicates the problem, it is useful to eliminate the time in (4.11) by taking the inverse Fourier transform.

Substituting

$$\begin{aligned} \psi(\boldsymbol{x}, t) &= \int_{-\infty}^{\infty} \Psi(\boldsymbol{x}, \omega)\, \mathrm{e}^{\mathrm{i}\omega t}\, d\omega \\ f(\boldsymbol{x}, t) &= \int_{-\infty}^{\infty} F(\boldsymbol{x}, \omega)\, \mathrm{e}^{\mathrm{i}\omega t}\, d\omega \end{aligned} \tag{4.13}$$

into (4.11) we find that $\Psi(\boldsymbol{x}, \omega)$ obeys the time-independent Helmholtz-wave equation

$$(\nabla^2 + k^2)\,\Psi(\boldsymbol{x}, \omega) = -F(\boldsymbol{x}, \omega) \tag{4.14}$$

where we have for brevity

$$k = \omega/v = \omega\sqrt{\varepsilon\mu} \ . \tag{4.15}$$

The left-hand side of (4.14) is linear in Ψ, but the function F on the right-hand side prevents the application of a superposition principle to the complete equation.

This should not be confused with the true proposition that an arbitrary linear combination of solutions of the *homogeneous* equation can always be added to any peculiar solution of (4.14). We are searching for a convenient method to construct a peculiar solution of (4.14) that fulfils the given initial or boundary conditions. This method is provided by Green's functions.

The Green's functions are solutions of an inhomogeneous differential equation in the form of (4.14), but with a special, convenient form on the right-hand side. This right-hand side is chosen such that the general function F can be expanded into a linear combination of these special functions. The solution Ψ of the general equation (4.14) is then formed by the same kind of linear superposition as F.

The Green's function $G(\boldsymbol{x}, \boldsymbol{x}')$ therefore is defined as the solution of

$$\boxed{(\nabla^2 + k^2)\, G(\boldsymbol{x}, \boldsymbol{x}') = -\,\delta(\boldsymbol{x} - \boldsymbol{x}')} \quad . \tag{4.16}$$

It has to be chosen such that it not only is a solution to (4.16) but also has the symmetries specific for the problem and obeys the initial or boundary conditions that are imposed. $G(\boldsymbol{x}, \boldsymbol{x}')$ is a function of two positions, the reason for this will become clear if we consider the corresponding problem in electrostatics. (4.14) then corresponds to the potential equation of an arbitrary charge distribution $F(\boldsymbol{x})$, while (4.16) is the equation for a unit point charge at the point $\boldsymbol{x}'$.

$$F(\boldsymbol{x}, \omega) = \int F(\boldsymbol{x}', \omega)\, \delta(\boldsymbol{x} - \boldsymbol{x}')\, d^3 x' \tag{4.17}$$

then corresponds to the representation of the arbitrary charge distribution by an ensemble of point charges, each with the charge $F(\boldsymbol{x}', \omega)$ at the position $\boldsymbol{x}'$, and the solution of (4.14) is then given by

$$\Psi(\boldsymbol{x}, \omega) = \int F(\boldsymbol{x}', \omega)\, G(\boldsymbol{x}, \boldsymbol{x}')\, d^3 x' \tag{4.18}$$

due to the linearity of the left-hand side of (4.14). In the electrostatic example, this corresponds to building up the total potential field from the fields of individual point charges, each situated at the position $\boldsymbol{x}'$.

The inverse Fourier transform of $G(\boldsymbol{x}, \boldsymbol{x}')$,

$$g(\boldsymbol{x}, t, \boldsymbol{x}', t') = \frac{1}{2\pi} \int G(\boldsymbol{x}, \boldsymbol{x}')\, e^{i\omega t}\, d\omega \ , \tag{4.19}$$

is then a solution of the inverse Fourier transform of Eq. (4.16), that is of

$$\left(\nabla^2 - \frac{1}{v^2} \frac{\partial^2}{\partial t^2}\right) g(\boldsymbol{x}, \boldsymbol{x}', t, t') = -\,\delta(\boldsymbol{x} - \boldsymbol{x}')\, \delta(t - t') \ . \tag{4.20}$$

This method of the Green's functions will now be applied to the case of spherical waves emitted from a point source. A spherical coordinate system (r, ϑ, φ) is appropriate in this case, so that (4.16) becomes [see Appendix (A 27)]

$$\frac{1}{r} \frac{d^2}{dr^2} (r\, G) + k^2 G = -\,\delta(r) \ . \tag{4.21}$$

For $r \neq 0$ this can be integrated immediately

$$G = \frac{1}{4\pi r} e^{\pm i k r} , \tag{4.22}$$

and it can be shown that this solution applies to $r \to 0$ as well. The corresponding Green's function for the time dependent problem is then obtained by the inverse Fourier transform of (4.22), that is, by

$$g(\boldsymbol{x}, \boldsymbol{x}'; t) = \frac{1}{4\pi |\boldsymbol{x} - \boldsymbol{x}'|} \frac{1}{2\pi} \int_{-\infty}^{\infty} e^{i(\omega t \pm k|\boldsymbol{x} - \boldsymbol{x}'|)} d\omega , \tag{4.23}$$

or, introducing a new, retarded (or advanced) time t'

$$t' = t \mp \frac{k}{\omega} |\boldsymbol{x} - \boldsymbol{x}'| = t \mp \frac{|\boldsymbol{x} - \boldsymbol{x}'|}{v} , \quad \text{by} \tag{4.24}$$

$$g(\boldsymbol{x}, \boldsymbol{x}', t, t') = \frac{\delta\left(t' + \frac{|\boldsymbol{x} - \boldsymbol{x}'|}{v} - t\right)}{4\pi |\boldsymbol{x} - \boldsymbol{x}'|} . \tag{4.25}$$

In (4.24) two choices of the sign are in principle possible; here the upper sign yielding the *retarded* potentials is selected because only this results in the proper causal relation. If advanced potentials were chosen, then the field at a given point at the time t would depend on charges and currents at other points at times later than t. That the retarded and not the advanced solution is selected should be taken as indication of the *arrow of time*.

The solution for the wave equation (4.11) is then, in the absence of boundaries,

$$\psi(\boldsymbol{x}, t) = \frac{1}{4\pi} \iint \frac{f(\boldsymbol{x}', t') \delta\left(t' + \frac{|\boldsymbol{x} - \boldsymbol{x}'|}{v} - t\right)}{|\boldsymbol{x} - \boldsymbol{x}'|} d^3 x' \, dt' . \tag{4.26}$$

If the integration over t' is performed we finally arrive at the result

$$\boxed{\psi(\boldsymbol{x}, t) = \frac{1}{4\pi} \int \frac{f\left(\boldsymbol{x}', t - \frac{|\boldsymbol{x} - \boldsymbol{x}'|}{v}\right)}{|\boldsymbol{x} - \boldsymbol{x}'|} d^3 x'} . \tag{4.27}$$

A short hand version of this is

$$\boxed{\psi(\boldsymbol{x}, t) = \frac{1}{4\pi} \int \frac{[f(\boldsymbol{x}', t')]_{\text{ret}}}{|\boldsymbol{x} - \boldsymbol{x}'|} d^3 x'} , \tag{4.28}$$

where the square brackets $[\,]_{\text{ret}}$ mean that the time t has to be taken at the retarded time $t' = t - |\boldsymbol{x} - \boldsymbol{x}'|/v$.

4.3 Antennas for Electromagnetic Radiation. The Hertz Dipole

Using the expression (4.27) or (4.28) for the retarded Green's function in the wave equation for the electrodynamic potential (4.9) and (4.10), we find that it should be possible to write any sensible solution of Maxwell's equation as

$$\boldsymbol{A}(\boldsymbol{x},t) = \frac{\mu}{4\pi}\iiint \frac{\boldsymbol{J}\left(\boldsymbol{x}', t - \frac{|\boldsymbol{x}-\boldsymbol{x}'|}{v}\right)}{|\boldsymbol{x}-\boldsymbol{x}'|}\, d^3x' \tag{4.29}$$

$$\Phi(\boldsymbol{x},t) = \frac{1}{4\pi\varepsilon}\iiint \frac{\varrho\left(\boldsymbol{x}', t - \frac{|\boldsymbol{x}-\boldsymbol{x}'|}{v}\right)}{|\boldsymbol{x}-\boldsymbol{x}'|}\, d^3x' \; . \tag{4.30}$$

To determine the electrodynamic potentials therefore we need to know the distribution of the currents $\boldsymbol{J}$ and electric charges ϱ in the whole space. To make things more complicated, we are not completely free in this, because a given field of $\boldsymbol{A}$ and Φ (or $\boldsymbol{E}$ and $\boldsymbol{H}$) will induce currents and charges, so that the question of self-consistency of this problem arises.

For realistic problems this situation can often be greatly simplified. The currents $\boldsymbol{J}$ (and charges, if they are given at all) are confined to a finite region, the antenna proper, and we postulate a given distribution for $\boldsymbol{J}$. Any reactions that the field has on these currents will be neglected. Thus the integrals in (4.29) and (4.30) extend over finite regions only and they can be computed.

The simplest case is that of the Hertz dipole. Here the volume over which the integrals have to be extended is that of an infinitesimal dipole with a length Δl and a cross-section q. The solution for this configuration was first given by H. Hertz in 1888 and applied in his famous experiments on electromagnetic waves.

If a current I is flowing in this dipole, the current density is $|\boldsymbol{J}| = I/q$ in the dipole, and $\boldsymbol{J} = 0$ outside. Then the integration volume is only that of the dipole

$$dV = q\,\Delta l$$

and the current density

$$\boldsymbol{J} = \boldsymbol{J}_0\,\mathrm{e}^{-\mathrm{i}\omega t} = \frac{I}{q}\,\mathrm{e}^{-\mathrm{i}\omega t}\boldsymbol{j} \quad \text{with} \quad |\boldsymbol{j}| = 1 \; .$$

If the rectangular coordinate system (x, y, z) is oriented such that the dipole extends from $z = -\,\Delta l/2$ to $z = +\,\Delta l/2$ on the z-axis, then $J_x = J_y = 0$, and

$$J_z = \frac{I}{q}\,\mathrm{e}^{-\mathrm{i}\omega t} \; .$$

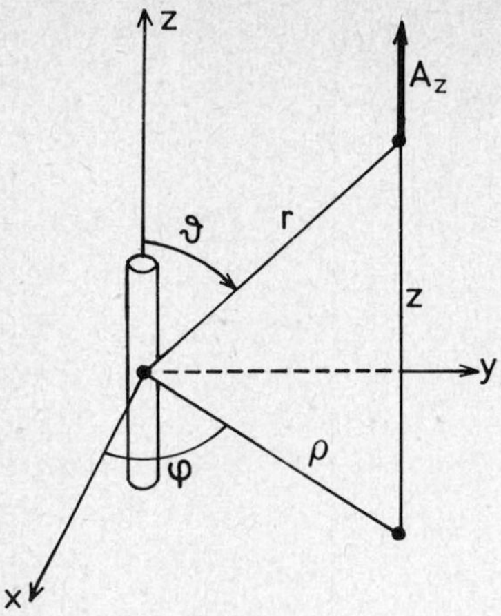

Fig. 4.1. The electric dipole (Hertz dipole) and the coordinate system used

The vector potential $\boldsymbol{A}$ according to (4.29) then has the single component

$$A_z = \frac{\mu}{4\pi} \int_{-\Delta l/2}^{\Delta l/2} \frac{I}{q} \exp\left[-\mathrm{i}\,\omega\left(t - \frac{r}{c}\right)\right] \frac{q}{r}\, dl$$

resulting in

$$\boxed{A_z = \frac{\mu}{4\pi} \frac{I\,\Delta l}{r} \exp\left[-\mathrm{i}\left(\omega t - \frac{2\pi}{\lambda} r\right)\right]} \quad . \tag{4.31}$$

Thus A_z is constant on concentric spheres $r^2 = x^2 + y^2 + z^2$. Introducing cyclindrical coordinates (ϱ, φ, z, see Fig. 4.1) we derive from (4.1)

$$B_\varphi = (\nabla \times \boldsymbol{A})_\varphi = \frac{\partial A_\varrho}{\partial z} - \frac{\partial A_z}{\partial \varrho} \quad .$$

But since

$$A_\varrho \equiv 0$$

we find that

$$B_\varphi = -\frac{\partial A_z}{\partial \varrho} = -\frac{\partial A_z}{\partial r} \frac{\partial r}{\partial \varrho} \quad .$$

Since

$$r^2 = \varrho^2 + z^2$$

$$\frac{\partial r}{\partial \varrho} = \frac{\varrho}{r} = \sin\vartheta \quad \text{and}$$

$$H_\varphi = \frac{1}{\mu} B_\varphi \; ,$$

we find that

$$H_\varphi = -\,\mathrm{i}\,\frac{I\,\Delta l}{2\,\lambda}\,\frac{\sin\vartheta}{r}\left[1-\frac{1}{\mathrm{i}\,\dfrac{2\,\pi}{\lambda}\,r}\right]\exp\left[-\mathrm{i}\left(\omega\,t-\frac{2\,\pi}{\lambda}\,r\right)\right] \tag{4.32}$$

where we have introduced λ by

$$\frac{\omega}{c}=\frac{2\,\pi}{\lambda}\ . \tag{4.33}$$

The other components of $\boldsymbol{H}$ are zero, because $A_\varrho \equiv A_\vartheta \equiv 0$. For the electrical field, we again make recourse to Maxwell's equations. According to (2.12)

$$\nabla\times\boldsymbol{H}=\dot{\boldsymbol{D}}+\boldsymbol{J}\ .$$

Outside the dipole, $\sigma = 0$ and therefore $\boldsymbol{J} = 0$ and for a harmonic wave according to (2.40), $\dot{\boldsymbol{D}} = -\,\mathrm{i}\,\omega\,\boldsymbol{D}$ so that for $\varepsilon \cong \varepsilon_0$

$$\boldsymbol{E}=\frac{\mathrm{i}}{\omega\,\varepsilon}\,(\nabla\times\boldsymbol{H})\ .$$

Returning to spherical coordinates (r, ϑ, φ) we find that

$$E_\vartheta=\frac{\mathrm{i}}{\omega\,\varepsilon}\,(\nabla\times\boldsymbol{H})_\vartheta\ .$$

Now $H_r \equiv 0$ and thus, from Appendix (A 26),

$$(\nabla\times\boldsymbol{H})_\vartheta=-\,\frac{1}{r}\,\frac{\partial\,(r\,H_\varphi)}{\partial r}\quad\text{so that}$$

$$E_\vartheta=-\,\mathrm{i}\,\sqrt{\frac{\mu_0}{\varepsilon_0}}\,\frac{I\,\Delta l}{2\,\lambda}\,\frac{\sin\vartheta}{r}\left[1-\frac{1}{\mathrm{i}\,\dfrac{2\,\pi}{\lambda}\,r}+\frac{1}{\left(\mathrm{i}\,\dfrac{2\,\pi}{\lambda}\,r\right)^2}\right]\exp\left[-\mathrm{i}\left(\omega t-\frac{2\,\pi}{\lambda}\,r\right)\right]\ . \tag{4.34}$$

Finally, since $H_\vartheta = 0$, we find that

$$(\nabla\times\boldsymbol{H})_r=\frac{1}{r\sin\vartheta}\,\frac{\partial\,(\sin\vartheta\,H_\varphi)}{d\vartheta}$$

and thus

$$E_r=\mathrm{i}\,\sqrt{\frac{\mu_0}{\varepsilon_0}}\,\frac{I\,\Delta l}{2\lambda}\,\frac{2\cos\vartheta}{r}\left[\frac{1}{\mathrm{i}\dfrac{2\pi}{\lambda}\,r}-\frac{1}{\left(\mathrm{i}\,\dfrac{2\pi}{\lambda}\,r\right)^2}\right]\exp\left[-\mathrm{i}\left(\omega t-\frac{2\pi}{\lambda}\,r\right)\right]\ . \tag{4.35}$$

$E_\varphi \equiv 0$ since $H_r \equiv 0$ and $H_\vartheta \equiv 0$. (4.32), (4.34) and (4.35) are therefore the only non-vanishing components of the electromagnetic field of an electric dipole. Forming the scalar product of $\boldsymbol{E}$ and $\boldsymbol{H}$ we find

$$\boxed{\boldsymbol{E} \cdot \boldsymbol{H} = 0} \tag{4.36}$$

just as in the case of plane electromagnetic waves. $\boldsymbol{E}$ and $\boldsymbol{H}$ of a radiating dipole are thus perpendicular everywhere, however, the expressions for $\boldsymbol{E}$ and $\boldsymbol{H}$ contain the radial distance r with different powers. Near- and far-field of an oscillating Hertz dipole are shown in Figs. 4.2 and 4.3.

The terms with $1/r^2$ in (4.32), (4.34) and (4.35) represent the *induction field* of a quasistationary electric dipole for slow oscillations; in $\boldsymbol{E}$ we find in addition the $1/r^3$ field of the *static dipole*. Most important is the *radiation field* which decays with only $1/r$. Its only components are

$$H_\varphi = -\,\mathrm{i}\,\frac{I\,\Delta l}{2\,\lambda}\,\frac{\sin\vartheta}{r}\exp\left[-\,\mathrm{i}\left(\omega\,t - \frac{2\,\pi}{\lambda}r\right)\right] \tag{4.37}$$

$$E_\vartheta = -\,\mathrm{i}\,\sqrt{\frac{\mu_0}{\varepsilon_0}}\,\frac{I\,\Delta l}{2\,\lambda}\,\frac{\sin\vartheta}{r}\exp\left[-\,\mathrm{i}\left(\omega\,t - \frac{2\,\pi}{\lambda}r\right)\right] \quad . \tag{4.38}$$

Just as in the case of plane electromagnetic waves

$$\frac{|\boldsymbol{E}|}{|\boldsymbol{H}|} = \sqrt{\frac{\mu_0}{\varepsilon_0}} \quad . \tag{4.39}$$

The Poynting vector for the radiation field is directed radially outwards and its average value is, according to (2.26),

$$\boxed{|\langle \boldsymbol{S} \rangle| = |\mathrm{Re}(\boldsymbol{E} \times \boldsymbol{H}^*)| = \sqrt{\frac{\mu_0}{\varepsilon_0}}\left(\frac{I\,\Delta l}{2\,\lambda}\right)^2 \frac{\sin^2\vartheta}{r^2}} \quad . \tag{4.40}$$

Thus the total radiated power is

$$P = \int_0^{2\pi}\int_0^{\pi} |\langle \boldsymbol{S} \rangle| r^2 \sin\vartheta \, d\vartheta \, d\varphi \ ,$$

and using

$$\int_0^{\pi} \sin^3\vartheta \, d\vartheta = \tfrac{4}{3}$$

this becomes

$$\boxed{P = \frac{8\,\pi}{3}\sqrt{\frac{\mu_0}{\varepsilon_0}}\left(\frac{I\,\Delta l}{2\,\lambda}\right)^2} \quad . \tag{4.41}$$

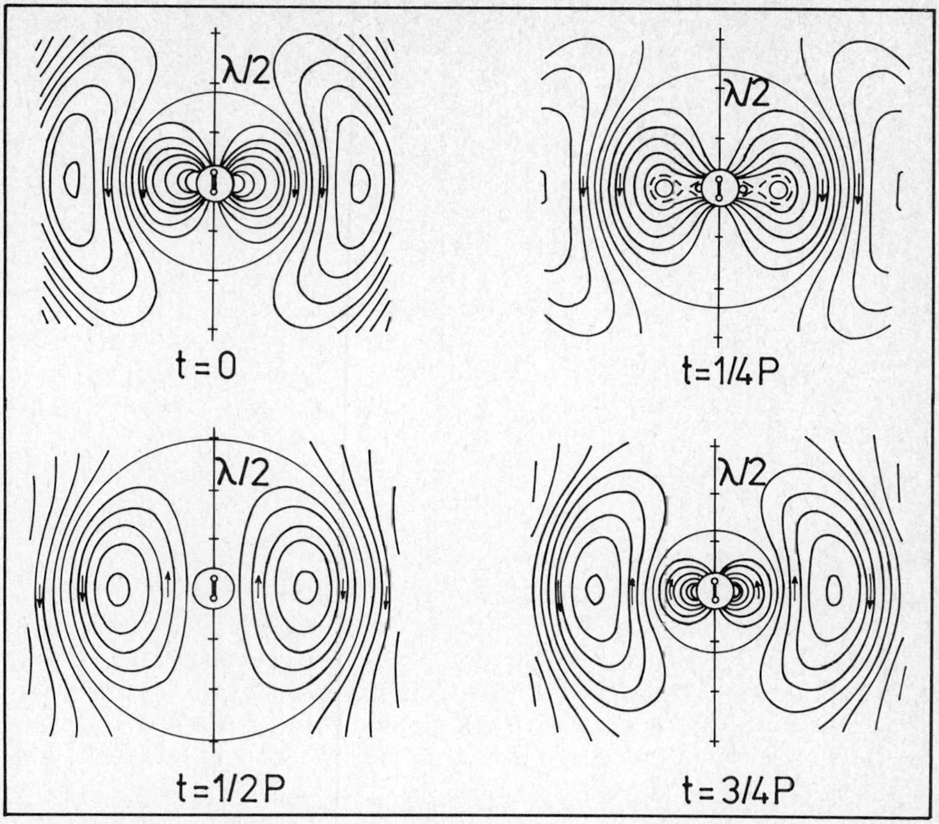

Fig. 4.2. The wavefield for an oscillating Hertz dipole: the region close to the dipole where P is the period of the oscillation

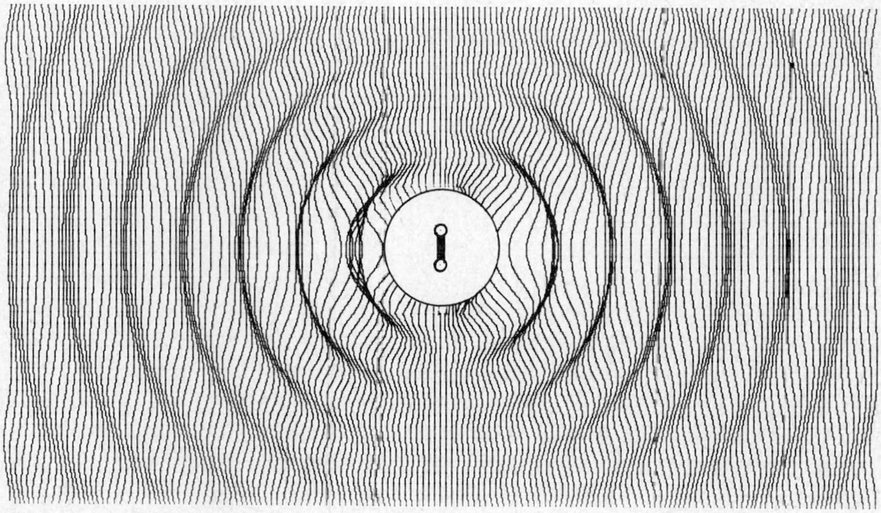

Fig. 4.3. The far-field of an oscillating Hertz dipole

This expression has the same general form as that giving the ohmic losses of a resistor

$$P = \tfrac{1}{2} R I^2$$

so that we are led to introduce a *radiation impedance* R_s of the Hertz dipole

$$R_s = \frac{4\pi}{3} \sqrt{\frac{\mu_0}{\varepsilon_0}} \left(\frac{\Delta l}{\lambda}\right)^2 . \tag{4.42}$$

Since $\sqrt{\mu_0/\varepsilon_0}$ is the intrinsic impedance of free space (vacuum), R_s is given in multiples of this where the controlling factor is the length of the dipole Δl in units of the wavelength. This formula is valid, however, only for very short dipoles; if $\Delta l \cong \lambda$ the current distribution along the dipole is no longer uniform so that the solutions (4.37) and (4.38) will not be applicable.

4.4 The Reciprocity Theorem

In the last section we have investigated the properties of a Hertz dipole as a transmitting device. It is possible, however, to use the same device as a receiving antenna. The obvious question is whether the antenna parameters for this purpose are the same or whether they differ. While it would be rather simple to check this for the Hertz dipole, there exists a general reciprocity theorem that answers this problem once and for all.

This theorem was stated first by Rayleigh and Helmholtz and it was later applied to the problem of antennas by Carson. Let us consider two antennas, 1 and 2, and let 1 be a transmitting antenna powered by the generator G while 2 is a receiving antenna that induces a certain current measured by M (Fig. 4.4). The receiving antenna is assumed to be oriented such that M shows a maximum deflection and we will assume that no ohmic losses occur in 2.

The reciprocity theorem now states that the current-measurement on M remains the same even if we exchange generator G and indicator M. Therefore it does not matter which antenna is transmitting and which is receiving. However, for the medium between 1 and 2, we must require that it is free of reciprocity

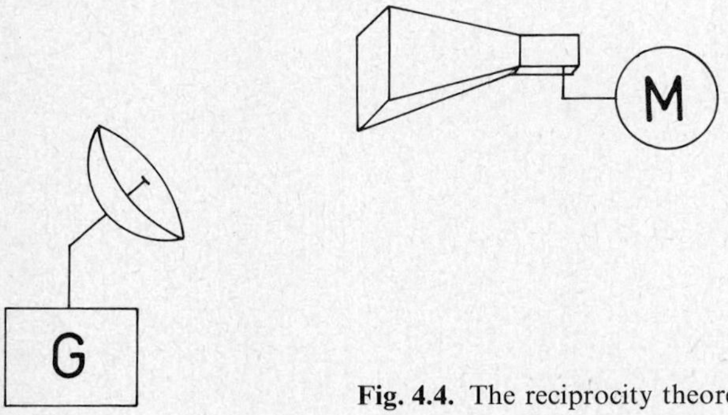

Fig. 4.4. The reciprocity theorem

failures; that is, that its transmission properties are the same from 1 directed towards 2 as from 2 directed towards 1. There are some materials, such as certain ferrities in a magnetic field, that do not fulfil this prerequisite; these are used in direction-sensitive devices such as circulators, isolators and the like. If such material is involved, the application of the reciprocity theorem needs special care.

To prove the reciprocity theorem consider Maxwell's equations for the two systems 1 and 2

$$\nabla\times\boldsymbol{H}_1 = \boldsymbol{J}_1 + \varepsilon\dot{\boldsymbol{E}}_1 \qquad \nabla\times\boldsymbol{H}_2 = \boldsymbol{J}_2 + \varepsilon\dot{\boldsymbol{E}}_2$$

$$\nabla\times\boldsymbol{E}_1 = -\mu\dot{\boldsymbol{H}}_1 \qquad \nabla\times\boldsymbol{E}_2 = -\mu\dot{\boldsymbol{H}}_2 \quad .$$

Forming

$$\begin{aligned}\nabla\cdot(\boldsymbol{E}_1\times\boldsymbol{H}_2) &= \boldsymbol{H}_2\cdot(\nabla\times\boldsymbol{E}_1) - \boldsymbol{E}_1\cdot(\nabla\times\boldsymbol{H}_2)\\ &= -\mu\dot{\boldsymbol{H}}_1\cdot\boldsymbol{H}_2 - \boldsymbol{E}_1\cdot\boldsymbol{J}_2 - \varepsilon\boldsymbol{E}_1\cdot\dot{\boldsymbol{E}}_2\end{aligned}$$

$$\begin{aligned}\nabla\cdot(\boldsymbol{E}_2\times\boldsymbol{H}_1) &= \boldsymbol{H}_1\cdot(\nabla\times\boldsymbol{E}_2) - \boldsymbol{E}_2\cdot(\nabla\times\boldsymbol{H}_1)\\ &= -\mu\boldsymbol{H}_1\cdot\dot{\boldsymbol{H}}_2 - \boldsymbol{E}_2\cdot\boldsymbol{J}_1 - \varepsilon\dot{\boldsymbol{E}}_1\cdot\boldsymbol{E}_2\end{aligned}$$

according to Appendix (A 9), the difference of these two lines is

$$\begin{aligned}\nabla\cdot(\boldsymbol{E}_1\times\boldsymbol{H}_2 - \boldsymbol{E}_2\times\boldsymbol{H}_1) = \mu(\boldsymbol{H}_1\cdot\dot{\boldsymbol{H}}_2 - \dot{\boldsymbol{H}}_1\cdot\boldsymbol{H}_2) - (\boldsymbol{E}_1\cdot\boldsymbol{J}_2 - \boldsymbol{E}_2\cdot\boldsymbol{J}_1)\\ - \varepsilon(\boldsymbol{E}_1\cdot\dot{\boldsymbol{E}}_2 - \dot{\boldsymbol{E}}_1\cdot\boldsymbol{E}_2)\ .\end{aligned}$$

If we now consider harmonic waves as in (2.40)

$$\dot{\boldsymbol{H}} = -\mathrm{i}\omega\boldsymbol{H}\ , \quad \dot{\boldsymbol{E}} = -\mathrm{i}\omega\boldsymbol{E} \quad \text{and thus}$$

$$\boldsymbol{H}_1\cdot\dot{\boldsymbol{H}}_2 - \dot{\boldsymbol{H}}_1\cdot\boldsymbol{H}_2 = 0\ , \quad \boldsymbol{E}_1\cdot\dot{\boldsymbol{E}}_2 - \dot{\boldsymbol{E}}_1\cdot\boldsymbol{E}_2 = 0$$

so that

$$\nabla\cdot(\boldsymbol{E}_1\times\boldsymbol{H}_2 - \boldsymbol{E}_2\times\boldsymbol{H}_1) = \boldsymbol{E}_2\cdot\boldsymbol{J}_1 - \boldsymbol{E}_1\cdot\boldsymbol{J}_2\ .$$

But according to the Gauss theorem [see Appendix (A 21)]

$$\int_V \nabla\cdot(\boldsymbol{E}_1\times\boldsymbol{H}_2 - \boldsymbol{E}_2\times\boldsymbol{H}_1)\,dV = \oint_S(\boldsymbol{E}_1\times\boldsymbol{H}_2 - \boldsymbol{E}_2\times\boldsymbol{H}_1)\,d\boldsymbol{S}\ .$$

If V is a sphere, the radius of which tends towards ∞, then $\boldsymbol{E}$ and $\boldsymbol{H}$ tend towards zero at the surface of this sphere and we suppose that the surface integral on the right-hand side vanishes. For a rigorous proof of this, we would have to show that $|\boldsymbol{E}_1\times\boldsymbol{H}_2 - \boldsymbol{E}_2\times\boldsymbol{H}_1|$ indeed is decreasing faster than $1/r^2$. This can be done considering that $\boldsymbol{E}\perp\boldsymbol{H}$ for spherical waves, but we will not give the details here; but adopting this we find

$$\int_V(\boldsymbol{E}_2\cdot\boldsymbol{J}_1 - \boldsymbol{E}_1\cdot\boldsymbol{J}_2)\,dV = 0\ . \tag{4.43}$$

Now, if the two antennas 1 and 2 are contained in different space regions V_1 and V_2, then obviously

$$\int_{V_1} \boldsymbol{E}_2 \cdot \boldsymbol{J}_1 \, dV = \int_{V_2} \boldsymbol{E}_1 \cdot \boldsymbol{J}_2 \, dV \tag{4.44}$$

since $\boldsymbol{J}_1 = 0$ in V_2 and $\boldsymbol{J}_2 = 0$ in V_1; the distance between the two antennas is of no concern here.

If an antenna is contained in an infinitesimal cylinder with the cross section q and the length dl

$$dV = q\,dl \ ;$$

the total current in the antenna is

$$I = q\,|\boldsymbol{J}|$$

and the emf

$$U = E\,dl \quad \text{and thus}$$

$$\boxed{U_2 I_1 = U_1 I_2} \ . \tag{4.45}$$

Here U_1 is the voltage induced by antenna 2 in antenna 1 and I_1 is the total current in antenna 1; U_2 and I_2 are similar quantities for antenna 2. Equation (4.45) is a quantitative formulation of the reciprocity theorem. It is therefore always possible to formulate the properties of an antenna either as a transmitting or a receiving device without distinguishing between the two.

4.5 Descriptive Antenna Parameters

The Hertz dipole is one of the very few kinds of antennas that can be investigated theoretically. Most other antenna systems, especially those of high gain and directivity which find application in radio astronomy, have to be treated by very much simplified methods or sometimes they can only be described empirically. But in order to be able to do this we need a set of characteristic parameters that can describe the properties of the given antenna. Some of these parameters have simple theoretical explanations; others are more difficult to put into a theoretical background. These descriptive antenna parameters will be outlined here, and usually the Hertz dipole will be used as an example. In the next chapter they will be applied to practical single-aperture antennas.

4.5.1 The Power Pattern $P(\vartheta, \varphi)$

The time-averaged Poynting flux of Hertz dipole is, according to (4.40),

$$|\langle \boldsymbol{S} \rangle| = S_0 \sin^2 \vartheta \ . \tag{4.40 a}$$

Such an antenna will thus radiate different amounts of power in different directions. The direction dependent part in (4.40 a) is given by the factor $\sin^2 \vartheta$; it depends in this case only on ϑ. In a more general case the shape of this power pattern $P(\vartheta, \varphi)$ will depend on both spherical coordinates ϑ and φ. According to the reciprocity theorem this applies to both transmitting and receiving antennas:

$$|\langle \boldsymbol{S} \rangle| = P(\vartheta, \varphi) \ . \tag{4.46}$$

Quite often not the power pattern but only the *normalized power pattern*

$$\boxed{P_{\rm n}(\vartheta, \varphi) = \frac{1}{P_{\rm max}} P(\vartheta, \varphi)} \tag{4.47}$$

is needed. The reciprocity theorem provides a method of measuring it. Suppose antenna A is connected to a generator and is transmitting. If now a second arbitrary antenna is moved on the surface of a sphere concentric with antenna A such that it always has a constant relative orientation to A, then the indications on the power meter M connected to this receiving antenna give a measure of $P(\vartheta, \varphi)$, and so $P_{\rm n}(\vartheta, \varphi)$ can be determined. The generator G and the meter M can be exchanged without a change in the results, because of the reciprocity theorem, and indeed it is this configuration which is usually applied in radio astronomy: the generator plus antenna B is formed by a small diameter radio source.

If the power pattern is measured using artificial transmitters, care should be taken that the distance from antenna A to antenna B is so large that B is in the radiation or far field of A. For large antennas like the 100 m telescope at Effelsberg, this distance can be 50 to 100 km!

Let us now consider the power pattern of the antenna as a transmitting device. If a total power of $W_\nu\,[\mathrm{W\,Hz^{-1}}]$ is fed into a lossless isotropic antenna that radiates equally well in all directions, it would transmit P power units per solid angle, the total radiated power being

$$W_\nu = 4\pi P \ .$$

In a realistic, but still lossless antenna, a power $P(\vartheta, \varphi)$ per unit solid angle is radiated in the direction (ϑ, φ). If we now call the factor by which $P(\vartheta, \varphi)$ supersedes P the directive gain $G(\vartheta, \varphi)$,

$$P(\vartheta, \varphi) = G(\vartheta, \varphi)\, P \quad \text{or}$$

$$\boxed{G(\vartheta, \varphi) = \frac{4\pi P(\vartheta, \varphi)}{\iint P(\vartheta, \varphi)\, d\Omega}} \ . \tag{4.48}$$

Thus the gain or directivity is also some kind of normalized power pattern similar to (4.47), but with the difference that the normalizing factor is $\int P(\vartheta, \varphi)\, d\Omega / 4\pi$. This can also be referred to as gain relative to a lossless isotropic source. But since such an isotropic source cannot be realized in practice, a measurable quantity is the gain relative to some standard antenna such as a half-wave dipole whose directivity is known from theoretical considerations.

As an example consider the Hertz dipole. According to (4.40 a)

$$P(\vartheta, \varphi) = S_0 \sin^2 \vartheta \quad \text{so that}$$

$$P_n(\vartheta, \varphi) = \sin^2 \vartheta \ . \tag{4.49}$$

Now

$$\int_0^{2\pi} \int_{-\pi/2}^{\pi/2} P_n(\vartheta, \varphi) \sin \vartheta \, d\vartheta \, d\varphi = 2\pi \int_{-\pi/2}^{\pi/2} \sin^3 \vartheta \, d\vartheta = \frac{8\pi}{3}$$

and, using (4.48),

$$G(\vartheta, \varphi) = \tfrac{3}{2} \sin^2 \vartheta \ . \tag{4.50}$$

4.5.2 The Main Beam

The *beam solid angle* Ω_A of an antenna is given by

$$\boxed{\Omega_A = \iint_{4\pi} P_n(\vartheta, \varphi) \, d\Omega = \int_0^{2\pi} \int_{-\pi/2}^{\pi/2} P_n(\vartheta, \varphi) \sin \vartheta \, d\vartheta \, d\varphi} \tag{4.51}$$

and it is measured in sr. The integration is extended over the full sphere 4π, such that Ω_A is the solid angle of an ideal antenna having $P_n = 1$ for all of Ω_A and $P_n = 0$ everywhere else. Such an antenna is an ideal model only and cannot be built in practice.

For most antennas the (normalized) power pattern has considerably larger values for a certain range of both ϑ and φ than for the remainder of the sphere. This range is called the main beam or main lobe of the antenna; the remainder is called the side lobes or back lobes (Fig. 4.5). The separation of the power pattern into main beam and side lobes is somewhat arbitrary and mainly governed by convention.

Quite analogously to (4.51) we then define the *main lobe solid angle* Ω_M by

$$\boxed{\Omega_M = \iint_{\substack{\text{main}\\\text{lobe}}} P_n(\vartheta, \varphi) \, d\Omega} \ . \tag{4.52}$$

Obviously the quality of an antenna as a direction measuring device depends on how well the power pattern is concentrated towards the main beam. If a large fraction of the received power comes from the side lobes it becomes rather difficult to tell where the radiation source is situated, and therefore it is appropriate to

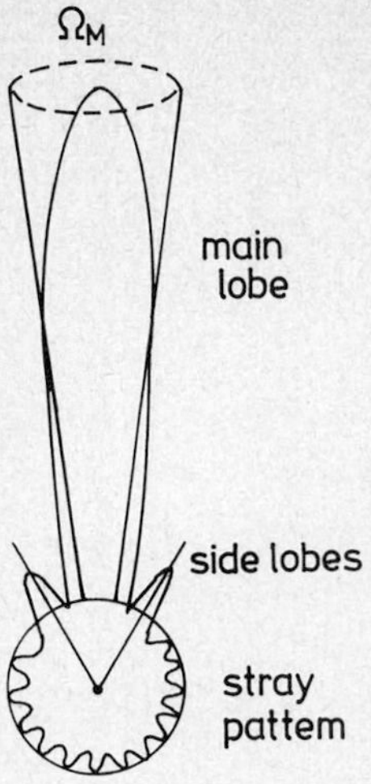

Fig. 4.5. A polar power pattern and its different parts

define a *main beam efficiency* or simply *beam efficiency* η_M by

$$\boxed{\eta_M = \frac{\Omega_M}{\Omega_A}} \quad . \tag{4.53}$$

This beam efficiency has nothing to do with the angular size of the main beam; it is quite conceivable that a small antenna with a wide main beam has a high beam efficiency. It only means that the power pattern is well concentrated in the main beam. In the next chapter we will show how this main beam efficiency can be controlled within certain limits for parabolic antennas by choosing the proper feed systems to provide the appropriate illumination of the dish.

Substituting (4.51) into (4.48) it is easy to see that the maximum directive gain G_{max} or *directivity* $\mathscr{D}$ can be expressed as

$$\boxed{\mathscr{D} = G_{max} = \frac{4\pi}{\Omega_A}} \quad . \tag{4.54}$$

The angular extent of the main beam is usually described by the *half power beam width* (HPBW), which is the angle between extreme points of the main beam where the normalized power pattern reaches 1/2 (Fig. 4.6). Other definitions such as the *beam width between first nulls* (BWFN) or the *equivalent width of the main beam* (EWMB) are sometimes used. The latter quantity is defined by

$$\mathrm{EWMB} = \sqrt{\frac{12}{\pi}\,\Omega_M} \quad . \tag{4.55}$$

For main beams with non-circular cross-sections two beam widths at orthogonal directions are needed.

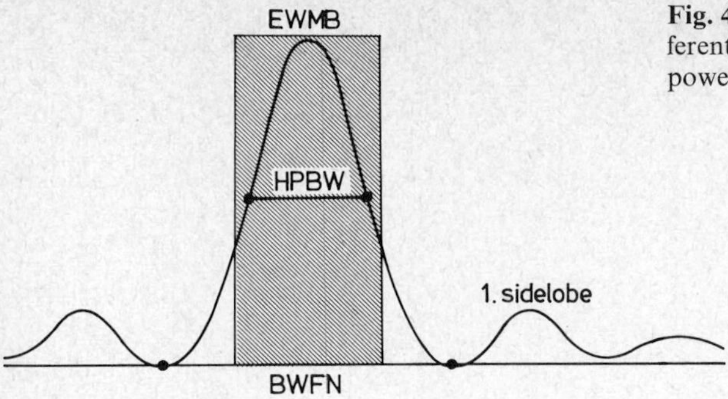

Fig. 4.6. The beam width and its different measures for a one-dimensional power pattern

The beam width is intimately connected with the geometric size of the antenna and with the wavelength used; the relevant formulae will be derived in Chap. 5.

4.5.3 The Effective Aperture

Let a plane wave with the power density $|\langle \boldsymbol{S} \rangle|$ be intercepted by an antenna. A certain amount of power is then extracted by the antenna from this wave, and let this amount of power be P_e. We will then call the fraction

$$A_e = P_e / |\langle \boldsymbol{S} \rangle| \tag{4.56}$$

the *effective aperture* of the antenna. A_e is a quantity very much like a cross-section in particle physics and it has the dimension of m^2.

Comparing this to the *geometric aperture* A_g we can define an aperture efficiency η_A by

$$\boxed{A_e = \eta_A A_g} \quad . \tag{4.57}$$

There are, however, some antennas like the Hertz dipole that do not permit an obvious definition of a geometric aperture. In such cases no obvious aperture efficiency η_A can be defined.

The effective aperture usually depends on the relative orientation of the antenna and the direction of wave propagation, and its peak value will be in the direction of the main beam. Effective aperture and directivity $\mathscr{D}$ are related by

$$\boxed{\mathscr{D} = \frac{4\pi A_e}{\lambda^2}} \tag{4.58}$$

which according to (4.54) is equivalent to

$$\boxed{A_e \Omega_A = \lambda^2} \quad . \tag{4.59}$$

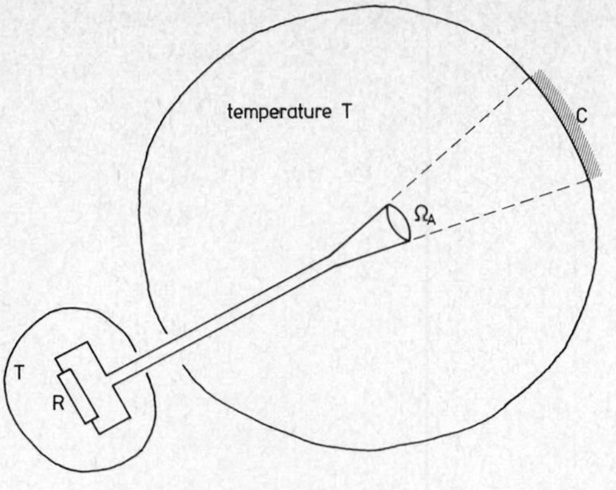

Fig. 4.7. Effective aperture and directivity

Quite often the proof of (4.58) or (4.59) is given by computing both $\mathscr{D}$ and A_e for some simple antennas and generalizing the result. But such a procedure is unsatisfactory if we want a result that is applicable to all antennas. Therefore we will report a derivation for (4.58) given by Pawsey and Bracewell in 1954 which uses thermodynamic considerations. Let antenna, receiver and a radiating surface C all be enclosed by a black body at the temperature T, and let us assume thermodynamic equilibrium for the whole system. Then the antenna will radiate power into the black enclosure, and this power will be absorbed there. On the other hand the black body will radiate too, and part of this radiation will be picked up by the antenna. Let the radiation surface C subtend the solid angle Ω_A as seen from the antenna (Fig. 4.7), its directivity be $\mathscr{D}$, the effective aperture A_e and the receiver bandwidth $\Delta\nu$. According to the Rayleigh-Jeans law, the surface C radiates with the specific power

$$P = \frac{2\,k\,T}{\lambda^2}\,\Delta\nu$$

per unit solid angle. Then the antenna collects a total power of

$$W\,\Delta\nu = A_e\,\frac{k\,T}{\lambda^2}\,\Delta\nu\,\Omega_A \tag{4.60}$$

according to (4.56) because the antenna can pick up only one polarisation component.

But if the whole system is assumed to be in thermal equilibrium the antenna must reradiate the same amount of power that it receives (detailed balance). If the antenna terminals are terminated by a matched resistor R, then the retransmitted power according to the Nyquist theorem (1.35) is

$$L\,\Delta\nu = k\,T\,\Delta\nu\ .$$

According to the definition (4.48) the fraction

$$L'\,\Delta\nu = k\,T\,\Delta\nu\,\mathscr{D}\,\frac{\Omega_A}{4\,\pi} \tag{4.61}$$

is intercepted by the surface C. As stated (4.60) and (4.61) should be equal if thermodynamic equilibrium prevails; thus

$$A_e \frac{k\,T}{\lambda^2}\,\Delta\nu\,\Omega_A = k\,T\,\Delta\nu\,\mathscr{D}\frac{\Omega_A}{4\,\pi}$$

and

$$\mathscr{D} = \frac{4\,\pi\,A_e}{\lambda^2}\ . \tag{4.58}$$

This relation has been proved here only for thermodynamic equilibrium, but since it is a relation between quantities that do not involve any thermodynamic quantities it should be always valid.

4.5.4 The Antenna Temperature

Consider a receiving antenna with a normalized power pattern $P_n(\vartheta,\varphi)$ that is pointed at a brightness distribution $B_\nu(\vartheta,\varphi)$ in the sky. According to the definitions of the effective area, a realistic antenna will then deliver at its output terminals a total power per unit bandwidth of

$$W = \tfrac{1}{2}\eta_R\,A_e \iint B_\nu(\vartheta,\varphi)\,P_n(\vartheta,\varphi)\,d\Omega \tag{4.62}$$

where η_R is the *radiation efficiency*, that is, the ratio of the available power at the antenna terminals to the total power absorbed by the antenna. It specifies the losses in the feed and the reflective surface due to its finite conductivity. η_R is difficult to measure, but a well-built antenna will have $\eta_R \cong 1$.

Using the Nyquist theorem (1.35) we can introduce an equivalent *antenna temperature* T_A by

$$W = k\,T_A\ . \tag{4.63}$$

Instead of the effective aperture A_e we can introduce the beam solid angle Ω_A according to (4.59), so that (4.62) becomes

$$T_A = \eta_R \frac{\lambda^2}{2\,k} \frac{\int B_\nu(\vartheta,\varphi)\,P_n(\vartheta,\varphi)\,d\Omega}{\int P_n(\vartheta,\varphi)\,d\Omega}\ .$$

Using now the Rayleigh-Jeans law we substitute here the brightness temperature (1.28)

$$T_b(\vartheta,\varphi) = \frac{\lambda^2}{2\,k} B_\nu(\vartheta,\varphi)$$

to obtain for the antenna temperature

$$\boxed{T_A(\vartheta_0,\varphi_0) = \eta_R \frac{\int T_b(\vartheta,\varphi)\,P_n(\vartheta-\vartheta_0,\varphi-\varphi_0)\sin\vartheta\,d\vartheta\,d\varphi}{\int P_n(\vartheta,\varphi)\,d\Omega}} \tag{4.64}$$

which is a convolution of the brightness temperature with the beam pattern of the telescope. The brightness temperature $T_b(\vartheta, \varphi)$ corresponds to the thermodynamic temperature of the radiating material only for thermal radiation of optically thick layers; in all other cases T_b is only an artificial but convenient quantity that in general will depend on the frequency.

As T_A is the quantity measured while T_b is the wanted one, (4.64) has to be inverted. Now (4.64) is an integral equation of the first kind, which in theory can be solved if the full range of $T_A(\vartheta, \varphi)$ and $P_n(\vartheta, \varphi)$ are known. But in practice this inversion is possible only approximately. Usually both $T_A(\vartheta, \varphi)$ and $P_n(\vartheta, \varphi)$ are known only for a limited range of ϑ and φ values, and the measured data are not free of errors. Therefore usually only an approximate deconvolution is performed. This will be described in the next chapter, when the steps necessary to calibrate an antenna are discussed.

5. Filled Aperture Antennas

5.1 The Radiation Field of Localized Sources of Finite Extent

When the radiation field of a system of oscillating charges and currents is investigated, we lose no generality by considering only such systems that vary sinusoidally with time. Therefore we will adopt

$$\begin{aligned} \varrho(\boldsymbol{x}, t) &= \varrho(\boldsymbol{x})\,\mathrm{e}^{-\mathrm{i}\omega t} \\ \boldsymbol{J}(\boldsymbol{x}, t) &= \boldsymbol{J}(\boldsymbol{x})\,\mathrm{e}^{-\mathrm{i}\omega t}\ . \end{aligned} \tag{5.1}$$

The amplitudes $\varrho(\boldsymbol{x})$ and $\boldsymbol{J}(\boldsymbol{x})$ can be complex quantities, so that the phases of the oscillations will be dependent on the position $\boldsymbol{x}$.

According to (4.29) the vector potential generated by these currents is

$$\boldsymbol{A}(\boldsymbol{x}, t) = \boldsymbol{A}(\boldsymbol{x})\,\mathrm{e}^{-\mathrm{i}\omega t} \quad \text{where} \tag{5.2}$$

$$\boldsymbol{A}(\boldsymbol{x}) = \frac{\mu}{4\pi} \iiint\limits_V \boldsymbol{J}(\boldsymbol{x}') \frac{\mathrm{e}^{\mathrm{i}k|\boldsymbol{x}-\boldsymbol{x}'|}}{|\boldsymbol{x}-\boldsymbol{x}'|}\, d^3x' \tag{5.3}$$

and

$$k = \frac{\omega}{c} = \frac{2\pi}{\lambda}\ . \tag{5.4}$$

V is the volume in which the current $\boldsymbol{J}$ flows.

But if $\boldsymbol{A}(\boldsymbol{x}, t)$ is known for a given position $\boldsymbol{x}$, the electromagnetic field for that position can be computed by standard methods. According to (4.1)

$$\boldsymbol{B} = \nabla \times \boldsymbol{A} \tag{5.5}$$

and from (2.12)

$$\dot{\boldsymbol{D}} = \varepsilon \dot{\boldsymbol{E}} = \nabla \times \boldsymbol{H} - \boldsymbol{J}\ .$$

As we are mainly interested in the electromagnetic field in free space, far away from the antenna structure, we can adopt $\sigma = 0$ and thus put $\boldsymbol{J} \equiv 0$, $\varepsilon = \varepsilon_0$, $\mu = \mu_0$. Using $\varepsilon_0 \mu_0 = 1/c^2$, we find for harmonic waves

$$\boldsymbol{E} = \frac{\mathrm{i}c}{k} \nabla \times \boldsymbol{B} = \frac{\mathrm{i}c}{k} \nabla \times (\nabla \times \boldsymbol{A})\ . \tag{5.6}$$

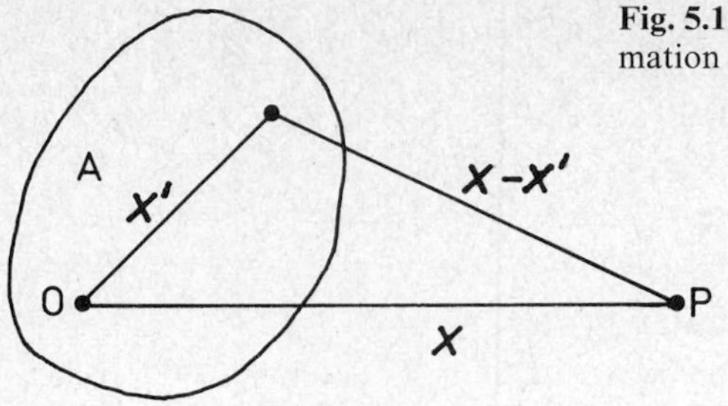

Fig. 5.1. The geometry of the Fraunhofer approximation

Therefore, given the current distribution $\boldsymbol{J}$ in the antenna, the electromagnetic field can be determined provided the integral (5.3) can be computed. For all but the most simple configurations like the Hertz dipole this is quite a formidable task which often can only be done using numerical methods. But in certain situations this is the only method that can be applied, and it has been successfully used in determining the radiation characteristics for satellites with complicated shapes.

Here we are interested in general results that can be stated analytically, even if they apply only approximately. A considerable simplification of (5.3) is already possible if we restrict the investigation to the radiation field far away from the antenna. Even more far-reaching simplifications are possible if we assume that the linear dimensions of the antenna structure are small compared to the distance of the point $\boldsymbol{x}$ from it. This then is the "Fraunhofer approximation", for which (Fig. 5.1)

$$\begin{aligned} |\boldsymbol{x} - \boldsymbol{x}'| &\cong r - \boldsymbol{n} \cdot \boldsymbol{x}' \\ r &= |\boldsymbol{x}| \end{aligned} \tag{5.7}$$

where $\boldsymbol{x}'$ is an arbitrary point inside the antenna structure and $\boldsymbol{n}$ is a unit vector in the direction of $\boldsymbol{x}$ ($\boldsymbol{n} = \boldsymbol{x}/|\boldsymbol{x}|$). For $r \gg \boldsymbol{n} \cdot \boldsymbol{x}'$ Eq. (5.3) then becomes

$$\boldsymbol{A}(\boldsymbol{x}) = \frac{\mu}{4\pi} \frac{\mathrm{e}^{ikr}}{r} \iiint_V \boldsymbol{J}(\boldsymbol{x}') \mathrm{e}^{-ik\boldsymbol{n} \cdot \boldsymbol{x}'} d^3 x' \ . \tag{5.8}$$

But this is the vector potential for an outgoing spherical wave.

If we now consider filled surface antennas such as parabolic dishes, the current $\boldsymbol{J}$ will be restricted to the surface of this dish with a corresponding simplification of (5.8). Even further simplifications are possible if the current distribution is such that all currents are parallel, that is, if

$$\boldsymbol{J}(\boldsymbol{x}') = g(\boldsymbol{x}')\boldsymbol{J}_0 \ ; \qquad \boldsymbol{J}_0 = \text{const} \ .$$

Then (5.8) can be reduced to a scalar equation for $g(\boldsymbol{x})$.

But note that this is quite a strong restriction which will not apply to many practical configurations. This simplification does not, strictly speaking, even apply to deep parabolic dishes with a small f/D ratio because the current $\boldsymbol{J}$ will always be parallel to the surface, and therefore $\boldsymbol{J}(\boldsymbol{x})/|\boldsymbol{J}(\boldsymbol{x})|$ will vary across the surface. Such an antenna therefore cannot be described by a scalar equation equivalent to (5.8) if full precision is required. This is particularly so if polarization properties are to be investigated. But many parameters of a filled aperture antenna can be described adequately by considering the scalar equivalent of (5.8).

5.2 Aperture Illumination and Antenna Pattern

For a two-dimensional antenna structure the current distribution $\boldsymbol{J}(\boldsymbol{x})$ has to be specified for the aperture and the volume integral in (5.8) can be simplified to a two-dimensional one. We will simplify the arguments here further by assuming this aperture to be plane. In addition, many complicated intermediate steps can be avoided by making direct use of the formulae derived for the Hertz dipole. We choose the coordinate system such that the aperture is a finite area of the plane $z' = 0$; this aperture is assumed to be a surface of constant wave phase and the unit vector of the current density is chosen to be $\boldsymbol{J}_0 = (0, J_0, 0)$.

The current in a surface element $dx'\,dy'$ at $\boldsymbol{x}'$ in the aperture is then $J_0\, g(\boldsymbol{x}')\, dx'$, and the only component of the electric field induced by this current element in $\boldsymbol{x}$ is then according to (4.38), if we restrict $\boldsymbol{x}$ to be approximately perpendicular to $\mathscr{A}$

$$dE_y = -\frac{\mathrm{i}}{2}\sqrt{\frac{\mu_0}{\varepsilon_0}}\,\lambda\, J_0\, g(\boldsymbol{x}')\,\frac{F_\mathrm{e}(\boldsymbol{n})}{|\boldsymbol{x}-\boldsymbol{x}'|}\,\mathrm{e}^{-\mathrm{i}(\omega t - k|\boldsymbol{x}-\boldsymbol{x}'|)}\,\frac{dx'}{\lambda}\,\frac{dy'}{\lambda}\;. \tag{5.9}$$

Here $F_\mathrm{e}(\boldsymbol{n})$ is the field pattern of the current element for the direction $\boldsymbol{n} = \boldsymbol{x}/|\boldsymbol{x}|$ which, for the Hertz dipole, is $\sin\vartheta$ where ϑ is the angle between $\boldsymbol{n}$ and $\boldsymbol{J}_0$. The total field in $\boldsymbol{x}$ is then obtained by integrating (5.9) over the full aperture.

As we are only interested in the far-field, and since we assumed the extent of the aperture to be small compared to the distance r, this integration can be simplified considerably by introducing the approximation (5.7). Because $r \gg |\boldsymbol{n}\cdot\boldsymbol{x}'|$, we can neglect $\boldsymbol{n}\cdot\boldsymbol{x}'$ compared to r everywhere except in the exponent. There the term $k\,\boldsymbol{n}\cdot\boldsymbol{x}'$ appears. But we assumed an aperture larger than λ, that is $|\boldsymbol{x}'| > \lambda$, so that

$$k(\boldsymbol{n}\cdot\boldsymbol{x}') = \frac{2\pi}{\lambda}(\boldsymbol{n}\cdot\boldsymbol{x}') \gtrsim 1\;.$$

Therefore

$$E_y = -\frac{\mathrm{i}}{2}\sqrt{\frac{\mu_0}{\varepsilon_0}}\,\lambda\, J_0\,\frac{F_\mathrm{e}(\boldsymbol{n})}{r}\,\mathrm{e}^{-\mathrm{i}(\omega t - kr)}\iint\limits_{\mathscr{A}} g(\boldsymbol{x}')\,\mathrm{e}^{-\mathrm{i}k\boldsymbol{n}\cdot\boldsymbol{x}'}\,\frac{dx'}{\lambda}\,\frac{dy'}{\lambda} \tag{5.10}$$

or

$$E_y = -\,\mathrm{i}\,\lambda\, J_0\,\pi\sqrt{\frac{\mu_0}{\varepsilon_0}}\,\frac{F_\mathrm{e}(\boldsymbol{n})}{r}\,f(\boldsymbol{n})\,\mathrm{e}^{-\mathrm{i}(\omega t - kr)} \tag{5.11}$$

where

$$\boxed{f(\boldsymbol{n}) = \frac{1}{2\pi} \iint_{-\infty}^{\infty} g(\boldsymbol{x}') \mathrm{e}^{-\mathrm{i}k\boldsymbol{n}\cdot\boldsymbol{x}'} \frac{dx'}{\lambda}\frac{dy'}{\lambda}} \quad . \tag{5.12}$$

The integral over the aperture $\mathscr{A}$ in (5.10) has been formally replaced by the two-dimensional Fourier integral (5.12) by putting $g(\boldsymbol{x}') = 0$ for $\boldsymbol{x}' \notin \mathscr{A}$. An expression similar to (5.11) can be derived for the magnetic field strength $\boldsymbol{H}$.

The normalized power pattern P_{n} as defined by (4.47) is then given by

$$P_{\mathrm{n}}(\boldsymbol{n}) = \frac{P(\boldsymbol{n})}{P_{\max}} = \frac{|\boldsymbol{E}\cdot\boldsymbol{E}^*|}{|\boldsymbol{E}\cdot\boldsymbol{E}^*|_{\max}} \quad \text{so that}$$

$$\boxed{P_{\mathrm{n}} = \frac{|f(\boldsymbol{n})|^2}{|f_{\max}|^2}} \quad . \tag{5.13}$$

The application of (5.12) and (5.13) is illustrated best by an *example*:

The Normalized Power Pattern of a Rectangular Aperture with Uniform Illumination. If the linear dimensions of the aperture are L_x and L_y, then the current grading can be written as

$$g(x, y) = \begin{cases} 1 & \text{for } |x| \leq L_x/2, \ |y| \leq L_y/2 \\ 0 & \text{else} \end{cases} \quad . \tag{5.14}$$

The components of the unit vector are $\boldsymbol{n} = (l, m, n)$ with $l^2 + m^2 + n^2 = 1$. If the aperture is part of the plane $z' = 0$, then (5.12) becomes

$$f(l, m) = \frac{1}{2\pi} \int_{-\infty}^{\infty} \int_{-\infty}^{\infty} g(x', y') \exp\left\{-\mathrm{i}\frac{2\pi}{\lambda}(l x' + m y')\right\} \frac{dx'}{\lambda}\frac{dy'}{\lambda} \quad .$$

with (5.14) this becomes

$$f(l, m) = \frac{\sin(\pi l L_x/\lambda)}{\pi l L_x/\lambda} \frac{\sin(\pi m L_y/\lambda)}{\pi m L_y/\lambda} \tag{5.15}$$

and the normalized power pattern is

$$\boxed{P_{\mathrm{n}}(l, m) = \left[\frac{\sin(\pi l L_x/\lambda)}{\pi l L_x/\lambda} \frac{\sin(\pi m L_y/\lambda)}{\pi m L_y/\lambda}\right]^2} \quad . \tag{5.16}$$

The main beam is the solid angle between the first nulls of P_{n} at

$$l_0 = \pm\lambda/2L_x \ ; \qquad m_0 = \pm\lambda/2L_y \tag{5.17}$$

and the full half power beam width (HPBW), i.e. the angle between those points of the main beam for which $P_n = 1/2$, is then

$$\mathrm{HPBW}_x = 0.88\,\frac{\lambda}{L_x}\ \mathrm{rad} = 50\overset{\circ}{.}3\,\frac{\lambda}{L_x} \tag{5.18}$$

with a similar expression for HPBW_y. The first sidelobes have a height of $P_n = 0.0472$ corresponding to an attenuation of 13.3 dB (relative to the main beam).

Obviously all these numbers depend on the shape of g, and they would be changed by changing the illumination or grading $g(x, y)$. This will be investigated in more detail for the case of a spherical aperture because this is of greater practical importance in radio astronomy. But antennas with a rectangular aperture have also been used in radio astronomy. Many of the antennas used for the first radio astronomical investigations consisted of rectangular arrays of dipoles which could be described best as a full rectangular aperture. The Kraus-type antenna consisting of a fixed vertical paraboloid and a tiltable plane reflector has a rectangular aperture, and this is also true for the individual arms of the Mills cross at Molonglo, Australia. Its East-West arm is a tiltable cylindrical paraboloid with $L_x = 1600$ m and $L_y = 12$ m resulting in a $\mathrm{HPBW}_x = 1\overset{\prime}{.}4$, $\mathrm{HPBW}_y = 3\overset{\circ}{.}06$ for $\lambda = 73$ cm. Combining this with a North-South arm of identical dimensions a resulting resolution of $1\overset{\prime}{.}4 \times 1\overset{\prime}{.}4$ can be achieved.

5.3 Circular Apertures

For a circular aperture it is convenient to introduce polar coordinates ϱ, φ by

$$\begin{aligned} x &= \lambda\,\varrho\cos\varphi \\ y &= \lambda\,\varrho\sin\varphi\ . \end{aligned} \tag{5.19}$$

If we now assume that the aperture is defined by $\varrho \leqq D/2\,\lambda$ and that the current grading g depends on ϱ only, then the resulting beam pattern will show circular symmetry too. Instead of two directional cosines l and m, a single number u giving the sine of the angle beween $\boldsymbol{n}$ and the direction of the main beam (0, 0, 1) is sufficient. Substituting (5.19) into (5.12) we obtain

$$f(u) = \frac{1}{2\pi}\int_0^{2\pi}\int_0^{\infty} g(\varrho)\,\mathrm{e}^{-\mathrm{i}\,2\pi u\varrho\cos\varphi}\,\varrho\,d\varrho\,d\varphi\ . \tag{5.20}$$

Considering the integral representation of the Bessel function of order zero

$$J_0(z) = \frac{1}{2\pi}\int_0^{2\pi} \mathrm{e}^{\mathrm{i}\,z\cos\varphi}\,d\varphi\ , \tag{5.21}$$

(5.20) can be written as

$$f(u) = \int_0^\infty g(\varrho)\, J_0(2\pi u \varrho)\, \varrho\, d\varrho \quad . \tag{5.22}$$

For the case of circular symmetry the electric and magnetic field strength is thus the Hankel-transform of the current grading. For the normalized beam pattern we then obtain

$$P_n(u) = \left[\frac{\int_0^\infty g(\varrho)\, J_0(2\pi u \varrho)\, \varrho\, d\varrho}{\int_0^\infty g(\varrho)\, \varrho\, d\varrho}\right]^2 \tag{5.23}$$

because $J_0(0) = 1$.

For a circular aperture with uniform illumination, that is for

$$g(\varrho) = \begin{cases} 1 & \text{for } \varrho \leqq D/2\lambda \\ 0 & \text{else} \end{cases} , \tag{5.24}$$

(5.23) then becomes

$$P_n(u) = \left[\frac{\int_0^{D/2\lambda} J_0(2\pi u \varrho)\, \varrho\, d\varrho}{\int_0^{D/2\lambda} \varrho\, d\varrho}\right]^2 = \left[\frac{2\lambda^2}{\pi^2 u^2 D^2} \int_0^{\pi u D/\lambda} J_0(z)\, z\, dz\right]^2 .$$

But for Bessel functions, the relation

$$\frac{d}{dz}\{z^n J_n(z)\} = z^n J_{n-1}(z) \tag{5.25}$$

[c.f. Abramowitz and Stegun (1964), Eq. (9.1.30)] applies so that

$$x^n J_n(x) = \int_0^x z^n J_{n-1}(z)\, dz \ ,$$

and we obtain for the normalized power pattern

$$P_n(u) = \left[\frac{2 J_1(\pi u D/\lambda)}{\pi u D/\lambda}\right]^2 \quad . \tag{5.26}$$

If the region up to and including the first nulls of $P_n(u)$ at $\pi u D/\lambda = 3.83171$ is included in the main beam region, a full beam width between the first nulls of

$$\text{BWFN} = 2.439 \frac{\lambda}{D} \text{ rad} \cong 139\overset{\circ}{.}8 \frac{\lambda}{D} \tag{5.27}$$

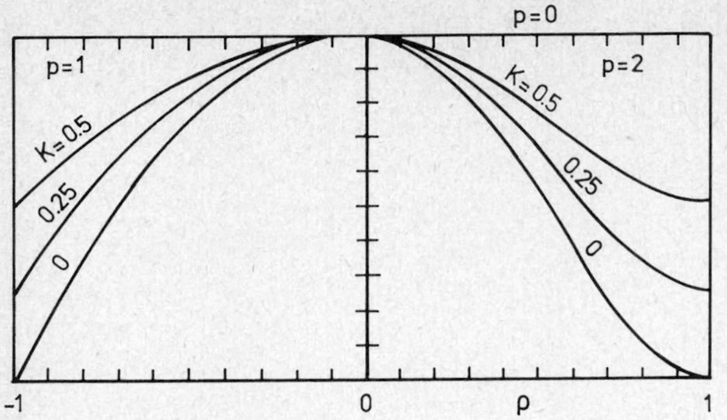

Fig. 5.2. A representative set of illumination tapers
$g(\varrho) = K + (1 - \varrho^2)^p$

and a full half power beam width

$$\mathrm{HPBW} = 1.02\,\frac{\lambda}{D} \cong 58.^\circ 4\,\frac{\lambda}{D} \tag{5.28}$$

is obtained.

For an aperture with a non-uniform illumination or grading the antenna pattern will be different from (5.26), see Fig. 5.2; the relation between grading and antenna pattern is given by (5.23). It can be quite difficult to evaluate this integral depending on the choice of $g(\varrho)$. But some ideas how P_n and the related parameters will change can be obtained by selecting a convenient interpolation formula for g which is chosen such that (5.23) can be evaluated.

Such a family of functions is

$$g(\varrho) = \left[1 - \left(\frac{2\lambda\varrho}{D}\right)^2\right]^p . \tag{5.29}$$

Because

$$\int_0^1 (1 - r^2)^p J_0(q\,r)\,r\,dr = \frac{2^p\,p!\,J_{p+1}(q)}{q^{p+1}} \tag{5.30}$$

[Gradshteyn and Ryzhik (1965), Eq. (6.567.1)], (5.23) can be evaluated in terms of J_p:

$$P_n(u) = \left[\frac{2^{p+1}\,p!\,J_{p+1}(\pi\,u\,D/\lambda)}{(\pi\,u\,D/\lambda)^{p+1}}\right]^2 . \tag{5.31}$$

All relevant parameters can now be computed from this. From the shape of P_n we find the half-power beamwidth HPBW and the beamwidth between the first nulls BWFN. Integrating (5.31) over the main beam area and the full sphere give the main beam solid angle Ω_M and total solid angle of the antenna Ω_A; their ratio then is the beam efficiency η_M (4.53), and the effective area of the antenna can be computed using (4.59), resulting in the aperture efficiency (see Fig. 5.3)

$$\eta_A = A_e \Big/ \frac{\pi D^2}{4} .$$

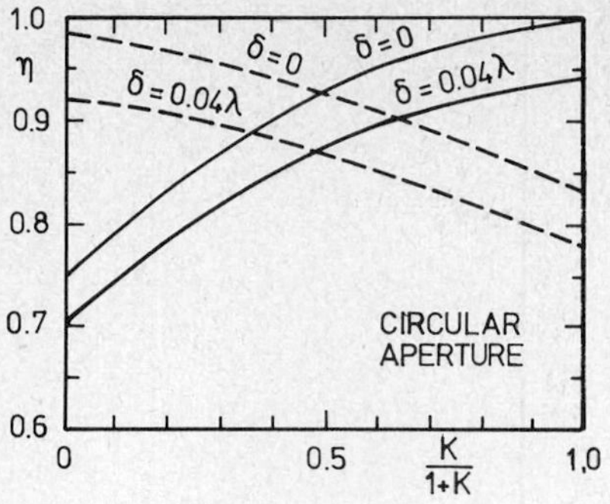

Fig. 5.3. Aperture efficiency η_A (——) and beam efficiency η_M (-----) for different values of K in (5.32). The values for both an ideal reflector ($\delta = 0$) and one that introduces random phase errors of $\delta = 0.04\,\lambda$ are given [after Nash (1964)]

Table 5.1. Normalized power pattern characteristics produced by an aperture illumination (5.32)

p	K	$\mathrm{HPBW} = k_1 \frac{\lambda}{D}$ rad	$\mathrm{BWFN} = k_2 \frac{\lambda}{D}$ rad	η_A	First sidelobe dB
0		1.02	1.22	1.00	−17.6
1		1.27	1.63	0.75	−24.6
2		1.47	2.03	0.56	−30.6
3		1.65	2.42	0.44	
4		1.81	2.79	0.36	
1	0.25	1.17	1.49	0.87	−23.7
2	0.25	1.23	1.68	0.81	−32.3
1	0.50	1.13	1.33	0.92	−22.0
2	0.50	1.16	1.51	0.88	−26.5

Table 5.1 gives numerical values for an aperture grading slightly more general than (5.29), for the two-parameter family

$$g(\varrho) = K + \left[1 - \left(\frac{2\lambda\varrho}{D}\right)^2\right]^p . \tag{5.32}$$

By selecting the tapering of the illumination, one can influence the properties of the antenna to some extent. A strong gradation results in a low sidelobe-level and a high beam efficiency but, simultaneously, the aperture efficiency becomes low. Therefore in practical antenna design some kind of compromise is needed and the choice will in general depend on the intended application. For antennas used for transmission it will differ from the choice selected for low-noise receiving antennas.

Since the far-off sidelobes are directed towards the hot (300 K) ground, low-noise antennas should have a beam pattern with low sidelobe level. This is achieved by using a strong gradation for $g(\varrho)$, while antennas used for high-power transmission in space mission must be designed for high effective aperture; that is, must use a low gradation resulting in a high sidelobe level.

5.4 Primary Feeds

In the preceding paragraphs we showed how the antenna pattern depends on the current grading across the aperture, but nothing was said about how this grading is achieved in a practical situation. A receiving antenna can be considered quite generally to be a device that transforms an electromagnetic wave in free space into a guided wave. The reflector transforms the plane wave into a converging spherical wave, and the primary feed then picks it up and transforms it into a wave in a waveguide.

For a successful antenna design many aims have to be met simultaneously and some of these are even contradictory, so that fulfilling one means that the other cannot simultaneously be met. Antenna design therefore is a subject that needs great experience. This is the more so since no general theory that covers all aspects is available.

It is quite true that, provided the current distribution $\boldsymbol{J}(\boldsymbol{x}')$ is given the vector potential $\boldsymbol{A}(\boldsymbol{x})$ and thus the electromagnetic fields $\boldsymbol{E}(\boldsymbol{x})$ and $\boldsymbol{H}(\boldsymbol{x})$ can be computed. But these then induce currents $\boldsymbol{J}$ so that we have the problem of self-consistency. These complications make rigorous solutions so difficult to obtain that Sommerfeld's 1896 rigorous solution of the diffraction of a plane wave by a perfectly conducting semi-infinite screen has not been significantly surpassed even today. Methods for approximate solutions are therefore nesessary.

Due to the reciprocity theorem the parameters of a given antenna are identical if used for transmission or for reception, but some concepts are more easily visualized if a receiving situation is assumed, while others are best understood in a transmission setup. Current grading as influenced by the feed is probably most simply discussed for transmission.

Let us therefore assume the primary feed to be a transmitting antenna. At point $\boldsymbol{x}'$ of the reflector this power will induce a surface current depending on the amplitude of the oscillating field strength. If the primary feed is far enough away from the reflector that far-field conditions can be adopted, the relative distribution of the field strength (both the electric and the magnetic one) can be computed from the normalized power pattern of the feed and, in most cases, this can be used at least as a first approximation.

By taking into account the various effects in which the real configuration deviates from the ideal far-field situation gradual improvements of the antenna theory can be obtained. But it is usually not necessary to know these details in order to be able to use an existing system; they are only needed if the antenna performance is to be improved in certain aspects. What is always needed, however, is a working knowledge of the advantages and inherent limitations of the most frequently used systems.

5.4.1 Dipole and Reflector

The most simple feed is formed by installing a short dipole at the focus, the dipole wings being parallel to the aperture. Because the beam pattern is torus-shaped with the wings of the dipole as axis, half of the beam pattern is directed away

Fig. 5.4. Dipole-disk feed assembly. An array of 2×2 dipoles forms two orthogonal beams concentric with a horn feed. The isolation between feeds is < -30 dB and the mutual influence of the radiation pattern is negligible (courtesy Max-Planck-Institut für Radioastronomie, Bonn)

from the reflector. This can be improved upon by putting a reflecting disk $\lambda/4$ away from the dipole.

Such dipole-disk (see, e.g., Fig. 5.4) feeds have frequently been used as primary feeds for deep parabolic dishes with a small f/D ratio (f = focal length, D = diameter). These are not very good primary feeds, but for focal plane reflectors ($f/D = 0.25$) that subtend the half-sphere surrounding the focus, very few other feed constructions able to illuminate such a great angle are available and therefore they are still used sometimes. For reflectors of greater f/D ratio simple dipole-disk feeds produce quite large spillover losses. Sometimes the illumination angle has been adapted to the reflector by using small dipole-arrays as prime-feeds, but usually this was done only if the central focus position was already occupied by a feed for some other frequency. A dipole-feed, which is sensitive to linear polarization only, has the electric field strength directed parallel to the dipole wings.

The greatest disadvantage of a dipole-feed, however, is that it produces a rather non-uniform illumination resulting in a non-circular mainbeam of the telescope (Fig. 5.5). And since the phases of the electromagnetic waves vary

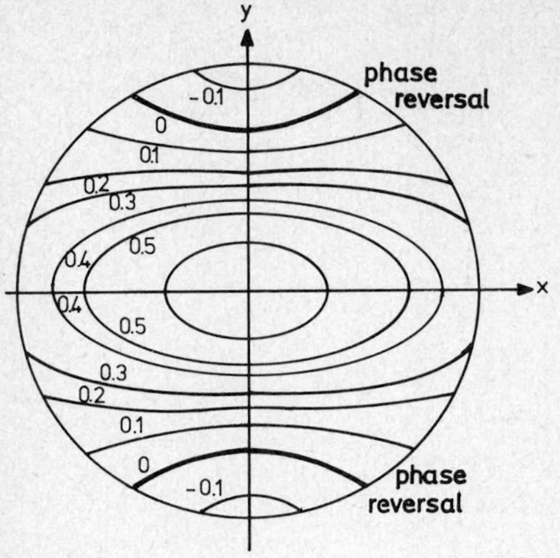

Fig. 5.5. Equivalent current distribution in the aperture plane of a dipole-feed paraboloid. The numbers give the absolute value of the current density, phases are not indicated, except for the locus of the phase reversal [after Heilmann (1970)]

rather strongly across the aperture, both the aperture efficiency and the beam efficiency of dipole-disk feeds are rather low.

5.4.2 Horn-Feeds

The electric and magnetic field strengths of a wave propagating along a waveguide will vary across its aperture, and therefore this aperture will radiate into free space; the power pattern of this radiation depends both on the dimension of the waveguide in units of the wavelength and on the mode of the wave. The greater the dimension of the waveguide in units of the wavelength, the greater is the directivity of this power pattern. However, the larger the cross-section of a waveguide becomes in terms of the wavelength, the more difficult it becomes to make certain that only a single wave mode is excited and therefore waveguides of a given size are used only for a limited frequency range. The aperture required for a selected directivity is then obtained by flaring the sides of a section of the waveguide so that the waveguide becomes a horn. These flaring sections act like waveguide filters preventing the unwanted higher modes from being propagated.

Such simple pyramidal horns (Fig. 5.6) are usually designed to transmit the TE_{10} (or H_{10}) mode. But for this mode, the electric and magnetic field strengths are unfortunately distributed rather differently along the sides of the horn aperture. The electric field strength varies as $\cos(\pi x/a)$ where x is the position in the aperture while the magnetic field vectors are independent of the position y.

Even if the mainlobe of the horn can be made to have a circular cross-section by a proper choice for the ratio a/b of the sides a and b of the horn aperture, the sidelobes will differ strongly due to the difference in the grading in the horn aperture in different directions.

A much better approximation of the sidelobe-structure for orthogonal directions can be achieved by a horn-cluster. In this the fields of three horns mounted side-by-side are coupled such that the electric fields are identical, while the magnetic field strength in the central horn is larger than in the two neighbouring ones. The resulting beam shape and the sidelobe structure is of fairly good circular

Fig. 5.6. A simple pyramidal horn feed (courtesy Max Planck-Institut für Radioastronomie, Bonn)

Fig. 5.7. A hybrid-mode feed for $f = 35$ GHz ($\lambda = 8$ mm) (courtesy Max Planck-Institut für Radioastronomie, Bonn)

symmetry, but this applies only to linearly polarized radiation of the proper polarization angle. If radiation of arbitrary polarization is to be measured different kinds of feeds have to be used.

Fortunately, great advances in the design of feeds have been made in the years since 1970, and most parabolic dish antennas now use hybrid mode feeds (Fig. 5.7). If a truly circular beam for an arbitrary polarization angle is wanted, obviously no TE-modes can be used; the electric field in the aperture has to be orientated in the direction of propagation. But then the conductivity of the horn in this direction has to be zero. This is achieved by making up the wall of the circular horn from rings that form $\lambda/4$-chokes in the z-direction of propagation.

Horns of this kind, the theory of which is rather complicated, can be adapted to practically all parabolic dishes, and they are suitable for any kind of polarization measurements.

5.4.3 Cassegrain and Gregorian Systems

When the size of a radio telescope as measured in units of the wavelength is large, quite often a design similar to that of optical telescopes is used, and so both Cassegrain and Gregorian systems have been built successfully. In a Cassegrain system a convex hyperbolic reflector is introduced into the converging beam just before the prime focus. This hyperbolic reflector transfers the converging rays to a secondary focus which, in most practical systems is situated closely to the apex of the main dish. A Gregorian system uses for the same purpose a concave reflector with an elliptical profile which, however, must be positioned behind the prime focus in the diverging beam.

There are several reasons why such systems are useful in radio astronomy. In small telescopes quite often the weight of the secondary reflector is less than that of bulky receiver frontends, especially if these have to be cooled. And in addition, they are usually more easily mounted and are more accessible at the apex. Because the effective focal ratio f/D of Cassegrain or Gregorian systems (Fig. 5.8) is usually 5 to 10 times larger than that of prime-focus systems, the usable field where coma distortions are negligible is much larger in such two mirror systems than in prime-focus configurations, so that often several receiving systems working at different frequencies can be accomodated in the secondary focus.

Finally, it is much easier to build low-noise systems using a folded design. High aperture efficiency requires a current grading with a good illumination up to the edge of the dish. If, however, in a prime-focus configuration, the power pattern of the feed extends beyond the edge of the dish, the feed will see the thermal radiation of the ground with $T_b \cong 300$ K in these spillover lobes. In a system with a secondary reflector the feed illumination beyond the edge of the secondary reflector sees radiation from the sky with only a few K, resulting in an overall system noise temperature significantly less than in prime-focus systems.

Quite naturally Cassegrain and Gregorian systems have disadvantages too. While the angle that the primary disk subtends as seen from the focus is usually between 100° and 180°, a secondary reflector usually subtends only 10°–15°, so that the primary feed horns must have much greater directivity and consequently greater dimensions. The dimensions go up in direct proportion to the wavelength

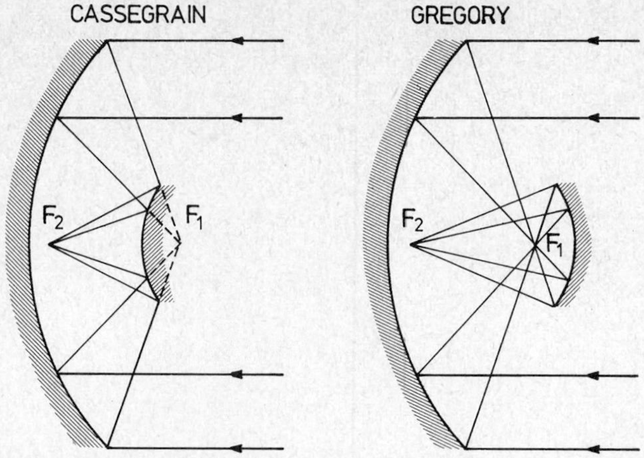

Fig. 5.8. The geometry of Cassegrain and Gregory systems

and therefore there is usually a lower limit for frequencies measured at the secondary focus. For the Effelsberg 100 m secondary focus this limit is near $f = 2.3$ GHz, for which frequency the primary feed horns have aperture diameters of 1.5 m and overall lengths of 3 m.

Another disadvantage of two-reflector systems is that they are quite often more susceptable to "baseline ripple" across the receiver band than prime-focus systems. Part of this ripple is caused by multiple reflections of radionoise in the antenna structure in front of the receiver. The effect is discussed in somewhat more detail in Sect. 10.6. But this is one of those features of a telescope than can only be handled by experience, and its influence on the observations can be reduced quite significantly by a proper arrangement of the observing procedure.

5.5 Antenna Tolerance Theory

When the relation between aperture illumination and antenna pattern was derived in Sect. 5.2, the aperture was assumed to be a plane of constant phase. If the conditions in a telescope deviate from this, some of the results will have to be modified. These modifications, due to phase variations across the aperture are the subject of this section but most results will only be stated qualitatively; a more detailed treatment can be found in textbooks on antenna theory.

It is convenient to distinguish several different kinds of phase errors in the current distribution across the aperture of a two-dimensional antenna.

A phase error that varies linearly along some direction across the aperture is treated most simply by defining a new aperture plane orientated such that the phase remains constant across it. All directions then have to be measured relative to this new aperture plane. A linear phase error therefore results only in a tilt of the direction of the main beam.

The investigation of a phase error which varies quadratically across the aperture is more difficult and leads to the introduction of Fresnel's integrals of the theory of diffraction which describe the conditions of the electromagnetic field in a slightly defocused state. We will not discuss this further here, but such errors can always be avoided by carefully focussing the telescope.

A third class of phase errors which is to some degree unavoidable is caused by the fabrication tolerances of the reflector. The theoretically requested shape of the reflector can be achieved only up to some finite limit of tolerance ε which will then cause a phase error

$$\delta = 4\pi \frac{\varepsilon}{\lambda} \tag{5.33}$$

of the equivalent current density $\boldsymbol{J}(\boldsymbol{x}')$ in the aperture plane. δ is measured in radians if ε is the displacement of the reflector surface element in the direction of the wave progagation.

The current grading in the aperture plane according to (5.9) can then be written as

$$g(\boldsymbol{x}) = g_0(\boldsymbol{x})\,\mathrm{e}^{\mathrm{i}\,\delta(\boldsymbol{x})}\,, \qquad g_0 \text{ real}\,. \tag{5.34}$$

The directivity gain of the reflector is, according to (4.48), (5.12) and (5.13)

$$G = \frac{4\pi}{\lambda^2} \frac{\left|\iint\limits_{\mathscr{A}} g_0(\boldsymbol{x})\,\mathrm{e}^{-\mathrm{i}[k\boldsymbol{n}\cdot\boldsymbol{x} - \delta(\boldsymbol{x})]}\,d^2x\right|^2}{\iint\limits_{\mathscr{A}} g_0^2(\boldsymbol{x})\,d^2x}\,. \tag{5.35}$$

Assuming δ to be a small quantity that is randomly distributed across the aperture with a given distribution function, the exponential function in (5.35) can be expanded in a power series including terms up to the second order

$$\mathrm{e}^{\mathrm{i}\,\delta} = 1 + \mathrm{i}\,\delta - \tfrac{1}{2}\delta^2 \ldots\,.$$

The ratio of the directivity gain of a system with random phase errors δ to that of an error-free system G_0 of identical dimensions then becomes

$$\frac{G}{G_0} = 1 + \bar{\delta}^2 - \overline{\delta^2} \quad \text{where} \tag{5.36}$$

$$\bar{\delta} = \frac{\iint\limits_{\mathscr{A}} g_0(\boldsymbol{x})\,\delta(\boldsymbol{x})\,d^2x}{\iint\limits_{\mathscr{A}} g_0(\boldsymbol{x})\,d^2x} \quad \text{and} \tag{5.37}$$

$$\overline{\delta^2} = \frac{\iint\limits_{\mathscr{A}} g_0(\boldsymbol{x})\,\delta^2(\boldsymbol{x})\,d^2x}{\iint\limits_{\mathscr{A}} g_0(\boldsymbol{x})\,d^2x}\,. \tag{5.38}$$

By selecting a suitable aperture plane, we can always force $\bar{\delta}$, the illumination weighted mean phase error, to be zero so that only the illumination weighted mean square phase error remains resulting in

$$\frac{G}{G_0} = 1 - \overline{\delta^2} \,. \tag{5.39}$$

For practical applications this series expansion has two drawbacks: it is valid only for small δ while phase errors of $\delta \gtrsim 1$ and even larger will occur if antennas are used near their short-wave limit.

The second reason why a more more sophisticated antenna tolerance theory is needed is that the phase errors $\delta(\boldsymbol{x})$ are not completely independent and randomly distributed across the aperture. If at some point $\delta < 0$, chances are great that $\delta < 0$, too, in an area surrounding this point. The reason for this is that the reflecting surface is smooth and has a certain stiffness, so that some kind of correlation distance for the phase errors has to be introduced. If this correlation distance d is of the same order of magnitude as the diameter of the reflector, part of the phase error can be treated like a systematic phase variation, either a linear one resulting only in a tilt of the main beam, or in a quadratic phase error which could be largely compensated for by refocussing. If $d \ll D$ the phase errors are almost independently distributed across the aperture. Ruze (1952, 1966) showed that a good estimate for the expected value of the rms phase error is given by

$$\boxed{\overline{\delta^2} = \left(\frac{4\pi\varepsilon}{\lambda}\right)^2 \left[1 - \exp\left\{-\frac{\Delta^2}{d^2}\right\}\right]} \tag{5.40}$$

where Δ is the distance between two points in the aperture that are to be compared and d is the correlation distance. The gain of the system now depends both on $\overline{\delta^2}$ and on d and, in addition in a complicated way both on the grading of the illumination and on the way δ is distributed across the aperture, even if d remains fixed. Ruze has given several approximations, the most simple being

$$G(u) = \eta\left(\frac{\pi D}{\lambda}\right)^2 \mathrm{e}^{-\overline{\delta^2}} + \left(\frac{2\pi d}{\lambda}\right)^2 (1 - \mathrm{e}^{-\overline{\delta^2}})\, \Lambda_1\left(\frac{2\pi d u}{\lambda}\right) \tag{5.41}$$

where

- η is the aperture efficiency
- $u = \sin \vartheta$
- $\Lambda_1(u) = \frac{2}{u} J_1(u)$ is the Lambda function
- D the diameter of the reflector, and
- d the correlation distance of the phase errors.

The surface irregularities thus have a twofold effect:

1) The axial gain is reduced (compare Fig. 5.9) to

$$G = \eta\left(\frac{\pi D}{\lambda}\right)^2 \exp\left[-\left(\frac{4\pi\varepsilon}{\lambda}\right)^2\right] \quad \text{and} \tag{5.42}$$

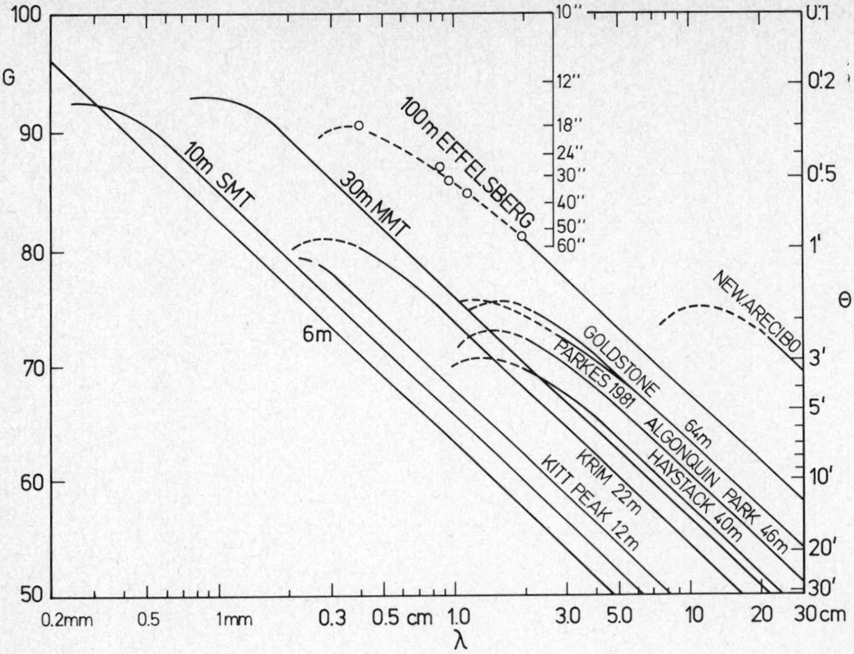

Fig. 5.9. Gain G in dB of some high-precision filled aperture radio telescopes. θ is the full half power beam width

2) the sidelobe level is increased in a way that depends both on $\overline{\delta^2}$ and on the correlation distance d. For large $\overline{\delta^2}$ and d, an error beam surrounding the main beam is formed, the width of which is governed by d: the HPBW of this error beam is approximately

$$\mathrm{HPBW_e} = 0.70\,\frac{\lambda}{d}\,[\mathrm{rad}] \cong 40^\circ\,\frac{\lambda}{d}\ . \tag{5.43}$$

Thus when Harten (1973) finds error beams with $\mathrm{HPBW_e} \cong 2^\circ$ and 6° at $\lambda = 21$ cm for the old surface of the Greenbank 300′ telescope, this corresponds to a correlation distance of $d \cong 105$ and 320 cm.

Therefore the directive gain of a filled surface antenna with phase irregularities δ does not increase indefinitely with increasing frequency but reaches a maximum at $\lambda_m = 4\pi\varepsilon$, and the gain thus achieved is 4.3 dB below that of an errorfree antenna of identical dimensions. If therefore the frequency can be determined at which the gain of a given antenna attains its maximum, the rms phase error and the surface irregularities ε can be measured electrically. Experience with the 100 m telescope at Effelsberg shows reasonably good agreement of such values for ε with direct measurements giving empirical support for the Ruze tolerance theory.

5.6 The Practical Design of Parabolic Reflectors

The discussion in the preceding sections showed how the radio frequency characteristics of parabolic reflectors depend on the electric properties of the design and the precision with which it can be build. This has still to be supplemented by some remarks about how far mechanical and practical limitations cause restrictions on what can be achieved.

Due to the diurnal rotation of the earth, all celestial objects rotate about the celestial pole, and therefore some kind of mounting has to be provided that permits the compensation for this if prolonged measurements on any celestial object are required.

For small reflectors a straight-forward adaptation of the classical equatorial mounting for optical telescopes has frequently been used. In this the telescope is turned with constant angular velocity around a polar axis which is parallel to the earth's axis of rotation, while different declinations can be reached by tilting the reflector about the declination axis perpendicular to the polar axis.

The advantage of this design is the simplicity of the resulting telescope control: as long as the telescope is aiming at a point with fixed celestial coordinates the telescope must rotate only about the polar axis with a constant angular velocity. On the other hand this kind of mounting has the strong disadvantage that the forces due to the weight of the construction act at an arbitrary angle on the bearings, and this angle is ever changing in the course of the diurnal motion for the case of the declination axis.

For these mechanical reasons equatorial mountings are usually restricted to small-diameter telescopes with $D \lesssim 25$ m; larger telescopes are generally supported by an alt-azimuthal mounting. The azimuthal axis is vertical, the elevation axis horizontal and both remain so even when the telescope is turning. The gravity load of the telescope acts either parallel or perpendicular to these axes and, when the dish is tilted, the resulting gravitational force vector will always remain in a plane as seen from the dish, while for the equatorial mounting this force vector can point to any direction within a hemisphere.

The price for this mechanical simplicity is a more complicated control needed for the angular velocity with which the telescope has to turn about the azimuthal and elevation axes, since both angular velocities have to be time-variable in order to compensate for the diurnal rotation of the earth.

For celestial positions which pass through the zenith, the azimuthal angular velocity becomes singular, so that no observations are possible in a region surrounding the zenith. The size of this region depends on the maximum possible speed for azimuth, but usually a field with a diameter of not more than $2-5°$ has to be avoided.

When considerations of the costs are of prime importance, savings are often possible by restricting the range of the motion for the telescope. A transit telescope can only be tilted in elevation and the radio sources are passed through beam by their diurnal motion. The possible observing time can be prolonged somewhat by moving the primary feed, but the range of this motion is limited by the off-axis aberrations (coma and astigmatism) of the paraboloid. The NRAO 90 m

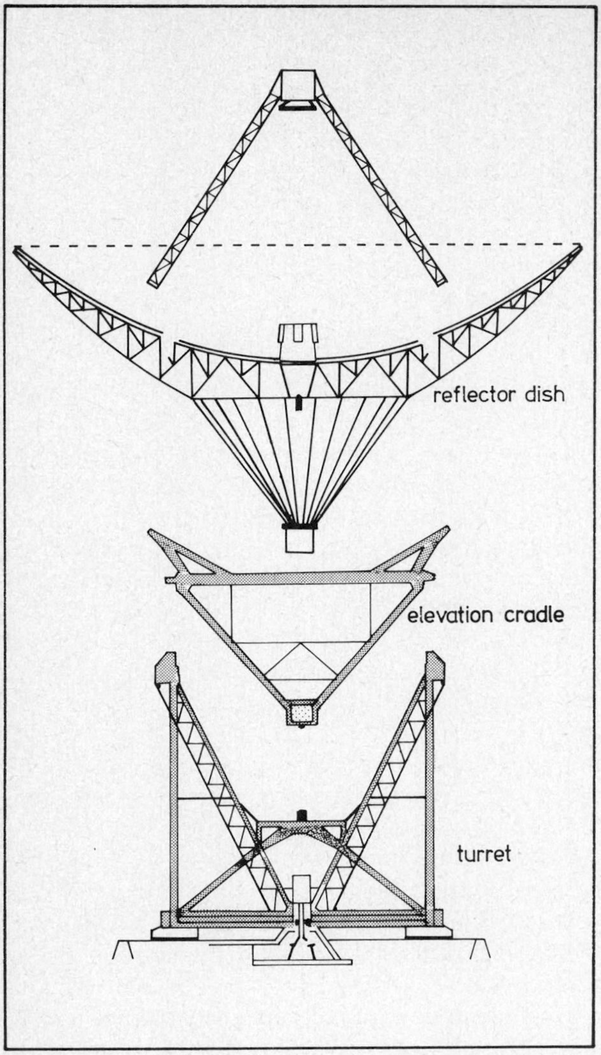

Fig. 5.10. The design principle of the 100 m-telescope at Effelsberg (after Grahl)

transit telescope in Greenbank, USA, is a well-known example of this type of telescope.

Transit telescopes of another kind are the Kraus standing parabola and tiltable flat reflectors at Ohio State University, USA, and at Nançay, France.

Even more restricted in telescope motion is the fixed 300 m spherical reflector at Arecibo, Puerto Rico. The dish is completely stationary in a suitable excavation in the ground. To avoid the large off-axis aberrations of parabolas the shape of the reflector was chosen to be spherical, since a sphere has no preferred axis. But a sphere has no true focus, only a focal line, and therefore line feeds have to be used to compensate for the spherical aberration. By moving this line feed towards the zenith along a truss rotating about the reflector axis, any direction within 20° of the zenith can be reached.

Fig. 5.11. The 64 m Parkes telescope (courtesy CSIRO)

In the antenna tolerance theory of the preceding section we showed that the shape of the reflecting surface has to be within $\lambda/20$ in order to achieve a gain (of the dish) that is within 67% of that of an ideal reflector. For a 30 m radio telescope usable down to $\lambda = 2$ mm this means a tolerance of 0.1 mm and what is more, this requirement must be met for all positions at which the telescope can be pointed.

Any structure that is built from existing material will show flexure due to its own weight if tilted. This is caused by the finite maximum stress that the material can transmit, the modulus of elasticity and the density of the material used for the construction. The geometric shape introduces only a numerical factor; it cannot completely suppress the deformation. In order to obtain some idea of the size of these deformations it should be noted that the rim of the dish of the 100 m telescope at Effelsberg (Fig. 5.10) is deformed by about 60 mm when the telescope is tilted from zenith to horizon. This should be compared to the required precision of about 0.5 mm rms if the telescope is to be fully usable for $\lambda = 1$ cm.

Compensation for these deflections with servos is complicated and expensive and has not yet been applied successfully to radio telescopes. However, closer scrutiny shows that what is needed is not the absence of deflections but only that the shape of the reflector should remain a paraboloid of revolution; changes in both the shape and the position (apex, focal point and axis) of this paraboloid could be tolerated. Such deformations are called "homologous", and it is ob-

Fig. 5.12. The 100 m-telescope at Effelsberg (courtesy Max-Planck-Institut für Radioastronomie, Bonn)

viously imperative in the telescope construction that only such deformations occur.

The first large radio telescope designed with these considerations in mind was the 100 m telescope at Effelsberg and the success of these ideas was such that this telescope is usable at wavelengths as short as $\lambda = 1$ cm, because the error due to telescope deformations always remains smaller than 0.4 mm. In the meantime the advance in telescope design is such that deformations due to gravity do not limit the performance of radio telescope at any wavelength. The practical limits are now given by thermal and wind deformations (Fig. 5.12).

As the dishes of radio telescopes usually are built from either steel tubes or beams, some kind of reflecting surface has to be installed to form the reflector proper. This surface may be perforated as long as the mesh size is only a fraction of the wavelength. For this reason the reflector surface is usually perforated sheet metal or wiremesh (see Fig. 5.11) if the intended wavelength is more than a few cm; for shorter wavelengths solid surfaces are preferred. Usually the surface is broken up into panels that can be adjusted individually to the desired paraboloid. At-

Fig. 5.13. A large modern mm wave telescope: the 30 m-telescope at Pico Veleta (courtesy Max Planck-Institut für Radioastronomie, Bonn)

Fig. 5.14. A small, versatile mm wave telescope: the 3 m-telescope at Cologne (courtesy *Winnewisser*, University of Cologne)

Fig. 5.13

Fig. 5.14 ▲

tempts to include the surface into the supporting structure of the dish have not been successful.

Usually some equipment has to be supported near the position of the prime focus. In a prime-focus telescope this is the primary feed and, in order to avoid the heavy losses of long transmission lines, the receiver fronted. If a secondary focus is used, as in Cassegrain or Gregory telescopes, a secondary reflector has to be supported. In symmetric telescopes both the supporting legs and the secondary reflector or the receiver box obstruct part of the aperture causing aperture blocking. Usually the loss of gain caused by the reduction of the effective aperture is of little significance. Of much greater importance is the influence on the sidelobe level. A strict theoretical treatment of this influence is rather difficult; therefore one usually has to fall back on empirical means. Supporting structures with three and four legs have been used; the resulting sidelobe structure shows a six-fold or a four-fold symmetry – at least at first approximation.

Considerations resulting in a high efficiency for the telescope are only part of the criteria in the design of radio telescopes. Of almost equal importance are features that result in an overall low-noise system. These refer mainly to the receiver design, but the telescope design is of importance, too.

Not all the radiation that enters the feed and thus eventually the receiver comes from the radio source at which the telescope is pointed. A large part of the signal, in practical systems up to 50%, comes from the immediate telescope vicinity. This could be radiation from the ground that is either leaking through the perforated reflecting surface or picked up by spillover lobes of the primary feed extending over the edge of the reflector dish. As pointed out already, the noise performance of Cassegrain or Gregorian systems is usually much better than that of systems with a primary feed because of such spillover lobes. The influence of spillover noise injection can be decreased by suitable screening jackets which have much the same purposes as screening tubes in optical telescopes (Fig. 5.13, 5.14).

5.7 Antenna Calibration

In order to be able to use a radio telescope as a research tool its relevant parameters as outlined in this and the preceding chapter have to be determined. Some of these, like the diameter of the reflector dish, are easily measured, while others, like the various efficiencies defined in Chap. 4, are more difficult to assess. For antennas used in radiocommunication this is usually done on an antenna test stand using transmitters with known power output. Since such a transmitter must be situated in the far-field

$$r > 2\,D^2/\lambda \tag{5.44}$$

if meaningful results are to be obtained, the required distance r will be in the order of $2 \cdot 10^3$ km if a telescope with $D = 100$ m is to be calibrated for $\lambda = 1$ cm. This is not possible to arrange on a routine basis and therefore radio telescopes must be calibrated using astronomical sources as standards.

Thus, in radio astronomy, we arrive at the the same two-step procedure that is characteristic for many astronomical applications: the measurements are calibrated using a set of celestial secondary calibration sources. To establish these calibration sources is a separate undertaking which is usually complicated and can only be done with much less precision than the relative calibration using the secondary standards.

The calibration standards are measured using special small telescopes, whose antenna parameters can either be computed theoretically with good confidence, or can be measured directly on an antenna test stand. As a result a list of radio sources with known flux densities for a wide range of frequencies is available; for convenience a list is given in Appendix E.

Practically all measurements made with a radio telescope are power measurements and therefore, as a first step in the calibration of a given combination of telescope, feed and receiver frontend, a calibrated noise power scale has to be established which refers to the receiver input. This is usually done by replacing the antenna signal by a noise signal of known strength as provided by a matched resistive load of known thermodynamic temperature. Since the output level for zero input power is not known (and cannot be achieved in a practical situation) the input of at least two different noise power levels as given by a hot (room temperature $T_n \cong 300$ K) and a cold load (either liquid N_2 or liquid He at $T_n = 78$ K or $T_n \cong 4.2$ K, respectively) have to be provided. The power output of the antenna can then be calibrated in K.

For the determination of the normalized power pattern $P_n(\vartheta, \varphi)$ of the telescope, this calibration is not even needed; P_n can be scanned directly using a strong point source of arbitrary strength.

However, if the antenna is pointed at a point source of known flux density S_ν, a power at the receiver input terminals

$$W_\nu \, d\nu = \tfrac{1}{2} A_e S_\nu \, d\nu = k \, T_A \, d\nu$$

is available. Thus

$$\boxed{T_A = \Gamma S_\nu} \tag{5.45}$$

where Γ is the *sensitivity* of the telescope measured in K Jy^{-1}. Introducing the aperture efficiency η_A according to (4.57) we find

$$\boxed{\Gamma = \eta_A \frac{\pi D^2}{8 k}} \quad . \tag{5.46}$$

Γ and η_A can thus be measured with the help of a calibrating source provided that the diameter D and the noise power scale in the receiving system are known.

The measurement of the main beam efficiency η_M of (4.53) can also be achieved using astronomical sources. In principle η_M could be computed by numerical integration of $P_n(\vartheta, \varphi)$ [cf. (4.51) and (4.52)], provided that $P_n(\vartheta, \varphi)$ could be measured for large range of ϑ and φ. Unfortunately this is not possible due to the confusion of the contribution of many independent sources. But again measurements of known astronomical objects can help.

Let us assume that there is a source with uniform brightness temperature over a certain solid angle Ω_s. The telescope then will provide an antenna temperature given by (4.64) which, for a constant brightness temperature across the source, simplifies to

$$T_A = \eta_R \frac{\int\limits_{\text{source}} P_n(\vartheta, \varphi)\, d\Omega}{\int\limits_{4\pi} P_n(\vartheta, \varphi)\, d\Omega} T_b$$

or, introducing (4.51)–(4.53),

$$T_A = \eta_R\, \eta_M \frac{\int\limits_{\text{source}} P_n(\vartheta, \varphi)\, d\Omega}{\int\limits_{\substack{\text{main}\\ \text{lobe}}} P_n(\vartheta, \varphi)\, d\Omega} T_b \ . \tag{5.47}$$

If the source diameter is of the same order of magnitude as the main beam the correcting factor of (5.47) can be determined with good precision from measurements of the normalized power pattern and thus (5.47) gives a direct determination of $\eta_R\, \eta_M$, the effective beam efficiency. A convenient source with constant surface brightness is the moon whose diameter of $\cong 30'$ is of the same order of magnitude as the beams of most large radio telescopes and whose brightness temperature

$$T_{b\,\text{moon}} \cong 225\ \text{K} \tag{5.48}$$

is of convenient magnitude.

Measurements of the mechanical properties of an antenna are of equal importance for its performance, especially if the telescope is of the kind which deforms homologously. These tend to be fairly soft in some of their mechanical parameters, so that it may become necessary to obtain information on how they vary when the dish is tilted.

In small telescopes with diameters up to 25 m the surface can often be adjusted with the help of templates. This is not possible for telescopes of larger diameter or if great precision is needed.

Surveying and adjustment of the surface panels can be done by measuring the position of fiducial marks with surveying techniques. Often the simplest method, using tapes along the surface and a theodolite from the apex of the dish, gives sufficient precision. Other methods use laser ranging, as well as the theodolite and, most recently, even holographic methods using a transmitter installed on a geostationary satellite as a source of coherent radiation. As a result the surface of a large reflector can be set with a precision of better than 0.1 to 0.5 mm depending on the diameter of the telescope and the effort put into the job.

The axial geometry of an alt-azimuth telescope is in principle identical to that of a theodolite, and thus both the theory of the errors for this kind of instrument and the methods for measuring them can be directly adapted. The pointing errors can be determined either by mounting a small optical telescope approximately parallel to the axis of the telescope or directly from radio measurements. Usually most of the errors remain fairly constant with time and can be included in the programme used for pointing and guiding the telescope. Other errors, like the collimation error, that depend on how the receiver currently in use is mounted in the telescope, will have to be determined more frequently.

Such efforts are usually supported by special pointing programmes for the control computer. If the control-programme includes such corrections, and if it compensates for known flexures of the telescope, changes in focus position, etc., then quite a considerable precision for the telescope pointing can be obtained. For the 100 m-telescope at Effelsberg in this way a rms pointing error of $\lesssim 10''$ in both coordinates is obtained, if no special efforts for obtaining positional accuracies are taken. If these efforts are made, an error of $\cong 4''$ is possible. Similar figures apply to other radio telescopes.

5.8 The Confusion Problem

Any antenna, irrespective of how it is constructed, will have only a finite angular resolution; it will receive radiation from a whole range of directions. According to (4.57, 4.62, 4.63) the antenna temperature at the receiver input is

$$T_{\mathrm{A}}(\boldsymbol{n}) = \eta_{\mathrm{R}}\,\eta_{\mathrm{A}}\left(\frac{\pi D^2}{8\,k}\right)\int_{4\pi} P_{\mathrm{n}}(\boldsymbol{n}-\boldsymbol{n}')\,B_\nu(\boldsymbol{n}')\,d\Omega'$$

or, with (5.46) and since $\eta_{\mathrm{R}} \approx 1$

$$\boxed{T_{\mathrm{A}}(\boldsymbol{n}) = \Gamma \int_{4\pi} P_{\mathrm{n}}(\boldsymbol{n}-\boldsymbol{n}')\,B_\nu(\boldsymbol{n}')\,d\Omega'} \quad . \tag{5.49}$$

This is the antenna convolution equation showing that the output of the antenna is always a weighted mean of the radiation for a whole range of directions. Taken literally, the radiation of the full sphere will enter into (5.49), but in a well-designed telescope only the radiaton in and close to the region of the main beam will be important. The signal measured is $T_{\mathrm{A}}(\boldsymbol{n})$. What is wanted though is $B_\nu(\boldsymbol{n})$ and therefore the convolution equation should be inverted. This is best done using the concepts of Fourier transforms.

Taking the Fourier transform of (5.49) the convolution theorem results in

$$t_{\mathrm{A}}(\boldsymbol{x}) = \Gamma\, p(\boldsymbol{x})\, b_\nu(\boldsymbol{x}) \tag{5.50}$$

where p and b_ν are the Fourier transforms of P_{n} and B_ν respectively. b_ν can thus be determined from t for those values of $\boldsymbol{x}$ where $p \neq 0$.

But taking the Fourier transform of (5.13) we find

$$p(\boldsymbol{x}) = \frac{1}{2\pi}\frac{1}{|f_{\max}|^2}\iint |f(\boldsymbol{n})|^2\, \mathrm{e}^{-ik\boldsymbol{n}\cdot\boldsymbol{x}}\, d\boldsymbol{n} \tag{5.51}$$

where, according to (5.12), $f(\boldsymbol{n})$ is the Fourier transform of the current grading across the antenna aperture. The autocorrelation theorem of Fourier analysis then shows that

$$p(\boldsymbol{x}) = p_0 \iint g(\boldsymbol{x}')\, g(\boldsymbol{x}-\boldsymbol{x}')\, dx'\, dy' \ . \tag{5.52}$$

The Fourier transform of the power pattern of an antenna is thus given by the autocorrelation function of the current grading. Therefore $p(\boldsymbol{x}) = 0$ for $\boldsymbol{x} \notin \mathscr{A}$, and (5.50) cannot be solved for such $\boldsymbol{x}$. The coordinate $\boldsymbol{x}$ here is a space-frequency as can be seen from (5.51), not a position in the aperture, and so we can state the result $p(\boldsymbol{x}) = 0$ for $\boldsymbol{x} \notin \mathscr{A}$ as: structures with space frequencies above certain limit – the resolution limit of the antenna – do not produce an output signal for the antenna. Therefore the intensity distribution as measured by a given telescope contains only a finite number of separate space frequencies. How many there are needed is given by the sampling theorem (Shannon):

> A function whose Fourier transform is zero for $|X| > X_c$ is fully specified by values at equal intervals not exceeding $\frac{1}{2}X_c^{-1}$ save for any harmonic terms with zeros at the sampling points [quoted from Bracewell (1965), p. 194].

For a two-dimensional distribution this theorem has to be applied to every direction so that as a practical rule-of-thumb, a grid-spacing of $\sqrt{2}/4$ HPBW $= 0.35$ HPBW is needed if the full information of the angular distribution is wanted.

Quite often the time needed for such a dense sampling is not available and only spacings of 1.0 HPBW are used. But note that this means undersampling with a possible resulting loss of resolution.

Sometimes efforts are made to compensate for the loss of resolution due to (5.49) by so called "restoration" of $B_\nu(\boldsymbol{n})$. Because $p(\boldsymbol{x}) = 0$ for $\boldsymbol{x} \notin \mathscr{A}$ this restoration is not unique. All we can hope for is the construction of the so-called "principal solution" $\bar{B}_\nu$ that differs from the true B_ν in that all components of its spatial frequencies above the sampling frequency are zero. Whether it is preferable to use $\bar{B}_\nu$ instead of T_A/Γ in discussing the physics of the objects is a matter that will have to be decided in each individual case. Today another approach is quite often taken in which the model B_ν is convolved with the beam P_n and the result is then compared directly with the measurements.

The whole problem of antenna convolution has been discussed extensively by Bracewell (1962) in his article on Radio Astronomy Techniques in Handbuch der Physik Vol. 54 where additional references can be found.

The finite angular resolving power of any antenna is the cause for another error source. When small diameter radio sources are measured, the finite angular width of the main beam not only limits the precision with which the source position can be determined, but also prevents the separation of closely neighbouring sources. Even a successful counting of sources is only possible if their density on the celestial sphere is low enough that the chances of two sources falling in the beam area can be neglected.

But even this may not be enough. In a source count survey only sources above a chosen flux limit will be counted; all fainter ones will be discarded. Due to the random distribution of these sources, it can happen that so many of these faint sources crowd by chance into a single beam area that their total flux is above the chosen limit and thus this agglomeration will be erroneously counted as a source. Mills and Slee first showed that this restricts the resolution limit more than the confusion of bright sources, and therefore this crowding of faint sources has to be considered.

Let us therefore determine how much the measured flux of a source with the true flux S_c will be affected by statistical crowding of the position for sources with a flux between S_L and S_c with $S_L \ll S_c$.

Let $p(S)$ be the probability density for the number of sources per steradian with a flux between S and $S + dS$; that is, on the average we will observe

$$\bar{\nu} = \Omega\, p(S)\, dS \tag{5.53}$$

sources with a flux in the interval $(S, S + dS)$ per beam Ω. Now ν, the number of sources per beam, must be by definition an integer number 0, 1,... and, if the sources are distributed at random, ν will be distributed according to a Poisson distribution

$$f(\nu) = \frac{\bar{\nu}^{\nu}}{\nu!}\,\mathrm{e}^{-\bar{\nu}} \tag{5.54}$$

as can be found in any textbook of mathematical statistics. The second moment μ_2 of this distribution is

$$\mu_2 = \sum_{\nu=0}^{\infty} \nu^2 f(\nu) = \bar{\nu}(\bar{\nu} + 1) \tag{5.55}$$

so that the dispersion becomes

$$\sigma_{\nu}^2 = \bar{\nu}\ . \tag{5.56}$$

For the sake of simplicity we will adopt the shape of a "pill-box" with perpendicular walls for the main beam, so that the total output of the antenna is the straight sum of the flux of all sources covered by the beam

$$S = \sum_k S_k\ . \tag{5.57}$$

Thus the average output $\bar{S}$ caused by sources $(S, S + dS)$ will be

$$\bar{S} = S\,\bar{\nu}\ . \tag{5.58}$$

Due to the Poisson statistics this output will have a second moment

$$\mu_2(S) = \sum_{\nu=0}^{\infty} S^2 f(\nu) = S^2 \sum_{\nu=0}^{\infty} \nu^2 f(\nu) = S^2\,\bar{\nu}(\bar{\nu} + 1) \tag{5.59}$$

and

$$\sigma_S^2 = S^2 \bar{\nu} \ . \tag{5.60}$$

If the signal is caused by sources with different fluxes, their dispersions (5.59) add quadratically. According to (5.53) and (5.59), sources with fluxes between S_c and S_L and a number density of $p(S)$ then result in a dispersion for the total flux of

$$\sigma_{S_c}^2 = \Omega \int_{S_L}^{S_c} S^2 p(S)\, dS \ . \tag{5.61}$$

Sources with a flux S_c therefore can be measured with a signal to noise ratio of

$$q = \frac{S_c}{\sigma_{S_c}} \quad \text{where} \tag{5.62}$$

$$\boxed{q^2 = \frac{S_c^2}{\Omega} \frac{1}{\int_{S_L}^{S_c} S^2 p(S)\, dS}} \ . \tag{5.63}$$

Practical experience with the distribution of faint point sources shows that

$$p(S) = n N_c S_c^n S^{-n-1} \tag{5.64}$$

for a fairly wide range of S with $n \sim 1.5$, so that

$$\boxed{q^2 = \begin{cases} \dfrac{1}{\Omega N_c} \dfrac{2-n}{n} \dfrac{1}{1-(S_L/S_c)^{2-n}} \ ; & n \neq 2 \\ \dfrac{1}{2} \dfrac{1}{\Omega N_c} \dfrac{1}{\ln (S_c/S_L)} \ ; & n = 2 \end{cases}} \ . \tag{5.65}$$

For $n < 2$ we find that

$$\lim_{S_L \to 0} q^2 = \frac{1}{\Omega N_c} \frac{2-n}{n}$$

while for $n > 2$, q^2 diverges if $S_L \to 0$. But in order to avoid an infinite total flux (5.57), either $n > 1$ or some finite lower flux limit S_L has to be adopted so that a finite q will result.

The signal-to-noise ratio of the faintest sources in a catalogue observed with a given instrument thus depends mainly on two parameters:

1) The average number of sources per beam ΩN_c at the limiting flux S_c, and
2) the power n of the source distribution density function.

Therefore, the larger the telescope, the fainter the lower flux limit S_c that can be chosen before the resulting signal-to-noise ratio q becomes excessive. This flux then is usually called the *confusion limit* for the telescope at the chosen frequency. For q a value between 3 and 5 is usually considered to be acceptable; it then corresponds to a probability of 10^{-3}–10^{-6} for an erroneous source identification.

In the derivation of the confusion limit of a telescope, we adopted a simplification that is not strictly necessary and can be dropped. We assumed a box-shaped main beam. If the true beam pattern is $P_{\rm n}(\boldsymbol{n})$, then the total flux measured is given by

$$S = \sum_k S_k P_{\rm n}(\boldsymbol{n}_k) \tag{5.66}$$

instead of (5.57). But according to Campbell's theorem (Rice 1954), the dispersion of this signal is

$$\sigma_S^2 = \int_{4\pi} [P_{\rm n}(\boldsymbol{n})]^2 \, d\Omega \int_{S_L}^{S_c} S^2 p(S) \, dS \tag{5.67}$$

so that a q^2 identical to (5.65) will result, provided that Ω is replaced by

$$\tilde{\Omega} = \int_{4\pi} [P_{\rm n}(\boldsymbol{n})]^2 \, d\Omega \; . \tag{5.68}$$

For a Gaussian beam $\tilde{\Omega} = \Omega_{\rm A}/2$, other beam shapes give similar results (see Höglund 1967 and Burns 1972).

6. Interferometers and Aperture Synthesis

6.1 The Quest for Angular Resolution

Using diffraction theory we have shown that the angular resolution of a radio telescope is $\delta \sim \lambda/D$ where δ is the smallest angular separation that two point sources can have in order to be recognized as separate objects, λ is the wavelength of the radiation received, and D is the diameter of the telescope. A numerical factor of order unity depends on the details of the telescope construction, such as the current grading, etc. Therefore, in order to improve this angular resolution for a given wavelength the diameter D of the telescope has to be increased.

Obviously there are practical limits for the telescope size which are in the range of 100–300 m depending on the technical and financial means available for the construction. Therefore radio astronomers have, right from the first days of radio astronomy, tried to improve the resolving power by linking two or more separate telescopes into an interferometer arrangement.

The principal idea behind this is that the direction to a point-like radio source is perpendicular to the planes of constant wave phase, that is, the planes of the crests or troughs of the electromagnetic waves. In this respect the individual telescopes of the interferometer act only as probes that sample the electromagnetic wave field; the coupling network determines the position of the crests (or troughs) at a selected moment. A source consisting of two separate point sources emits two separate wave fields that do not interact – they are incoherent, and this should be recognizable by the interferometer and be distinguished from the wavefield of a single point source.

This is only a simplistic sketch of the working principle of an interferometer but it permits to point out two important limitations:

1) An interferometer will work properly only if the electromagnetic field received can be well described by a wave with a definite wavelength, frequency, phase etc. All this is simple and straightforward for a strictly monochromatic wavefield. In the real situation, however, we have to deal with a superposition of wave trains of finite duration, with waves of different wavelength, and therefore questions of the coherence of the radiation come into play. As it turns out an interferometer is actually measuring the complex degree of coherence, and therefore the principle of partially coherent electromagnetic radiation fields have to be discussed first.

2) The precision with which the directions of the planes of constant phase can be determined will depend on λ/B, where B is the length of the interferometer base. Since values for B up to 10–20 km are technically feasible if the telescopes are

linked by cables, angular resolution of ~ 1 mas (10^{-3} arc seconds) can be achieved in this way. In VLBI (Very Large Baseline Interferometer) configurations in which the separate telescopes are not directly connected either by cables or by radio-links but are registering the received signals independently on magnetic tapes together with precise time marks, worldwide separations for the individual telescopes are possible, and angular resolutions of $\delta < 0.01$ mas have been achieved. But due to the difficulties in determining the phase of the mutual correlation function, VLBI interferometers have problems not encountered by phase-stable interferometers linked by cables. These problems will be discussed in some detail in this chapter.

6.2 The Mutual Coherence Function

It was shown in Chap. 3 on wave polarization that four parameters are generally needed to describe the full state of polarization of a quasi-monochromatic electromagnetic wavefield even at a single point. Therefore, when the wavefields at two separate points are to be compared, obviously a multitude of correlations can be formed. We will simplify the discussion by considering only wavefields in a simple state of polarization, so that one scalar quantity is all that is needed to describe the wavefield at a single point. This quantity could be any of the four Stokes parameters or any other component of an orthogonal representation of the wavefield. We will now consider the distribution of this one parameter over the whole wavefield.

Considering the simplest wavefield that can be imagined, that of a plane monochromatic wave, the field intensity at P_2 for all times can be calculated from that at P_1 once the phase-difference for these two points has been determined. An arbitrary polychromatic wavefield is in some respects the other extreme: even a full knowledge of the field and its time variation at P_1 does not give the slightest hint of the field at P_2 provided P_1 and P_2 are not too close together. The monochromatic plane wavefield is said to be fully *coherent*, while the second example is that of an *incoherent* wavefield; others will have properties that fall between these two extremes. Obviously a measure of this coherence is needed that can be determined with some practical set-up even for fields where the instantaneous values of the field strength at a chosen point cannot be measured. A useful measure of coherence must be based on some kind of time-average while making use of the fact that we will only consider stationary fields. Therefore the *mutual coherence function* of the (complex) wavefield $U(P_1, t_1)$ and $U(P_2, t_2)$ will be defined as

$$\boxed{\begin{aligned} \Gamma(P_1, P_2, \tau) &= \lim_{T \to \infty} \frac{1}{2T} \int_{-T}^{T} U(P_1, t)\, U^*(P_2, t+\tau)\, dt \\ &= \langle U(P_1, t)\, U^*(P_2, t+\tau) \rangle \end{aligned}} \quad . \qquad (6.1)$$

The sharp brackets are used to indicate the time-averaging as introduced in (3.49) and we will assume that the limits do exist. It will be recognized that the intensity of the wave (3.51) is a special case of this definition

$$I(P) = \Gamma(P, P, 0) = \langle U(P, t)\, U^*(P, t)\rangle \tag{6.2}$$

and that all quantities in the coherence matrix (3.55) are formed similarly, except that the vector properties of the wavefield enter into this case.

For a plane monochromatic wavefield propagating into the z-direction, Γ is easily computed.

Since

$$U(P, t) = U_0\, e^{i(kz - \omega t)} \quad \text{where}$$

$$P = (x, y, z)\ , \qquad k = 2\pi/\lambda = \text{const}\ , \qquad \omega = 2\pi\nu = \text{const}\ ,$$

we get

$$\Gamma(P_1, P_2, \tau) = |U_0|^2\, e^{i[k(z_1 - z_2) + \omega\tau]}\ . \tag{6.3}$$

The mutual coherence function of the travelling monochromatic wavefield is thus periodic with a constant amplitude and a wavelength equal to that of the original wavefield. But the coherence function is not propagating; it is a standing wave with a phase such that, for $\tau = 0$, $\Gamma = \Gamma_{max}$ for $z_1 = z_2$.

It is often useful to normalize Γ by referring it to a wavefield of intensity 1. Thus

$$\gamma(P_1, P_2, \tau) = \frac{\Gamma(P_1, P_2, \tau)}{\sqrt{I(P_1)\, I(P_2)}}\ . \tag{6.4}$$

For this complex degree of coherence, obviously

$$|\gamma(P_1, P_2, \tau)| \leqq 1\ . \tag{6.5}$$

6.3 The Coherence Function of Extended Sources: The van Cittert-Zernike Theorem

Taking Eqs. (6.3) and (6.4) together we see that for a monochromatic plane wave the complex degree of coherence has the constant amplitude 1, while an arbitrary polychromatic wavefield will result in $\gamma = 0$ for $P_1 \neq P_2$. It will be of interest to see which properties of a wavefield result in a partial loss of coherence. Therefore we will gradually reduce the restrictions imposed on the wavefield in order to see the effect this has on the degree of coherence.

A wavefield that is only slightly more complex than a monochromatic plane wave is formed by the (incoherent) superposion of two such wavefields with identical wavelengths (and frequencies) but propagating into different direc-

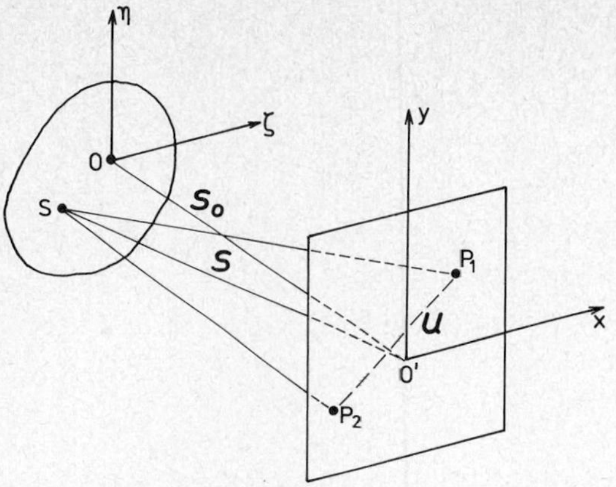

Fig. 6.1. Coordinate systems and designations used in the van Cittert-Zernike theorem

tions (Fig. 6.1):

$$\begin{aligned} U_a &= U_{0a}\,\mathrm{e}^{\mathrm{i}(k\boldsymbol{s}_a\cdot\boldsymbol{x}-\omega t)} \\ U_b &= U_{0b}\,\mathrm{e}^{\mathrm{i}(k\boldsymbol{s}_b\cdot\boldsymbol{x}-\omega t)}\ . \end{aligned} \tag{6.6}$$

Here $\boldsymbol{s}_a$ and $\boldsymbol{s}_b$ are unit vectors describing the propagation direction, and both $k = 2\pi/\lambda$ and $\omega = 2\pi\nu$ are assumed to be equal for both waves. The total wave field is then formed by

$$U = U_a + U_b$$

and the mutual coherence function (6.1) is

$$\begin{aligned} \langle U(P_1,t_1)\,U^*(P_2,t_2)\rangle &= \langle\{U_a(P_1,t_1) + U_b(P_1,t_1)\}\,\{U_a(P_2,t_2) + U_b(P_2,t_2)\}^*\rangle \\ &= \langle U_a(P_1,t_1)\,U_a^*(P_2,t_2)\rangle + \langle U_b(P_1,t_1)\,U_b^*(P_2,t_2)\rangle \\ &\quad + \langle U_a(P_1,t_1)\,U_b^*(P_2,t_2)\rangle \\ &\quad + \langle U_b(P_1,t_1)\,U_a^*(P_2,t_2)\rangle\ . \end{aligned} \tag{6.7}$$

If we now assume the two wavefields U_a and U_b to be *incoherent*, we adopt the fieldstrengths U_a and U_b to be uncorrelated even when taken at the same point so that

$$\langle U_a(P_1,t_1)\,U_b^*(P_2,t_2)\rangle = \langle U_b(P_1,t_1)\,U_a^*(P_2,t_2)\rangle \equiv 0\ . \tag{6.8}$$

Such incoherence is not possible for strictly monochromatic waves of identical polarization consisting of a wave train of infinite duration and length. Only if the wave is made up of sections of finite duration between which arbitrary phase jumps occur, can Eqs. (6.8) be correct. But then the waves are not strictly monochromatic but have a finite, although small bandwidth. The relation between coherence and bandwidth will be investigated more closely in the next chapter when the autocorrelation spectrometer is described.

Substituting (6.8) into (6.7) we obtain

$$\Gamma(P_1, P_2, \tau) = \langle U(P_1, t)\, U^*(P_2, t+\tau)\rangle$$
$$= \langle U_a(P_1, t)\, U_a^*(P_2, t+\tau)\rangle + \langle U_b(P_1, t)\, U_b^*(P_2, t+\tau)\rangle$$

or, using (6.6)

$$\Gamma(P_1, P_2, \tau) = |U_{0a}|^2\, \mathrm{e}^{\mathrm{i}(k\boldsymbol{s}_a\cdot\boldsymbol{u}+\omega\tau)} + |U_{0b}|^2\, \mathrm{e}^{\mathrm{i}(k\boldsymbol{s}_b\cdot\boldsymbol{u}+\omega\tau)} \tag{6.9}$$

where

$$\boldsymbol{u} = \boldsymbol{x}_1 - \boldsymbol{x}_2\ . \tag{6.10}$$

Thus only the difference of the two positions P_1 and P_2 enters into the problem. For the case of two waves of equal amplitude,

$$|U_{0a}| = |U_{0b}| = |U_0|\ ,$$

(6.9) can be simplified using the identities

$$\boldsymbol{s}_a = \tfrac{1}{2}(\boldsymbol{s}_a + \boldsymbol{s}_b) + \tfrac{1}{2}(\boldsymbol{s}_a - \boldsymbol{s}_b)$$
$$\boldsymbol{s}_b = \tfrac{1}{2}(\boldsymbol{s}_a + \boldsymbol{s}_b) - \tfrac{1}{2}(\boldsymbol{s}_a - \boldsymbol{s}_b)$$

resulting in

$$\Gamma(\boldsymbol{u}, \tau) = 2|U_0|^2 \cos\left(\frac{1}{2}k(\boldsymbol{s}_a - \boldsymbol{s}_b)\cdot\boldsymbol{u}\right)\exp\left[i\left(\frac{k}{2}(\boldsymbol{s}_a + \boldsymbol{s}_b)\cdot\boldsymbol{u} + \omega\tau\right)\right]\ , \tag{6.11}$$

or, if normalized,

$$\gamma(\boldsymbol{u}, \tau) = \cos\left(\frac{k}{2}(\boldsymbol{s}_a - \boldsymbol{s}_b)\cdot\boldsymbol{u}\right)\exp\left[\mathrm{i}\left(\frac{k}{2}(\boldsymbol{s}_a + \boldsymbol{s}_b)\cdot\boldsymbol{u} + \omega\tau\right)\right]\ . \tag{6.12}$$

For two waves propagating into directions that differ only slightly, $|\boldsymbol{s}_a - \boldsymbol{s}_b|/2$ is a small quantity, while $(\boldsymbol{s}_a + \boldsymbol{s}_b)/2$ differs only little from either $\boldsymbol{s}_a$ or $\boldsymbol{s}_b$. The normalized coherence function (6.12) therefore is similar to that of a single plane wave, but with an amplitude that varies slowly with position. We will have complete loss of coherence for

$$\frac{k}{2}(\boldsymbol{s}_a - \boldsymbol{s}_b)\cdot\boldsymbol{u} = (2n+1)\frac{\pi}{2}\ , \qquad n = 0, 1, 2\ldots\ . \tag{6.13}$$

This principle of superposition of simple monochromatic plane waves can be extended to an arbitrary number of plane waves, and the result will be an obvious generalization of (6.9) if we again assume these fields to be mutually incoherent. The signals at P_1 and P_2 are then the sum of the components $U_n(P, t)$

$$U(P, t) = \sum_n U_n(P, t) \tag{6.14}$$

and, if the different waves are incoherent, then

$$\langle U_m(P_1,t)\, U_n^*(P_2,t+\tau)\rangle = 0 \quad \text{for all } m \neq n \tag{6.15}$$

while

$$\langle U_n(P_1,t)\, U_n^*(P_2,t+\tau)\rangle = |U_{0n}|^2\, \mathrm{e}^{\mathrm{i}(k\boldsymbol{s}_n\cdot \boldsymbol{u}+\omega\tau)} \tag{6.16}$$

so that

$$\Gamma(\boldsymbol{u},\tau) = \langle U(P_1,t)\, U^*(P_2,t+\tau)\rangle = \sum_n |U_{0n}|^2\, \mathrm{e}^{\mathrm{i}(k\boldsymbol{s}_n\cdot \boldsymbol{u}+\omega\tau)} .$$

Or, if we go to the limit $n \to \infty$

$$\boxed{\Gamma(\boldsymbol{u},\tau) = \iint I(\boldsymbol{s})\, \mathrm{e}^{\mathrm{i}(k\boldsymbol{s}\cdot \boldsymbol{u}+\omega\tau)}\, d\Omega} \quad \text{where} \tag{6.17}$$

$$\boxed{I(\boldsymbol{s}) = \iint U(\boldsymbol{s})\, U^*(\boldsymbol{s})\, \mathrm{d}\Omega} \tag{6.18}$$

is the total intensity at the position P if the integral is taken over the angular extent of the directions $\boldsymbol{u}$, and the generalization of (6.14) is

$$\boxed{U(\boldsymbol{x},t) = \iint U(\boldsymbol{s})\, \mathrm{e}^{\mathrm{i}(k\boldsymbol{s}\cdot \boldsymbol{x}-\omega t)}\, d\Omega} \; . \tag{6.19}$$

(6.17) is the monochromatic version of the van Cittert-Zernike-theorem stating how the mutual coherence function of an arbitrary monochromatic wavefield (6.19) that is built up from plane waves is related to the intensity distribution (6.18).

Provided that $\Gamma(\boldsymbol{u},\tau)$ can be measured and Eq. (6.17) can be solved for $I(\boldsymbol{s})$, this gives an opportunity to measure $I(\boldsymbol{s})$.

As stated earlier the possible resolution for $I(\boldsymbol{s})$ depends on the size of the telescope used. Using (6.17) the spacing $|\boldsymbol{u}|$ is introduced and, since it turns out to be possible to measure $\Gamma(\boldsymbol{u},\tau)$ for values of $|\boldsymbol{u}|$ much larger than the largest telescope diameters possible, the resolution of $I(\boldsymbol{s})$ obtainable from the inversion of (6.17) is much greater than that which can be achieved by using telescopes that form direct images $I(\boldsymbol{s})$.

The principle, the methods and the limitations for such a construction of $I(\boldsymbol{s})$ from $\Gamma(\boldsymbol{u},\tau)$ will be discussed in the remaining parts of this chapter. But as both the theory of this method and the practical details of it are fairly complicated this should only be considered an introduction. For a thorough understanding, the specialized books given in the reference should be consulted.

6.4 Two-Element Interferometers

The coherence function $\Gamma(\boldsymbol{u},\tau)$ can be measured most simply by a two-element interferometer. Let two telescopes T_1 and T_2 be separated by the distance $\boldsymbol{B}$

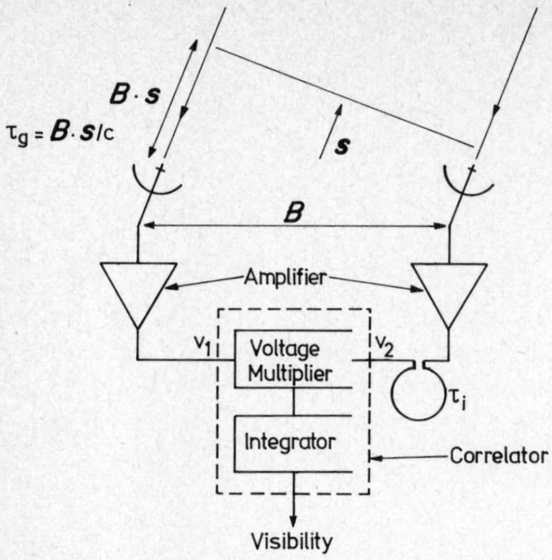

Fig. 6.2. A two-element correlation interferometer

(directed from T_2 to T_1), both telescopes being sensitive only for radiation of the same state of polarization (Fig. 6.2).

An electromagnetic wave induces the voltage U_1 at the output of antenna T_1

$$U_1 \propto E \cos \omega t \tag{6.20}$$

while at T_2 we obtain

$$U_2 \propto E \cos [\omega (t - \tau)] \tag{6.21}$$

where τ is the geometric delay caused by the relative orientation of the interferometer base $\boldsymbol{B}$ and the direction of the wave propagation. These outputs (6.20) and (6.21) will now be processed further in a way depending on the type of interferometer used.

In a *correlation interferometer* the signals are input to a multiplying device followed by a low pass filter such that the output is proportional to

$$R(\tau) \propto \frac{E^2}{T} \int_0^T \cos \omega t \cos \omega (t - \tau)\, dt \ .$$

If T is a time much longer than the time of a single full oscillation $T \gg 2\pi/\omega$ then the average over the time T will not differ much from the average over a single full period; that is

$$R(\tau) \propto \frac{\omega}{2\pi} E^2 \int_0^{2\pi/\omega} \cos \omega t \cos \omega (t - \tau)\, dt$$

$$\propto \frac{\omega}{2\pi} E^2 \int_0^{2\pi/\omega} \cos \omega t \left[\cos \omega t \cos \omega \tau + \sin \omega t \sin \omega \tau\right] dt$$

$$\propto \frac{\omega}{2\pi} E^2 \left[\cos \omega \tau \int_0^{2\pi/\omega} \cos^2 \omega t\, dt + \sin \omega \tau \int_0^{2\pi/\omega} \sin \omega t \cos \omega t\, dt\right]$$

resulting in

$$\boxed{R(\tau) \propto \tfrac{1}{2} E^2 \cos \omega \tau} \quad . \tag{6.22}$$

The output of the correlator + integrator thus varies periodically with τ, the delay-time; this output is simply the mutual coherence function (6.3) of the received wave. If the relative orientation of interferometer baseline $\boldsymbol{B}$ and wave propagation direction $\boldsymbol{s}$ remains invariable, τ remains constant too and so does $R(\tau)$. But if $\boldsymbol{s}$ is slowly turning due to the diurnal rotation of the earth, then τ will vary, and we will measure interference fringes as a function of time.

In an *adding or total power interferometer* the signals from the individual antennas are combined in a slightly different way. Let us describe the antenna outputs in the complex form, so that we have instead of (6.20)

$$U_1 \propto E\,\mathrm{e}^{\mathrm{i}\,\omega t}$$

and instead of (6.21)

$$U_2 \propto E\,\mathrm{e}^{\mathrm{i}\,\omega (t-\tau)} \quad .$$

These two signals are simply added and then the total intensity of the resulting signal is measured; that is, we form

$$\begin{aligned} I &= \langle (U_1 + U_2)(U_1 + U_2)^* \rangle \\ &= \langle U_1 U_1^* \rangle + \langle U_2 U_2^* \rangle + \langle U_1 U_2^* \rangle + \langle U_2 U_1^* \rangle \ . \end{aligned}$$

But

$$\langle U_1 U_1^* \rangle = I_1 \ , \qquad \langle U_2 U_2^* \rangle = I_2$$

$$\langle U_1 U_2^* \rangle = \sqrt{I_1 I_2}\,\mathrm{e}^{\mathrm{i}\,\omega\tau}$$

$$\langle U_2 U_1^* \rangle = \sqrt{I_1 I_2}\,\mathrm{e}^{-\mathrm{i}\,\omega\tau} \qquad \text{so that}$$

$$I = I_1 + I_2 + 2\sqrt{I_1 I_2}\cos \omega\tau \ . \tag{6.23}$$

For identical telescopes $I_1 = I_2$ and

$$I = 2 I_1 (1 + \cos \omega\tau) \ .$$

Therefore both the correlation interferometer and the total power or adding interferometer measure the mutual correlation function (Fig. 6.3).

In order to understand a working interferometer we will consider a two-element interferometer in somewhat more detail. The basic constituents are shown in Fig. 6.2. If the radio brightness distribution is given by $I_\nu(\boldsymbol{s})$, the power received per bandwidth $d\nu$ from the source element $d\Omega$ is $A(\boldsymbol{s})\, I_\nu(\boldsymbol{s})\, d\Omega\, d\nu$, where $A(\boldsymbol{s})$ is the effective collecting area in the direction $\boldsymbol{s}$; we will assume the same $A(\boldsymbol{s})$ for each of the antennas. The amplifiers introduce a constant gain factor which we will omit for simplicity.

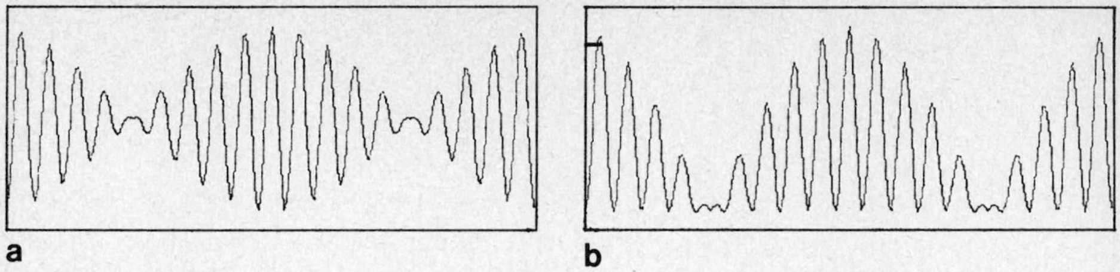

Fig. 6.3 (a) Output of a correlation interferometer. (b) Output of an adding interferometer

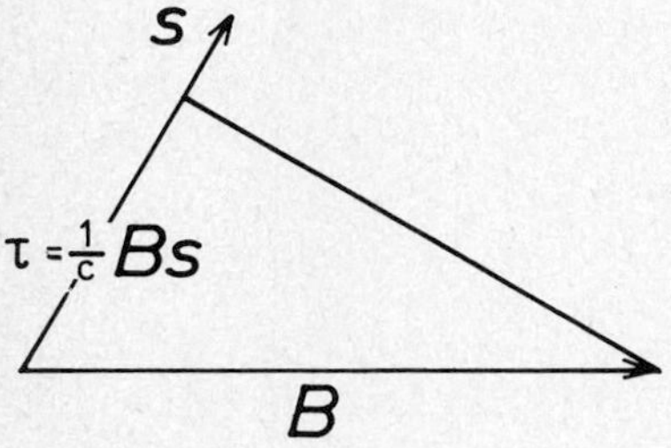

Fig. 6.4. The delay geometry for a two-element interferometer

The output of the correlator for radiation from the direction $\boldsymbol{s}$ (Fig. 6.4) then is

$$r_{12} = A(\boldsymbol{s})\, I_\nu(\boldsymbol{s})\, \mathrm{e}^{\mathrm{i}\omega\tau}\, d\boldsymbol{s}\, d\nu \tag{6.24}$$

where τ is the difference between the geometrical and instrumental delays τ_g and τ_i. If $\boldsymbol{B}$ is the baseline vector for the two antennas

$$\tau = \tau_g - \tau_i = \frac{1}{c}\boldsymbol{B}\cdot\boldsymbol{s} - \tau_i \tag{6.25}$$

and the total response is obtained by integrating over the source S

$$\boxed{R(\boldsymbol{B}) = \iint_S A(\boldsymbol{s})\, I_\nu(\boldsymbol{s}) \exp\left[\mathrm{i}\,\omega\left(\frac{1}{c}\boldsymbol{B}\cdot\boldsymbol{s} - \tau_i\right)\right] d\boldsymbol{s}\, d\nu} \quad . \tag{6.26}$$

This function $R(\boldsymbol{B})$ is obviously closely related to the mutual coherence function of the source (6.17) but, due to the power pattern $A(\boldsymbol{s})$ of the individual antennas, it is not identical to $\Gamma(\boldsymbol{B}, \tau)$. For well-made antennas $A(\boldsymbol{s}) \neq 0$ for only a limited region of the main beam, and (6.26) is integrated only over this region.

6.5 Aperture Synthesis

Aperture synthesis is a method of solving the interferometer equation (6.26) for $I(\boldsymbol{s})$ by measuring $R(\boldsymbol{B})$ at suitable values for the orientation $\boldsymbol{B}$. To do this effectively, convenient coordinate systems for the two vectorial quantities $\boldsymbol{s}$ and $\boldsymbol{B}$ will

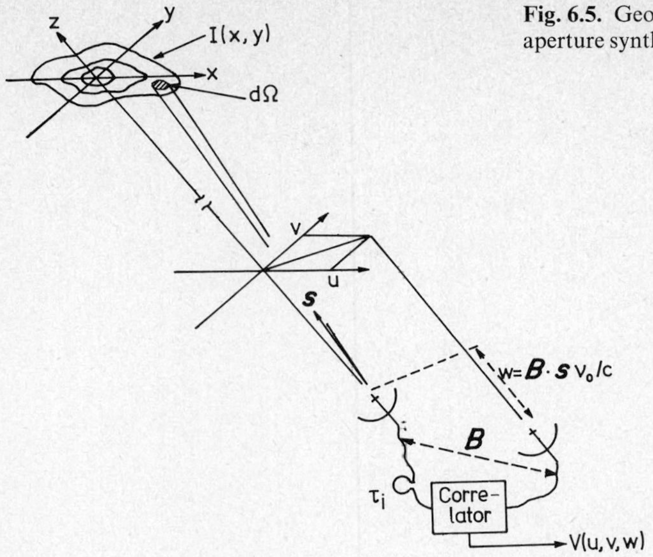

Fig. 6.5. Geometry and coordinates for aperture synthesis [after Thompson (1982)]

be introduced. $\boldsymbol{s}$ is a unit vector pointing towards the point on the celestial sphere where the radiation being investigated originates (Fig. 6.5).

$$\boldsymbol{s} = \boldsymbol{s}_0 + \boldsymbol{\sigma} \ , \qquad |\boldsymbol{\sigma}| = 1$$

where $\boldsymbol{s}_0$ is a conveniently chosen position close to the center of the region investigated.

Substituting this, (6.26) can be written as

$$R(\boldsymbol{B}) = \exp\left[\mathrm{i}\,\omega\left(\frac{1}{c}\boldsymbol{B}\cdot\boldsymbol{s}_0 - \tau_{\mathrm{i}}\right)\right] d\nu \iint_S A(\boldsymbol{\sigma})\, I(\boldsymbol{\sigma}) \exp\left(\mathrm{i}\frac{\omega}{c}\boldsymbol{B}\cdot\boldsymbol{\sigma}\right) d\boldsymbol{\sigma} \ . \tag{6.27}$$

The exponential factor extracted from the integral is describing a plane wave which defines the phase of $R(\boldsymbol{B})$ for the map center. The remaining integral is called the visibility function V of the intensity distribution $I(\boldsymbol{\sigma})$, viz.

$$\boxed{V(\boldsymbol{B}) = \iint_S A(\boldsymbol{\sigma})\, I(\boldsymbol{\sigma}) \exp\left(\frac{\mathrm{i}\,\omega}{c}\boldsymbol{B}\cdot\boldsymbol{\sigma}\right) d\boldsymbol{\sigma}} \ . \tag{6.28}$$

The visibility function for various brightness distribution models is given in Fig. 6.7.

Now if the coordinate systems are chosen such that

$$\frac{\omega}{2\pi c}\boldsymbol{B} = (u, v, w)$$

where u, v, w are measured in units of the wavelength $\lambda = 2\pi c/\omega$ and the direction $(0, 0, 1)$ is parallel to $\boldsymbol{s}_0$, u is pointing into the local direction east while v is pointing

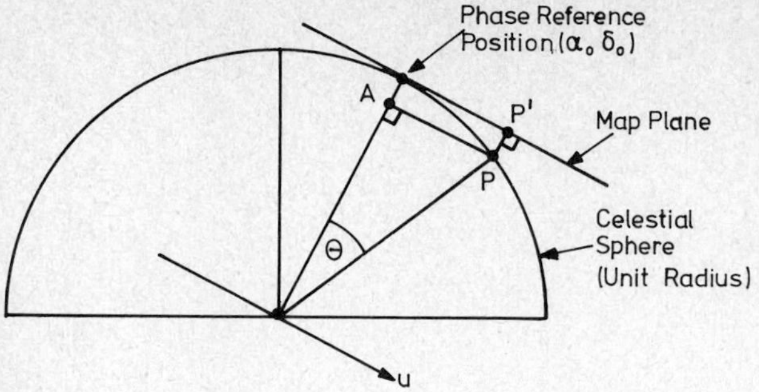

Fig. 6.6. Mapping of points on the celestial sphere onto a plane surface [after Thompson (1982)]

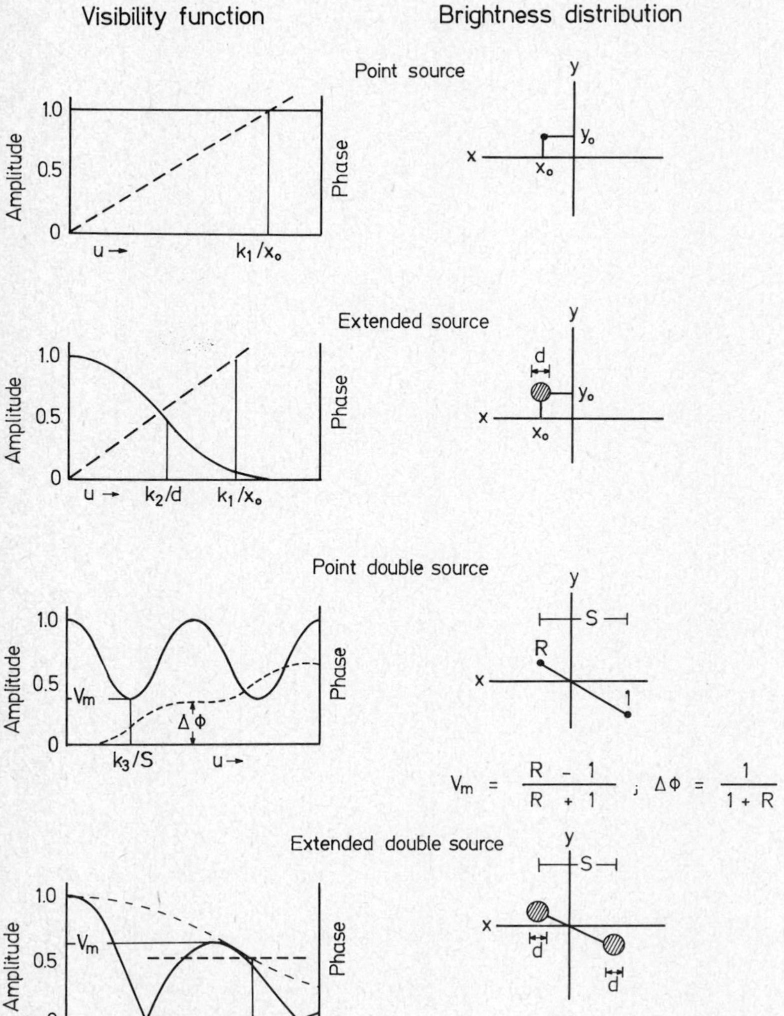

Fig. 6.7. The visibility function for various brightness distribution models [after Fomalont and Wright (1974)]

north; the vector $\boldsymbol{\sigma} = (x, y, z)$ is defined such that x and y are the direction cosines with respect to the u and v axes. Then the xy-plane represents a projection of the celestial sphere onto a tangent plane with the tangent point (and origin) at $\boldsymbol{s}_0$ (Fig. 6.6). In these coordinates (6.28) becomes

$$V(u, v, w) = \int_{-\infty}^{\infty} \int_{-\infty}^{\infty} A(x, y) I(x, y) \times \exp[\mathrm{i}\, 2\pi (ux + vy + w\sqrt{1 - x^2 - y^2})] \frac{dx\, dy}{\sqrt{1 - x^2 - y^2}} \; . \tag{6.29}$$

The integration limits have been formally extended to $\pm\infty$ by adopting $A(x, y) = 0$ for $x^2 + y^2 > l^2 < 1$; where l is the full width of the main beams. The integral closely resembles a two-dimensional Fourier integral; it would be one, if the term $w\sqrt{1 - x^2 - y^2}$ could be removed from under the integral sign.

If only a small region of the sky is to be mapped then $\sqrt{1 - x^2 - y^2} \cong \text{const} \cong 1$ and (6.29) becomes

$$\boxed{V(u, v, w)\, \mathrm{e}^{-\mathrm{i}\, 2\pi w} = \int_{-\infty}^{\infty} \int_{-\infty}^{\infty} A(x, y) I(x, y)\, \mathrm{e}^{\mathrm{i}\, 2\pi (ux + vy)}\, dx\, dy} \; . \tag{6.30}$$

The factor $\exp(-\mathrm{i}2\pi w)$ is the approximate conversion required to change the observed phase of V to the value that would be observed with antennas in the uv-plane:

$$V(u, v, w)\, \mathrm{e}^{-\mathrm{i}\, 2\pi w} \cong V(u, v, 0) \; . \tag{6.31}$$

Substituting this into (6.30) and performing the inverse Fourier transform we obtain

$$\boxed{I'(x, y) = A(x, y) I(x, y) = \int_{-\infty}^{\infty} \int_{-\infty}^{\infty} V(u, v, 0)\, \mathrm{e}^{-\mathrm{i}\, 2\pi (ux + vy)}\, du\, dv} \; , \tag{6.32}$$

the intensity $I(x, y)$ as modified by the primary beam shape $A(x, y)$.

The conversion factor $\exp(\mathrm{i}\, 2\pi w)$ for the phase of V is precise for the map center $x = y = 0$ only, and it results in a phase error of $\pi(x^2 + y^2)w$ for radiation at the point (x, y). This will produce a distortion of the map.

The problem of aperture synthesis is now handled by sampling the visibility function $V(u, v, 0)$ with separate telescopes distributed in the uv-plane. Many configurations are possible, because all that is needed is a reasonably dense covering of the plane. And to make things even simpler, it is unnecessary for all values of V to be measured simultaneously. So, in principle, all that is needed is a single pair of telescopes that are moved about in the uv-plane and are thus synthesizing the telescope aperture.

Obviously, if more than two telescopes are available, $V(u, v, 0)$ can be measured faster. Telescope arrays of quite different shapes range from arrays arranged in a circle (Culgoora solar telescope in Australia with 96 telescopes on a circle of

3 km diameter), arrays in the form of a cross (Christiansen cross antenna at Fleurs, Australia, or the Mills cross at Molonglo, Australia), or the Y-shaped very large array (VLA) of (movable) 25 m telescopes at Socorro, New Mexico, USA (Fig. 6.9). All these telescopes can produce an almost instantaneous "snapshot" value for $V(u, v, 0)$ which can then be Fourier-transformed into an estimate for $I(x, y)$.

There is, however, another way of covering the UV-plane effectively without bodily moving the individual antennas, making use of the rotation of the earth to cause the variation of the position-vectors $\boldsymbol{B}$ of the individual telescope pairs. Sampling of the *uv*-plane by earth rotation is sometimes called supersynthesis.

One problem met in using rotation synthesis is caused by the fact that the vectors $\boldsymbol{B}_i$ of the correlation pairs will generally not lie in a plane if the whole telescope array is rotated along with the earth, but will fill part of a volume. We have therefore a three-dimensional problem with all the complications connected with it. Only if the telescopes are arranged along a straight east-west line (Fig. 6.8) will they remain in a plane perpendicular to the axis of rotation. To make $w = 0$ we choose the w-axis in the direction of the pole. Then it is possible to map a whole hemisphere using the inverse Fourier transform

$$\frac{A(x, y)\, I(x, y)}{\sqrt{1 - x^2 - y^2}} = \int_{-\infty}^{\infty} \int_{-\infty}^{\infty} V(u, v)\, \mathrm{e}^{-\mathrm{i}\, 2\pi (ux + vy)}\, du\, dv \; . \tag{6.33}$$

For an east-west interferometer u and v are given by

$$\begin{aligned} u &= \frac{\nu L}{c} \cos t \\ v &= \frac{\nu}{c} L \sin t \sin \delta_0 \end{aligned} \tag{6.34}$$

where L is the length of the interferometer baseline, t the hour angle of the field centre, and δ_0 the declination of the field centre.

The coordinates x, y at the celestial sphere have their origin at the field center (t, δ_0) too, and they are parallel to u and v.

If such an interferometer tracks a source for some time, the projected interferometer baseline will trace out an ellipse in the *uv*-plane; if the synthesis telescope consists of several telescopes aligned on an east-west line, several concentric ellipses will be traced out simultaneously.

From (6.34) we see that these *uv*-ellipses are compressed in the v-direction by the factor $\sin \delta_0$: for $\delta_0 = \pm 90°$ the traces are circles; for $\delta_0 = 0$ they degenerate into straight lines.

Therefore the *uv*-plane of the visibility function for sources near the celestial equator is sampled insufficiently if a linear east-west array is used with earth-rotation aperture synthesis. Such sources will have to be measured using arrays that also extend in the north-south direction, even if the fact that the $\boldsymbol{B}$'s of the individual interferometer pairs do not remain coplanar causes complications in measurements extended over longer time intervals (Fig. 6.9).

Fig. 6.8. The Westerbork aperture synthesis telescope (courtesy WRST)

Fig. 6.9. The Very Large Array at Socorro, New Mexico. Aerial view of the VLA looking down the southwest arm; the prominent structure is the Antenna Assembly Building. The north arm branches off to the lower right; the southeast arm extends to the left. The array is shown in its concentrated configuration (courtesy NRAO/AUI)

6.6 The Transformation of the Visibility Function

Provided that the visibility function $V(u, v)$ is known for the full uv-plane both in amplitude and in phase, the (modified) intensity distribution $I'(x, y)$ can be determined from it by performing the Fourier transformation (6.32). But in a realistic situation $V(u, v)$ is only sampled at discrete points within a radius $\cong u_{\max}$ along elliptical tracks, and some regions of the uv-plane are not measured at all due to either missing short spacings or a missing angular wedge in the orientation of the baseline vector $\boldsymbol{B}$.

One way to determine $I'(x, y)$ is to compute the direct Fourier transform (DFT) in which (6.32) is summed only for the measured samples of the visibility function

$$I''(x, y) = \sum_k g(u_k, v_k)\, V(u_k, v_k)\, \mathrm{e}^{-\mathrm{i}\, 2\pi (u_k x + v_k y)} \,, \tag{6.35}$$

where $g(u, v)$ is a weighting function called the grading or apodization. It is to a large extent arbitrary and can be used to change the effective beam shape and side lobe level of the array. This reconstructed image I'' is often not even a particularly good representation of I', but it is certainly related to it because (6.35) has the property

$$I''(x, y) = P_D(x, y) \otimes I'(x, y) \quad \text{where} \tag{6.36}$$

$$P_D(x, y) = \sum_k g(u_k, v_k)\, \mathrm{e}^{-\mathrm{i}\, 2\pi (u_k x + v_k y)} \tag{6.37}$$

is the response to a point source (point spread function or dirty beam), and where $\otimes$ means convolution. The sum in (6.37) extends over the same positions u_k, v_k as in (6.35), and the sidelobe structure of the beam depends on the distribution of these points.

Grating Response. If there is some regularity in the distribution of the sample points in the uv-plane, the resulting sidelobe structure called the *grating response* or *grating lobes* becomes particularly strong. For an east-west array in which the telescopes are distributed at regular intervals, the uv-plane will be sampled in concentric elliptical tracks and between the tracks we implicitly set $V(u, v) = 0$. Now the Fourier transform of such a system of elliptical rings is again a system of elliptical rings with semi-axes $k\, c/\nu\, \Delta L$ and $k\, c/\nu\, \Delta L \sin\delta_0$ where ΔL is the interval in the baseline and k is an integer (Fig. 6.10).

So, to minimize the disturbance caused by the grating ellipses, the true baseline intervals. ΔL should be small. But if the interferometer consists of telescope of diameter D, the smallest interval possible is $\sim D$, and so the grating rings are unavoidable. But in many practical situations, especially if the intensity distribution $I(x, y) > 0$ only for part of the map, the grating lobes can often be almost completely eliminated by using a restoration procedure like CLEAN.

Fast Fourier Transform Inversion, Aliasing. The computation of (6.35) even for a modest data set is quite time consuming and therefore other methods for inverting (6.30) that are based on the Cooley-Tukey Fast Fourier Transform algorithm

Fig. 6.10. A "dirty" map of the Virgo cluster at $\lambda = 49$ cm taken with the Westerbork synthesis telescope from two 12-hour observations. The strong grating rings correspond to a 36 m sampling interval [from Hamaker (1979)]

(FFT) are usually used. In order to use the FFT, the visibility function must be obtained on a regular grid with sides that are a power of two if we restrict ourselves to the simplest version of the FFT.

Since the observed data seldom lie on such a grid, some interpolation scheme must be used. If the measured points are randomly distributed this interpolation is best by a convolution procedure.

The "gridded" visibility function may be represented by

$$V''(u, v) = \text{Ш}(u, v)\{G(u, v) \otimes V'(u, v)\} \tag{6.38}$$

where $V'(u, v)$ is the measured visibility function sampled on the irregular u_i, v_i grid, $G(u, v)$ is a convolving function, and $\otimes$ is again the convolution operator by which a value for an interpolated visibility function $V''(u, v)$ is defined for every u, v. Finally, the Sha-function

$$\text{Ш}(u, v) = \Delta u\, \Delta v \sum_{j,k=-\infty}^{\infty} \delta(u - j\,\Delta u)\, \delta(v - k\,\Delta v) \tag{6.39}$$

defines the regular grid on the uv-plane with Δu and Δv giving the cell size.

The intensity distribution $I'(x, y)$ is now obtained by substituting (6.38) into (6.32). The Fourier transform (6.32) can be computed by the FFT because $V(u, v)$ is now given on a regular grid. However, (6.32) cum (6.38) have certain unpleasant properties that are most easily seen by slightly rewriting these equations using some of the fundamental properties of Fourier transforms. Recalling that the Fourier transform of the Sha-function is another Sha-function with a grid spacing $1/\Delta u$ and $1/\Delta v$

$$\text{Ш}(x, y) = \sum_{i,j=-\infty}^{\infty} \delta(x - i/\Delta u)\,\delta(y - j/\Delta v) \tag{6.40}$$

and, applying the convolution theorem to (6.38), we obtain

$$I(x, y) = \text{Ш}(x, y) \otimes [g(x, y)\, I'(x, y)] \tag{6.41}$$

where $g(x, y)$ is the Fourier transform of $G(u, v)$, that is, the grading that results in the beam $G(u, v)$.

The important property of (6.41) is the fact that $g I'$ is convolved with a Sha-function that extends over the full x, y space. If the grading $g(x, y)$ does not remain equal to zero outside the map area, radiation from outside is aliased into the map. This will happen for most practical convolving functions. Adopting a "pill box beam"

$$G(u, v) = \begin{cases} 1 & \text{for } u^2 + v^2 \leqq u_{\max}^2 \\ 0 & \text{else} \end{cases}, \tag{6.42}$$

the corresponding grading will be a $(1/x) \sin x$ function which has non-zero values extending over the full xy-plane. Other convolving functions will be slightly less aliasing, but the effect as such can never be completely avoided. By a proper choice of the convolution function, aliasing can be suppressed by 10^2 to 10^3 at 2 to 3 map radii, and this usually is sufficient. Map structures caused by aliasing can, however, always be recognized by regridding the visibility function $V(u, v)$ to another grid size Δu, Δv, since the new aliases will then shift their position in the resulting map.

CLEANing the Map. The reconstructed image $I''(x, y)$ is thus always a deformed version of the true intensity distribution but, with the exception of the problems caused by aliasing, these deformations can be described by a convolution process (6.36). What is needed, therefore, is some kind of deconvolution algorithm to improve the resulting image.

One particular method, developed originally by Högbom and called CLEAN, has been applied by so many researchers that it has become a kind of standard with which all other methods are compared.

CLEAN tries to approximate the true but unknown intensity distribution $I(x, y)$ by the superposition of the radiation of a finite number of point sources of the flux A_i placed at the position (x_i, y_i). It is the aim of CLEAN to determine the A_i, x_i, y_i such that

$$I''(x, y) = \sum_i A_i\, P_D(x - x_i, y - y_i) + I_R(x, y) \tag{6.43}$$

where I'' is the "dirty" map obtained from the inversion of the visibility function and P_D is the "dirty" beam (6.37). $I_R(x, y)$ is the residual brightness distribution after decompositon. The approximation (6.43) is deemed successful if I_R is of the order of the error of the measured intensities. As this decomposition cannot be done analytically, an iterative technique has to be used.

Usually a cleaned map is obtained by convolving the δ-function with some suitable beam P_c. This beam is arbitrary, but is usually taken to be a Gaussian function with an elliptical beamwidth matching that of the "dirty" beam.

A great deal of literature exists on such image restoration methods, their advantages and drawbacks. The theory is complicated, and it is often difficult to estimate how large the influence of the method chosen is on the result obtained. But due to the obvious existence of side lobe noise etc. on the "dirty" maps, some efforts of this kind are necessary and unavoidable. For a more thorough discussion, however, we must refer to the literature given in the reference.

Calibration. Large antenna arrays and the electronics connecting the individual antennas combining their outputs are a complicated machinery that obviously need minute adjustments and calibrations in order to produce results as intended by the theory. The details of how this is done and checked will obviously depend on the actual instrument used, and usually a radio astronomer making measurements with this instrument will not be involved in the calibration procedure.

But a responsible scientist should always make certain that his research tool is working properly, and should check how his results are affected by possible calibration errors. To do this, some basic understanding of the calibration procedure is needed.

The basic and fundamental purpose of any array is to measure the mutual correlation function of the received radiaton field. For an array consisting of n identical antennas, the output for a strong point source of flux S_ν for any two antennas obviously should be

$$V_{jk} = S_\nu \exp\left[\mathrm{i}\frac{\omega}{c}\boldsymbol{B}_{jk}\cdot(\boldsymbol{s}-\boldsymbol{s}_0)\right] \tag{6.44}$$

where S_ν is the flux density, $\boldsymbol{B}_{jk}$ the base vector of antennas j and k, $\boldsymbol{s}$ the position of the source, and $\boldsymbol{s}_0$ the phase reference position. If the measured output does not agree with (6.44), complex calibration factors $C_{jk}(t)$ have to be determined such that $C_{jk}(t)\,V_{jk}$ is equal both in amplitude and in phase to the theoretically expected mutual correlation function. Usally for each aperture synthesis telescope a systematic procedure is available to determine these $C_{jk}(t)$. If they are applied, the calibration point source should produce a reasonable map.

New map fields can then be calibrated in terms of this calibration point source. The intensities are then given in units of "Jansky per beam", and this is indeed the unit in which most synthesis maps are calibrated. But if the measurements are to be discussed in astrophysical terms, these instrumental units have to be converted into proper physical ones. This can be done by using the descriptive antenna parameters outlined in Sect. 4.5.

The power that the array is intercepting from a source with the flux S_ν is, according to (4.59) to (4.60),

$$W = \frac{1}{2} A_e\, S_\nu = \frac{\lambda^2}{2}\left(\frac{S_\nu}{\Omega_A}\right) . \tag{6.45}$$

Using Nyquist's theorem (1.35) this can be expressed in terms of the antenna temperature (4.63) or, introducing the brightness temperature (1.28),

$$W = k\, T_A = \eta_M\, k\, T_b . \tag{6.46}$$

Taking (6.45) and (6.46) together gives

$$\boxed{T_b = \frac{1}{\eta_M} \frac{\lambda^2}{2\,k}\left(\frac{S_\nu}{\Omega_A}\right)} \tag{6.47}$$

or, using practical units,

$$\boxed{T_b = 3.62 \cdot 10^{-2} \frac{1}{\eta_M}\left(\frac{\lambda}{\mathrm{m}}\right)^2\left(\frac{S_\nu/\mathrm{Jy}}{\Omega_A/\mathrm{sr}}\right)} . \tag{6.48}$$

If the solid angle of the array is measured in the more appropriate unit of $1\,\square' \cong 84.62 \cdot 10^{-9}$ sr (6.48) becomes

$$\boxed{T_b = 4.28 \cdot 10^5 \frac{1}{\eta_M}\left(\frac{\lambda}{\mathrm{m}}\right)^2\left(\frac{S_\nu/\mathrm{Jy}}{\Omega_A/\square'}\right)} . \tag{6.49}$$

Although it is quite true that it is difficult for any observer to determine the effective beam solid angle appropriate to his measurements, he should still be in a better position to do so than the poor theoretician who later tries to conform his model to these measurements. And I would find it definitely bad style if this habit of quoting only in terms of Jy/beam were ever extended to measurements made with filled aperture antennas that permit a rather easy determination of an appropriate Γ connecting T_A with S_ν (see 5.45).

6.7 Aperture Synthesis with Limited Phase Information

For a given wavelength the angular resolving power of an interferometer depends only on the length of the interferometer base $\boldsymbol{B}$. But the need to provide a phase-stable coaxial link of the individual telescopes with the correlation unit limits $|\boldsymbol{B}|$ for practical and economical reasons to about 20–25 km. If longer baselines are wanted radio-links have been used, but it becomes more and more difficult to guarantee the phase stability, since transient irregularities in the transmission path will have detrimental effects.

The construction of atomic clocks with extremely phase stable oscillators opened up the possibility of avoiding altogether the transmission of a phase-stable local oscillator signal and observing independently at the two sites of the interferometer, registering the data on magnetic tapes together with precise time marks and determining the correlations later when the two tapes are brought together. The tapes contain true copies of the time-variation of the electrical field strength so that the time-averaged product of the playback signal of two such tapes gives directly the mutual correlation function.

Local oscillators with extreme phase stability are needed at each station for two reasons. The tapes – videotapes or video cartridges are normally used – usually permit signals in a band reaching from zero to at most a few MHz; the received signal therefore must be mixed down into this band, and for this a phase stable local oscillator is needed, since all phase jumps of the oscillator are reflected in the IF signal. The second use of the phase stability is to provide the extremely precise time marks needed to align the two tapes. Again phase jumps would destroy the coherence. For the local oscillators, different systems have been used with varying success, ranging from Rb-clocks, free running quartz oscillators and, most successfully, hydrogen masers. The phase stability of hydrogen masers results in a frequency stability

$$\frac{\Delta\nu}{\nu} = \begin{cases} 10^{-12}\,(t/\mathrm{s})^{-1} & \text{for } 10^{-1}\,\mathrm{s} < t < 10^{2}\,\mathrm{s} \\ 10^{-14} & \phantom{\text{for }} 10^{2}\,\mathrm{s} < t < 10^{4}\,\mathrm{s} \end{cases}, \tag{6.50}$$

where t is the time for which the measurement is extended. If we permit a random phase error of $\cong 1$ radian due to such irregular phase jumps at either of the two antennas of the interferometer, the signals will stay coherent for

$$\frac{t_c}{\mathrm{s}} = \begin{cases} 1.9 \cdot 10^{4} \left(\dfrac{\nu}{\mathrm{GHz}}\right)^{-1} & \text{for } 0 < \nu < 190\ \mathrm{GHz} \\ 0 & \phantom{\text{for } 0 <{}} \nu > 190\ \mathrm{GHz} \end{cases} \tag{6.51}$$

seconds. This is the maximum time over which the visibility function can be coherently integrated, that is, where we can determine the amplitude and phase of the visibility function. VLBI interferometers therefore are only able to measure the amplitude of the fringe-visibility function; the phase is not known.

Radio observatories all over the earth are regularly linked together by VLBI work and, since the telescope sites are not usually planned with VLBI work in mind, a three-dimensional interferometer geometry using u, v and w must be used. Most often the uv-plane is covered only in a very spotty way, and therefore the information eventually obtained on $I(x, y)$ is very restricted indeed.

Often the measured variation of the amplitude of the visibility function can only be compared with computed amplitudes for simple model distributions like that of a double point source, sources of finite extent or core-halo structures. From this then typical angular dimensions can be derived.

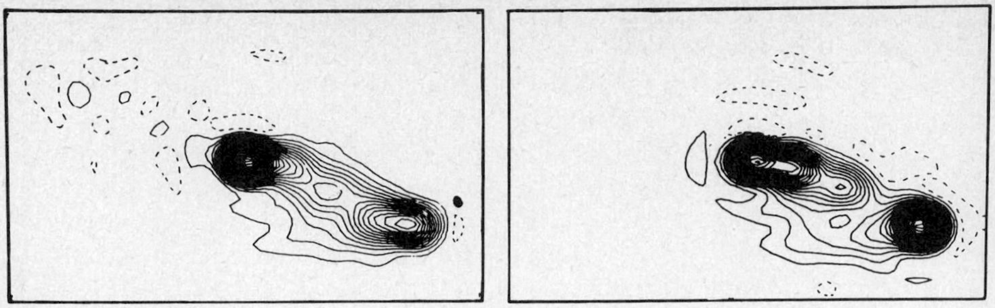

Fig. 6.11. An attempt to derive the structure of 3C382 from data without phase information. *Left*: image obtained without phase information, *right*: map obtained with observed amplitudes and phases [after Baldwin and Warner (1979)]

This comparison with model-fringe visibilities can be extended to fairly complicated structures for $I(x, y)$ if the uv-plane is covered reasonable well by a multi-station experiment. It is, however, still not quite clear which conditions have to be met in order that these image reconstructions are unique because they contain only the amplitude information on the visibility function, not the phase information which is unavailable (Figs. 6.11, 6.12).

In many cases it turned out to be possible to recover at least part of the lost phase information even when the coherence time caused by phase jumps of the local oscillators, time variability of the phase propagation properties of the atmosphere and the like is shorter than the time interval between calibration runs. The reason is that the calibration factors C_{ik} introduced in Sect. 6.6 or their inverse, the complex gain factors G_{ik} relating the complex mutual coherence function Γ_{ik} to the measured visibility function V_{ik}

$$V_{ik} = G_{ik}\,\Gamma_{ik} + \text{noise} \quad , \tag{6.52}$$

depend in a rather simple way on the gain factors of the individual antennas. Considering the definition of Γ in Sect. 6.3 we realize that

$$G_{ik} = g_i\,g_k^* \quad . \tag{6.53}$$

Introducing the phases φ, θ and ψ by

$$\begin{aligned} V_{ik} &= |V_{ik}|\exp\{i\,\varphi_{ik}\} \\ G_{ik} &= |g_i||g_k|\exp\{i\,\theta_i\}\exp\{-\,i\,\theta_k\} \\ \Gamma_{ik} &= |\Gamma_{ik}|\exp\{i\,\psi_{ik}\} \end{aligned} \tag{6.54}$$

Eqs. (6.52) and (6.53) then postulate that the visibility phase ψ_{ik} on the baseline $i\,k$ will be related to the observed phase φ_{ik} by

$$\varphi_{ik} = \psi_{ik} + \theta_i - \theta_k + \varepsilon_{ik} \tag{6.55}$$

where ε_{ik} is the phase noise. Then the *closure phase* Ψ_{ikl} around a closed triangle of baseline $i\,k$, $k\,l$, $l\,i$

$$\Psi_{ikl} = \varphi_{ik} + \varphi_{kl} + \varphi_{li} = \psi_{ik} + \psi_{kl} + \psi_{li} + \varepsilon_{ik} + \varepsilon_{kl} + \varepsilon_{li} \tag{6.56}$$

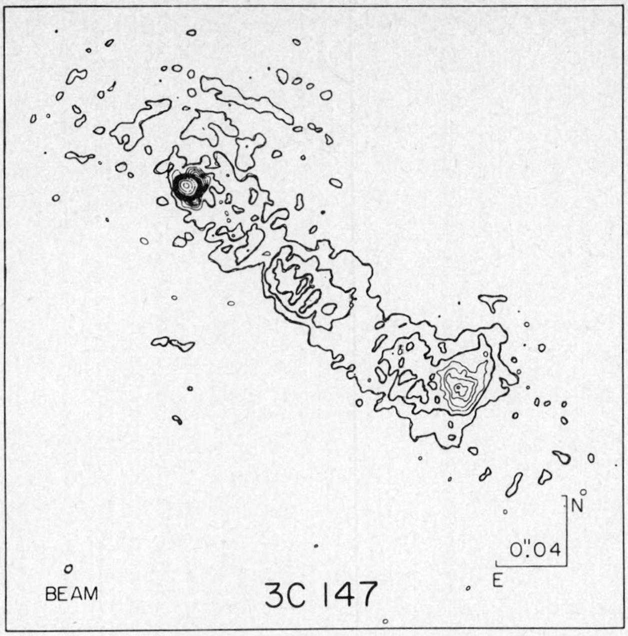

Fig. 6.12. A VLBI map of the quasar 3C147 at 1.7 GHz made using global fringe-fitting technique of 10 antennas. The resolution is 0".004 [from Pearson and Readhead (1984)]

will be independent of the phase shifts θ introduced by the individual antennas and of their time variations. There are certain residual errors that do not cancel, but these "closure errors" usually are small. Figure 6.13 shows an example.

If four antennae are used simultaneously ratios can be formed that are independent of the antenna gain factors as well:

$$A_{klmn} = \frac{|V_{kl}||V_{mn}|}{|V_{km}||V_{ln}|} = \frac{|\Gamma_{kl}||\Gamma_{mn}|}{|\Gamma_{km}||\Gamma_{ln}|} . \qquad (6.57)$$

These ratios are called *closure amplitudes*.

Both kinds of quantities can be used to improve the reconstructed maps by forming constraints for the visibility function computed from the reconstructed maps.

If each antenna introduces an unknown complex gain factor g with amplitude and phase, the total number of unknown factors in the array can be reduced significantly by measuring closure phases and amplitudes. If four antennas are available 50% of the phase information and 33% of the amplitude information can thus be recovered, in a 10 antennas configuration as in a typical VLBI arrangement these ratios are 80% and 78% respectively.

There exist different methods by which these closure informations can be used in the image reconstruction process. The most straightforward is to form certain special, highly redundant array configurations so that the closure equations can be solved directly to compute the visibilities of all baselines. But more efficient

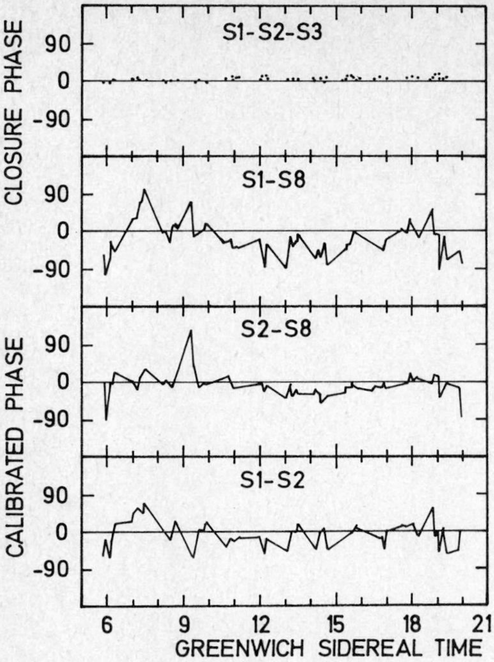

Fig. 6.13. Elimination of antenna-based phase errors by determination of the closure phase. The lower frames show measured phase of the VLA, Socorro, on three baselines in bad weather, the upper one shows the closure phase [after Readhead et al. (1980)]

methods have been developed which use the closure information in a complicated sequence of image restoration and convolution sequence. The discussion of this would go beyond the aim of this text, it is outlined and discussed in the review article by Pearson and Readhead (1984) where references are given too.

7. Receivers

7.1 Stationary Stochastic Processes

The concept of spectral power density was introduced in Chap. 1 in a purely phenomenological way. Radio receivers are devices to measure this spectral power density, but the working of certain receiver types – namely the autocorrelation spectrometers – as well as the discussion of the limiting receiver sensitivity can only be understood properly if some implications of this concept are discussed more thoroughly.

In the preceding chapters the signals considered were periodic functions of the time which could be conveniently expressed as the superposition of simple harmonic functions of time. It is now convenient to permit a more general class of time-variable functions; that is, to represent the signals as *stationary random processes* $x(t)$. The signal $x(t)$ is a function of the time t, but it is not fully determined by it: if the random process is acting at different times, different realizations of $x(t)$ will result. All that is fixed by the random process are certain statistical properties of the actual signal.

7.1.1 Probability Density, Expectation and Ergodicy

Perhaps the most important of these statistical quantities is the probability density function $p(x)$ giving the probability that at any arbitrary moment of time the value of the process $x(t)$ falls within an interval $(x - \frac{1}{2}dx, x + \frac{1}{2}dx)$.

For a stationary random process, $p(x)$ will be independent of the time t.

The *expected value* $E\{x\}$ or *mean value* of the random variable x is given by the integral

$$\boxed{E\{x\} = \int_{-\infty}^{\infty} x\,p(x)\,dx} \tag{7.1}$$

and, by simple analogy, the expectation value $E\{f(x)\}$ of a function $f(x)$ is given by

$$\boxed{E\{f(x)\} = \int_{-\infty}^{\infty} f(x)\,p(x)\,dx} \quad . \tag{7.2}$$

This is different from the expected value of the transformation $y = f(x)$

$$E\{y\} = \int_{-\infty}^{\infty} y\,p_y(y)\,dy = \int_{-\infty}^{\infty} f(x)\,p_x(x)\frac{dx}{|f'(x)|} \quad . \tag{7.3}$$

Expected values that are frequently used are the *mean value*

$$\mu = E\{x\} \tag{7.4}$$

and the *dispersion*

$$\sigma^2 = E\{x^2\} - E^2\{x\} \ . \tag{7.5}$$

Another kind of average that can be formed for a stationary random process is the *time average* of the values of the function $f(x(t))$ along a single realization of the process. This average will be designated (like in earlier chapters) by acute brackets:

$$\boxed{\langle f(x)\rangle = \lim_{T\to\infty} \frac{1}{2T} \int_{-T}^{T} f(x(t))\,dt} \ . \tag{7.6}$$

This limit may or may not exist, but conditions can be formulated [the ergodic theorem of Birkoff, see Khinchin (1949)] so that the results of the definitions (7.2) and (7.6) agree. We will assume this to be the case here.

7.1.2 Autocorrelation and Power Spectrum

The concept of the Fourier transform plays a fundamental role in many branches of physics and engineering, and therefore it would be convenient to apply it to the discussion of noise signals as well. There are, however, difficulties in doing this, because a stationary time series does not decay to zero for $t \to \pm\infty$, and therefore the obvious definition for the Fourier transform

$$X(\nu) = \lim_{T\to\infty} \int_{-1/2\,T}^{1/2\,T} x(t)\,\mathrm{e}^{-2\pi i \nu t}\,dt$$

does not exist; the integral varies irregularly as T increases. As Wiener first showed, the concept of the Cesaro sum of an improper integral helps. It is defined as

$$\int_{-\infty}^{\infty} A(x)\,dx = \lim_{N\to\infty} \frac{1}{N} \int_0^N \left[\int_{-r}^{r} A(x)\,dx\right] dr \ , \tag{7.7}$$

that is, as the limit of the average over the finite integrals. This limit will exist for a wide class of functions where the ordinary improper integral does not exist. For those cases where the ordinary limit exists, the Cesaro sum will be equal to it as can be seen if the sequence of the integrations in (7.7) is interchanged using Dirichlet's theorem on repeated integrations (see Whittaker and Watson, § 4.3):

$$\frac{1}{N} \int_0^N \left[\int_{-r}^{r} A(x)\,dx\right] dr = \int_{-N}^{N} \left(1 - \frac{|x|}{N}\right) A(x)\,dx \ . \tag{7.8}$$

For any finite section of a stochastic time series we can define the Fourier transform

$$X_T(\nu) = \int_{-1/2T}^{1/2T} x(t)\,\mathrm{e}^{-2\pi i\nu t}\,dt \ ,$$

and its mean-squared expected value then is

$$E_T\{|X(\nu)|^2\} = E\left\{\int_{-1/2T}^{1/2T}\int_{-1/2T}^{1/2T} x(s)\,x(t)\,\mathrm{e}^{-2\pi i\nu(t-s)}\,ds\,dt\right\} . \tag{7.9}$$

But, because $x(t)$ is assumed to be stationary, we must have

$$R_T(\tau) = E_T\{x(s)\,x(s+\tau)\} = E_T\{x(t-\tau)\,x(t)\} \tag{7.10}$$

where $R_T(\tau)$ is the autocorrelation function (ACF). Introducing this into the above expression and performing the integration with respect to s, we find

$$E_T\{|X(\nu)|^2\} = T\int_{-T}^{T}\left(1-\frac{|\tau|}{T}\right)R_T(\tau)\,\mathrm{e}^{-2\pi i\nu\tau}\,d\tau \ . \tag{7.11}$$

But the right-hand side is a Cesaro sum, and therefore by defining the power spectral density (PSD) $S(\nu)$ as

$$S(\nu) = \lim_{T\to\infty}\frac{1}{T}E_T\{|X(\nu)|^2\} \ , \tag{7.12}$$

we obtain from (7.11)

$$\boxed{S(\nu) = \int_{-\infty}^{\infty} R(\tau)\,\mathrm{e}^{-2\pi i\nu\tau}\,d\tau} \ . \tag{7.13}$$

This is the *Wiener-Khinchin theorem* stating that the ACF $R(\tau)$ and the PSD $S(\nu)$ of an ergodic random process are Fourier transform pairs (Fig. 7.1). Taking the inverse Fourier transform of (7.13) we obtained similarly

$$\boxed{R(\tau) = \int_{-\infty}^{\infty} S(\nu)\,\mathrm{e}^{2\pi i\nu\tau}\,d\nu} \ . \tag{7.14}$$

Thus the total power transmitted by the process is given by

$$R(0) = \int_{-\infty}^{\infty} S(\nu)\,d\nu = E\{x^2(t)\} \ . \tag{7.15}$$

The limit $T\to\infty$ of the autocorrelation function (ACF) $R_T(\tau)$ can be found using the Cesaro sum resulting in

$$R(\tau) = E\{x(s)\,x(s+\tau)\} = \lim_{T\to\infty}\int_{-T}^{T}\left(1-\frac{|s|}{2T}\right)x(s)\,x(s+\tau)\,ds \ . \tag{7.16}$$

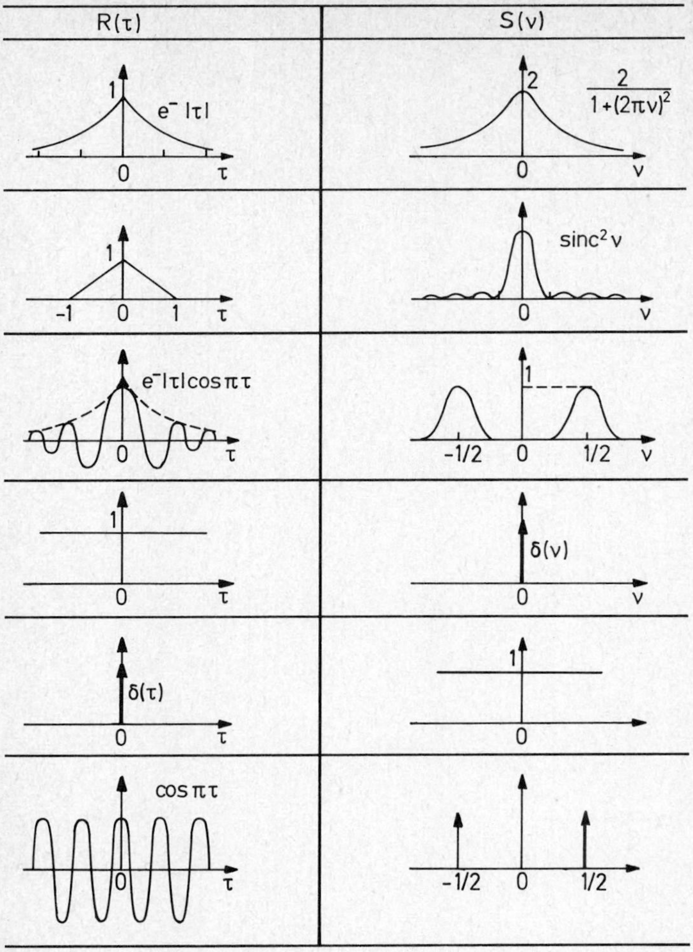

Fig. 7.1. Pictorial representation of some correlation functions $R(\tau)$ and corresponding spectral densities $S(\nu)$ [from Papoulis (1965)]

Using the concept of ensemble average, this can also be written as

$$R(\tau) = \iint_{-\infty}^{\infty} x_1(s)\, x_2(s+\tau)\, p(x_1, x_2; \tau)\, dx_1\, dx_2 \tag{7.17}$$

where $p(x_1, x_2; \tau)$ is the joint probability density function for observing the values x_1 and x_2 which are separated by the time τ. Again, for ergodic stationary processes, (7.16) and (7.17) lead to identical results, but sometimes one or the other definition may be easier to apply.

The application of these concepts will be shown in the following two sections where the influence that linear systems and a square law detector have on the signal of a random process will be disussed. This may serve as an example; the results will, however, be used later on in the discussion of the limiting sensitivity of radio receivers.

7.1.3 Linear Systems

Let the signal $x(t)$ be passed through a fixed linear filter whose time-response to a unit impulse $\delta(t)$ is $h(t)$, see (Fig. 7.2). The output of the system is then the convolution of $x(t)$ with $h(t)$, that is,

$$y(t) = \int_{-\infty}^{\infty} x(t-\tau)\,h(\tau)\,d\tau = \int_{-\infty}^{\infty} x(\tau)\,h(t-\tau)\,d\tau \;. \tag{7.18}$$

In physical systems the impulse response $h(t) = 0$ for $t < 0$. This permits a corresponding change of the integration bounds in (7.18). However, in the analysis here, it will not be necessary to make this assumption.

The Fourier transform of the filter response is

$$H(\nu) = \int_{-\infty}^{\infty} h(t)\,\mathrm{e}^{-2\pi i\nu t}\,dt \;. \tag{7.19}$$

Taking the expectation value of (7.18) and exchanging again the order of expectation value and integration we find

$$\boxed{E\{y(t)\} = \int_{-\infty}^{\infty} E\{x(t-\tau)\}\,h(\tau)\,d\tau}$$

or using (7.4)

$$\boxed{E\{y(t)\} = \mu_y = E\{x(t)\}\int_{-\infty}^{\infty} h(\tau)\,d\tau} \;. \tag{7.20}$$

With (7.19) this can be written as

$$\boxed{\mu_y = H(0)\,\mu_x} \tag{7.21}$$

showing how the mean value of a stochastic process will be affected if passed through a linear system. If the mean value of the input signal is zero, this will also be true for the output signal.

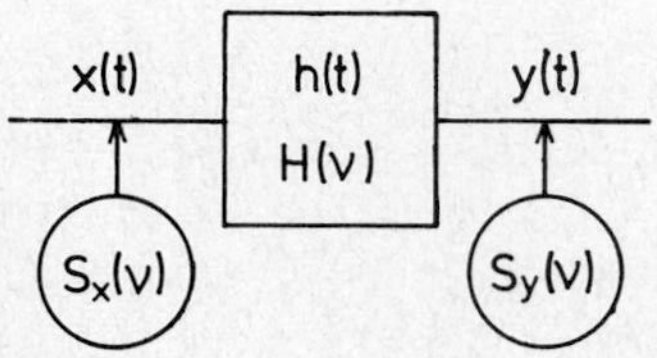

Fig. 7.2. Noise in a linear system

The autocorrelation $R_{yy}(\tau)$ of the output $y(t)$ is most easily determined by first considering the cross-correlation $R_{xy}(\tau)$ between $x(t)$ and $y(t)$. Multiplying both sides of (7.18) by $x(t-\vartheta)$ we have

$$y(t)\,x(t-\vartheta) = \int_{-\infty}^{\infty} x(t-\tau)\,x(t-\vartheta)\,h(\tau)\,d\tau \tag{7.22}$$

and

$$E\{x(t-\tau)\,x(t-\vartheta)\} = R_{xx}((t-\tau)-(t-\vartheta)) = R_{xx}(\vartheta-\tau)\ .$$

Taking the expectation value of both sides of (7.22) and again exchanging integration and expectation value, we get

$$E\{y(t)\,x(t-\vartheta)\} = \int_{-\infty}^{\infty} R_{xx}(\vartheta-\tau)\,h(\tau)\,d\tau\ .$$

This integral is obviously time-independent and equal to the convolution of $R_{xx}(\tau)$ with $h(\tau)$; the left side is the cross-correlation of $y(t)$ and $x(t)$, so that

$$R_{yx}(\tau) = R_{xx}(\tau) \otimes h(\tau)\ . \tag{7.23}$$

Multiplying (7.18) by $y(t+\vartheta)$ we have

$$y(t+\vartheta)\,y(t) = \int_{-\infty}^{\infty} y(t+\vartheta)\,x(t-\tau)\,h(\tau)\,d\tau$$

and

$$R_{yy}(\vartheta) = \int_{-\infty}^{\infty} R_{yx}(\vartheta+\tau)\,h(\tau)\,d\tau = R_{yx}(\vartheta) \otimes h(-\vartheta)\ . \tag{7.24}$$

Now obviously

$$R_{xy}(\tau) = R_{yx}(\tau)\ ,$$

so that, combining (7.23) and (7.24), we finally obtain

$$\boxed{R_{yy}(\tau) = R_{xx}(\tau) \otimes h(\tau) \otimes h(-\tau)} \tag{7.25}$$

which is written in extended form

$$R_{yy}(\tau) = \int_{-\infty}^{\infty} R_{xx}(\tau-t)\left[\int_{-\infty}^{\infty} h(\vartheta+t)\,h(\vartheta)\,d\vartheta\right]dt\ . \tag{7.25 a}$$

Therefore, in order to be able to compute only a single value of the output autocorrelation function of a linear filter, the whole input ACF has to be known.

Taking the Fourier transform of (7.25) we obtain for the input and output PSD the relation

$$\boxed{S_y(\nu) = S_x(\nu)\,|H(\nu)|^2} \quad . \tag{7.26}$$

7.1.4 Square-Law Detector

In radio receivers, the noise signal is passed through a device (Fig. 7.3) that produces an output signal $y(t)$ which is proportional to the power of a given input signal $x(t)$:

$$y(t) = a\,x^2(t) \quad . \tag{7.27}$$

The analysis of such a nonlinear system is quite naturally more difficult than that of a linear system. Therefore we will restrict the discussion to a special class of random-noise signals which, however, is wide enough to describe well the behaviour of real noise signals: we will discuss here stationary *normally distributed random variable* or *gaussian noise* for which the probability density distribution function is a gaussian function with the mean $\mu = 0$. Thus

$$p(x) = \frac{1}{\sigma\sqrt{2\pi}}\,\mathrm{e}^{-x^2/2\sigma^2} \quad . \tag{7.28}$$

For this random variable we obviously have

$$E\{x\} = \mu = 0 \quad \text{and} \quad E\{x^2\} = \sigma^2 \quad .$$

Let the ACF of $x(t)$ be

$$E\{x(t+\tau)\,x(t)\} = R(\tau) \quad ;$$

then the random variables $x(t+\tau)$ and $x(t)$ are jointly normal with the same variance $E\{x^2\} = R(0)$ and the correlation coefficient

$$r = \frac{E\{x(t+\tau)\,x(t)\}}{\sqrt{E\{x^2(t+\tau)\}\,E\{x^2(t)\}}} = \frac{R(\tau)}{R(0)} \tag{7.29}$$

so that their joint density distribution function is given by

$$p(x_1, x_2; \tau) = \frac{1}{2\pi\sqrt{R^2(0) - R^2(\tau)}}\exp\left[-\frac{R(0)\,x_1^2 - 2R(\tau)\,x_1 x_2 + R(0)\,x_2^2}{2\,[R^2(0) - R^2(\tau)]}\right] \tag{7.30}$$

$$= \frac{1}{2\pi\sigma_1\sigma_2\sqrt{1-r^2}}\exp\left[\frac{-1}{2(1-r^2)}\left(\frac{x_1^2}{\sigma_1^2} - \frac{2r\,x_1 x_2}{\sigma_1\sigma_2} + \frac{x_2^2}{\sigma_2^2}\right)\right] \quad . \tag{7.31}$$

If such gaussian noise is input to a square law detector (7.28) the output noise will no longer be gaussian (Fig. 7.4). The distribution function $p_y(y)$ can, however,

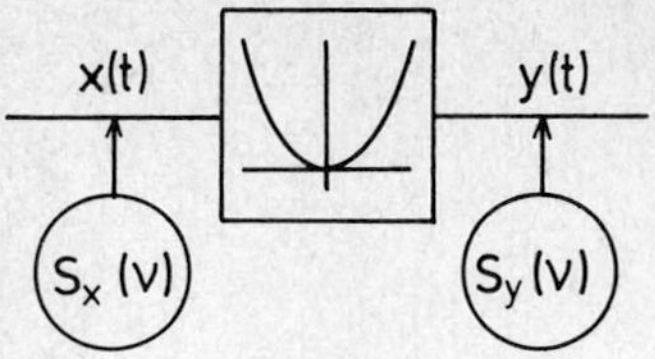

Fig. 7.3. Input and output signal for a square-law detector

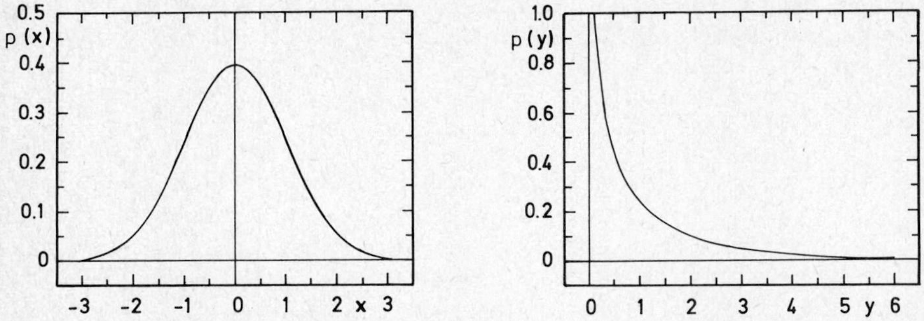

Fig. 7.4. Gaussian input probability density function for a square-law detector and the resulting output probability density function

be rather easily computed. It can be shown [see e.g. Papoulis (1965), p. 125] that if the random variable x is transformed into y according to

$$y = \varphi(x) \; ,$$

then the probability density $p(y)$ is given by

$$p_y(y) = \frac{p_x(x_1)}{|\varphi'(x_1)|} + \frac{p_x(x_2)}{|\varphi'(x_2)|} + \dots + \frac{p_x(x_n)}{|\varphi'(x_n)|} \tag{7.32}$$

where $x_1, x_2 \dots, x_n$ are all real roots of

$$y = \varphi(x_1) = \varphi(x_2) = \dots = \varphi(x_n) \; .$$

The relation (7.27) has two solutions

$$x_1 = \sqrt{\frac{y}{a}} \, , \qquad x_2 = -\sqrt{\frac{y}{a}} \quad \text{and}$$

$$\varphi' = \frac{dy}{dx} = 2\sqrt{a\,y} \; , \qquad \text{so that}$$

$$p_y(y) = \frac{p_x\left(\sqrt{\frac{y}{a}}\right) + p_x\left(-\sqrt{\frac{y}{a}}\right)}{2\sqrt{a\,y}}$$

which for (7.28) results in

$$p_y(y) = \begin{cases} \dfrac{1}{\sigma_x \sqrt{2\pi a y}} \exp\left(-\dfrac{y}{2a\sigma_x^2}\right) & \text{for } y \geqq 0 \\ 0 & \text{for } y < 0 \end{cases} . \tag{7.33}$$

Thus we find that

$$\begin{aligned} E\{y(t)\} &= E\{x^2(t)\} = a\sigma_x^2 = a R_x(0) \\ E\{y^2(t)\} &= 3a^2\sigma_x^4 = 3a^2 R_x^2(0) \end{aligned} \tag{7.34}$$

and hence

$$\sigma_y^2 = E\{y^2(t)\} - E^2\{y(t)\} = 2a^2 R_x^2(0) = 2E^2\{y(t)\} . \tag{7.35}$$

Contrary to linear systems, the mean value of the output signal of a square-law detector is not equal to zero, even if the input signal has this expected average.

In order to determine the shape of the output PSD of a square law detector we will compute the autocorrelation function of the output signal $y(t)$. From the definitions (7.10) and (7.27) we find that

$$R_y(\tau) = E\{y(t+\tau)\,y(t)\} = a^2 E\{x^2(t+\tau)\,x^2(t)\} . \tag{7.36}$$

Analogously to (7.17)

$$E\{x_1^2 x_2^2\} = \iint\limits_{-\infty}^{\infty} x_1^2 x_2^2\, p(x_1, x_2; \tau)\, dx_1\, dx_2$$

with $p(x_1, x_2; \tau)$ given by (7.31). By direct integration [the details will be skipped here; a more elegant, but still rather complicated proof can be found in Papoulis (1965), p. 221] we can show that

$$E\{x_1^2 x_2^2\} = E\{x_1^2\}\, E\{x_2^2\} + 2E^2\{x_1 x_2\} .$$

Therefore,

$$E\{x^2(t+\tau)\,x^2(t)\} = E\{x^2(t+\tau)\}\, E\{x^2(t)\} + 2E^2\{x(t+\tau)\,x(t)\} . \tag{7.37}$$

Combining (7.36) and (7.37) we obtain

$$\boxed{R_y(\tau) = a^2\,[R_x^2(0) + 2R_x^2(\tau)]} . \tag{7.38}$$

From this we obtain the PSD of the output signal by taking the Fourier transform resulting in

$$S_y(\nu) = a^2 R_x^2(0)\,\delta(\nu) + 2a^2 \int\limits_{-\infty}^{\infty} R_x^2(\tau)\, e^{-2\pi i \nu\tau}\, d\tau .$$

Now

$$\int_{-\infty}^{\infty} R_x^2(\tau)\,e^{-2\pi i\nu\tau}\,d\tau = \int_{-\infty}^{\infty} S_x(\nu')\,d\nu' \int_{-\infty}^{\infty} R_x(\tau)\,e^{-2\pi i(\nu-\nu')\tau}\,d\tau$$

$$= \int_{-\infty}^{\infty} S_x(\nu')\,S_x(\nu-\nu')\,d\nu' \ .$$

Hence

$$\boxed{S_y(\nu) = a^2 R_x^2(0)\,\delta(\nu) + 2\,a^2 \int_{-\infty}^{\infty} S_x(\nu')\,S_x(\nu-\nu')\,d\nu'} \qquad (7.39)$$

expressing the output PSD in terms of the input PSD (see also Fig. 7.6 b).

7.2 Limiting Receiver Sensitivity

Radio receivers are, as stated earlier, devices that measure the spectral power density and, to be able to do this, even an elementary receiver consisting only of the most essential parts must contain the following basic units (Figs. 7.5 and 7.6):

1) The reception filter with the power gain transfer function $G(\nu)$ defining the spectral range to which the receiver responds. For practical receivers this frequency range may be situated anywhere from the MHz range up to several hundered GHz with bandwidths from less than 1 kHz to several GHz.

2) The square-law detector used to produce an output signal that is proportional to the average power in the reception band.

3) The smoothing filter with the power gain transfer function $W(\nu)$ that determines the time response of the output. Its properties must correspond to the intended purpose of the receiver: the filter for a radio astronomical receiver used to measure faint radio sources by a on-off procedure will be different from that used for one that measures fast solar burst activities or for an audio receiver.

For a receiver to be useful, it should be able to detect as faint signals as possible. Just as with any other measuring device there are limits for this sensitivity which come from the fact that the signal that is to be detected is noisy. We will outline here how this limiting sensitivity depends on the receiver parameters.

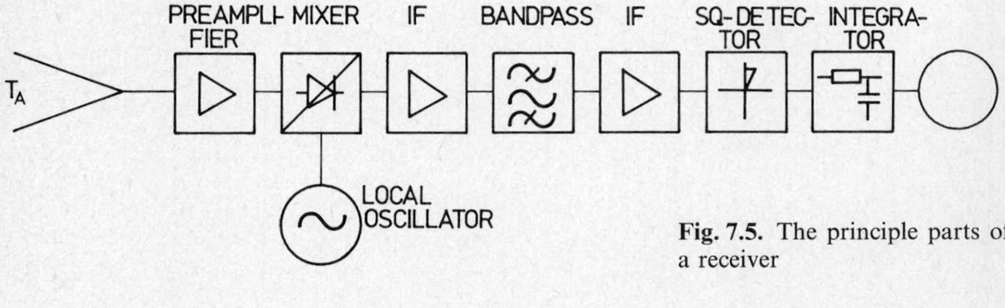

Fig. 7.5. The principle parts of a receiver

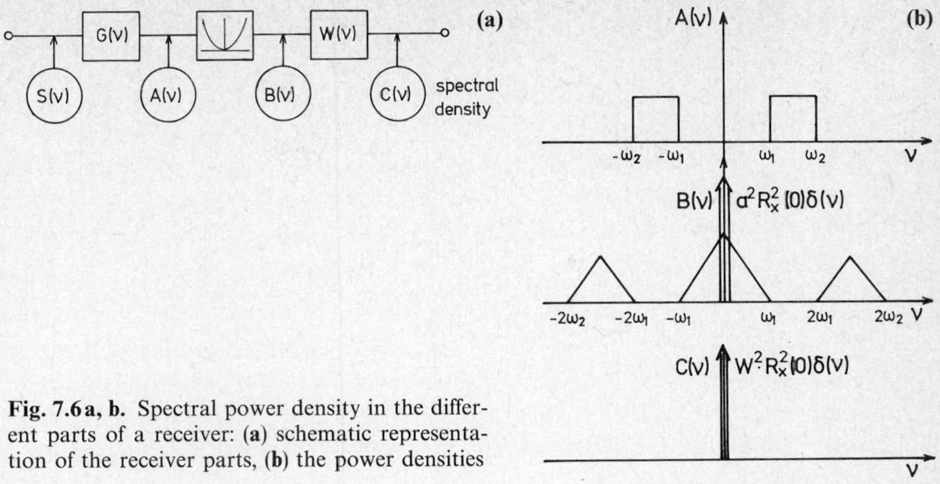

Fig. 7.6 a, b. Spectral power density in the different parts of a receiver: (**a**) schematic representation of the receiver parts, (**b**) the power densities

Even when no input source is connected to a receiver we will observe an output signal. The reason for this is that any receiver is built from parts that emit thermal noise. This noise is amplified along with any external signal and cannot in principle be distinguished from it.

If we want to analyze the performance of a real receiver we can replace it by an ideal receiver producing no internal noise and being connected simultaneously to two noise sources, one producing the external source and a second virtual source simulating the receiver noise.

While the spectral power density of the external source is arbitrary, experience shows that the virtual input PSD of the receiver noise is well represented by a white spectrum. Please note that we never measure this PSD directly; we only see an output signal which behaves as if produced by such a white PSD. Using the Nyquist theorem we can describe such a white PSD by an equivalent noise temperature [cf. Eq. (1.35)] that would produce just such a thermal PSD. Instead of specifying the power level of the receiver noise PSD we therefore can characterize the receiver quality by its system noise temperature T_{sys}.

In practical systems there will always be several sources of the system noise. Besides the receiver proper, the atmosphere, the spillover noise of the antenna consisting of ground radiation that enters into the system through far sidelobes looking at the ground, and noise produced by the unavoidable attenuation in the waveguides and coaxial cables connecting the feed to the receiver input all contribute to the total receiver output. These noise contributions are additive and therefore

$$T_{\mathrm{sys}} = \sum T_i$$

where T_i is the noise temperature of the individual contributors. The PSD of the system noise at the receiver input is then

$$S_{\mathrm{sys}}(\nu) = k\, T_{\mathrm{sys}} \quad . \tag{7.40}$$

The input to the square-law detector for the normal random noise signal $x(t)$ has the PSD

$$A(\nu) = k\,T_{\text{sys}}\,G(\nu)\ . \tag{7.41}$$

According to (7.33) the signal $y(t)$ does not have a Gaussian distribution at the output of the square-law detector; its PSD will be called $B(\nu)$. The output of the smoothing filter with the power transfer function $W(\nu)$ is $z(t)$; its PSD is $C(\nu)$ (Figs. 7.6 a, b).

We will call a signal detectable if the mean output increment is greater than or equal to the dispersion of z; i.e. if $\langle \Delta z\rangle \geqq \sigma_z$.

If z had a Gaussian probability density distribution, this would mean a detection significance of 68%; if other limits for Δz are chosen, higher confidence limits can be achieved.

Using (7.5) the detection limit therefore is

$$\Delta z = \sqrt{E\{z^2(t)\} - E^2\{z(t)\}}\ . \tag{7.42}$$

But according to (7.15)

$$\langle z^2\rangle = E\{z^2(t)\} = \int_{-\infty}^{\infty} C(\nu)\,d\nu$$

while (7.21) and (7.34) result in

$$\langle z\rangle = E\{z(t)\} = \sqrt{W(0)}\,E\{y(t)\} = \sqrt{W(0)}\,a\,R_x(0)\ . \tag{7.43}$$

But since

$$C(\nu) = W(\nu)\,B(\nu)$$

while according to (7.39)

$$B(\nu) = 2\,a^2\,[A(\nu) \otimes A(\nu)] + a^2\,R_x^2(0)\,\delta(\nu)$$

for a Gaussian noise input to the square-law detector, we obtain by substituting all this into (7.42):

$$\Delta z = \sqrt{2\,a^2 \int_{-\infty}^{\infty} W(\nu)\,[A(\nu) \otimes A(\nu)]\,d\nu}\ . \tag{7.44}$$

Generally, the smoothing filter output power transfer function is only a few Hertz wide, while $A(\nu) \otimes A(\nu)$ will have a width of kHz to MHz. Then $A(\nu) \otimes A(\nu)$ can be considered to be constant over the total width of $W(\nu)$, so that

$$\int W(\nu)\,[A \otimes A]\,d\nu \cong [A \otimes A]_0 \int W(\nu)\,d\nu\ .$$

From (7.43) and (7.44) expressing $R_x(0)$ in terms of $A(\nu)$ and using (7.15), we then find

$$\frac{\Delta z}{\langle z\rangle} = \sqrt{\frac{2\,[A \otimes A]_0}{\left(\int\limits_{-\infty}^{\infty} A(\nu)\,d\nu\right)^2}\;\frac{\int\limits_{-\infty}^{\infty} W(\nu)\,d\nu}{W(0)}}$$

Table 7.1. Noise bandwidth of some filters

Reception filter	$G(\nu)$	Δf
Rectangular pass band	$\begin{cases} 1 & \text{for } \nu_0 - \frac{1}{2}\Delta < \lvert\nu\rvert < \nu_0 + \frac{1}{2}\Delta \\ 0 & \text{elsewhere} \end{cases}$	Δ
Triangular pass band	$\begin{cases} 1 - 2\lvert\nu - \nu_0\rvert/\Delta & \text{for } \nu_0 - \frac{1}{2}\Delta < \lvert\nu\rvert < \nu_0 + \frac{1}{2}\Delta \\ 0 & \text{elsewhere} \end{cases}$	$\frac{3}{4}\Delta$
Single tuned circuit	$[1 + (\lvert\nu\rvert - \nu_0)^2/\Delta^2]^{-1}$	$2\pi\Delta$
Two isolated tandem tuned circuits	$[1 + (\lvert\nu\rvert - \nu_0)^2/\Delta^2]^{-2}$	$\frac{4}{5}\pi\Delta$
Gaussian pass band	$\exp[-(\lvert\nu\rvert - \nu_0)^2/2\Delta^2]$	$2\sqrt{\pi}\Delta$
Smoothing filter	$W(\nu)$	τ
Running mean over time T	$(\pi T \nu)^{-2}\sin^2 \pi T \nu$	T
Single RC circuit	$[1 + (2\pi RC\nu)^2]^{-1}$	$2RC$
Tandem RC circuit	$[1 + (2\pi RC\nu)^2]^{-2}$	$4RC$
Rectangular pass band	$\begin{cases} 1 & \lvert\nu\rvert < \nu_0 \\ 0 & \lvert\nu\rvert > \nu_0 \end{cases}$	$\frac{1}{2}\nu_0$
Gaussian pass band	$\exp\{-\nu^2/2\nu_0^2\}$	$\sqrt{2\pi}\,\nu_0$

or, if (7.41) is substituted,

$$\frac{\Delta z}{\langle z\rangle} = \sqrt{\frac{2\int_{-\infty}^{\infty} |G(\nu)|^2\, d\nu}{\left(\int_{-\infty}^{\infty} G(\nu)\, d\nu\right)^2} \frac{\int_{-\infty}^{\infty} W(\nu)\, d\nu}{W(0)}} \,. \tag{7.45}$$

The first term in (7.45) depends only on the pre-detector bandwidth of the receiver, while the smoothing filter time-constant governs the second. Thus defining

$$\boxed{\Delta f = \frac{1}{2}\frac{\left(\int_{-\infty}^{\infty} G(\nu)\, d\nu\right)^2}{\int_{-\infty}^{\infty} |G(\nu)|^2\, d\nu}} \quad \text{and} \tag{7.46}$$

$$\boxed{\tau = \frac{W(0)}{\int_{-\infty}^{\infty} W(\nu)\, d\nu}} \tag{7.47}$$

and remembering that $\Delta z \propto \Delta T$ and $\langle z \rangle \propto T_{sys}$

$$\boxed{\frac{\Delta T}{T_{sys}} = \frac{1}{\sqrt{\Delta f \, \tau}}} \, , \qquad (7.48)$$

a result first obtained by Dicke (1946). Δf and τ are different from the parameters that are usually employed to characterize filter passbands, although they naturally are related to them. In Table 7.1 values for some typical filters are given [taken from Bracewell (1962)].

7.3 Radiometers

The system noise temperature T_{sys} is one of the parameters on which the "quality" of a radiometer depends: of two receivers identical in all respects with the exception of the system noise, the one with the lower T_{sys} is the better one. Therefore it is of interest to discuss how electric noise is produced, and how a system has to be arranged for an optimal result.

7.3.1 Thermal Noise of an Attenuator

Attenuators appear at many positions in the circuit of a radiometer, either purposely in order to attenuate a signal or simply as a "lossy" piece of connecting cable, connector, switch etc. The equation of radiative transfer together with Kirchhoff's law can be used to determine the noise power emitted by such a device if we can assume it to be in local thermodynamic equilibrium. The PSD at the output of the attenuator is obtained by integrating the transfer equation (1.9) along the signal path. If isothermal conditions are assumed and no external signal is input, the signal at the output of the attenuator is according to (1.18)

$$I_{\nu\,\mathrm{noise}} = B_\nu(T)(1 - \mathrm{e}^{-\tau}) \cong (1 - \mathrm{e}^{-\tau}) \frac{2\,\nu^2}{c^2} \, k\, T$$

where T is the thermodynamic temperature of the attenuator and τ its total attenuation in Neper. We now introduce the concept of "noise temperature of an attenuator" T_N by defining it as the brightness temperature of thermal radiation such that this radiation sent through a noiseless attenuator would equal the output of a real attenuator of the same τ and kept at the thermodynamic temperature T. Therefore

$$\mathrm{e}^{-\tau} \frac{2\,\nu^2}{c^2} \, k\, T_N = (1 - \mathrm{e}^{-\tau}) \frac{2\,\nu^2}{c^2} \, k\, T \, , \quad \text{or}$$

output of ideal attenuator = output of real attenuator. If we abbreviate the loss of the attenuator by

$$L = \mathrm{e}^{\tau} \quad \text{then}$$

$$T_N = (L - 1)\, T \, . \qquad (7.49)$$

An example may illustrate the use of (7.49). In the input lines of a low noise receiver connectors, switches etc. introduce an attenuation of 0.3 dB corresponding to $L = 10^{0.3/10} \cong 1.072$. If these parts are at an ambient temperature of $T = 293$ K the resulting $T_N = 21$ K, but if they were cooled down to $T = 20$ K they would contribute only 1.4 K.

Expressing L in dB, that is

$$d/\mathrm{dB} = 10 \log L$$

and using the series expansion for $\log L$ up to and including the linear term we obtain

$$T_N = 0.230 \left(\frac{T_0}{\mathrm{K}} \right) (d/\mathrm{dB}) \tag{7.49 a}$$

or, with $T_0 = 293$ K

$$T_N = 67.5 (d/\mathrm{dB}) \ . \tag{7.49 b}$$

7.3.2 Cascading of Amplifiers

The power amplification needed for a practical receiver is of the order of 80–100 dB. Such a large amplification can only be obtained by cascading (Fig. 7.7) several amplification stages each with the gain G_i resulting in the total gain

$$G = \prod_{i=1}^{n} G_i \ .$$

The question is now: what is the total noise temperature of the cascaded system if each individual stage contributes the noise temperature T_{Si}.

If the input PSD of stage 1 is

$$P_0 = k\, T_A \tag{7.50}$$

then stage i produces an output PSD

$$P_i(\nu) = [P_{i-1}(\nu) + k\, T_{Si}]\, G_i(\nu) \ . \tag{7.51}$$

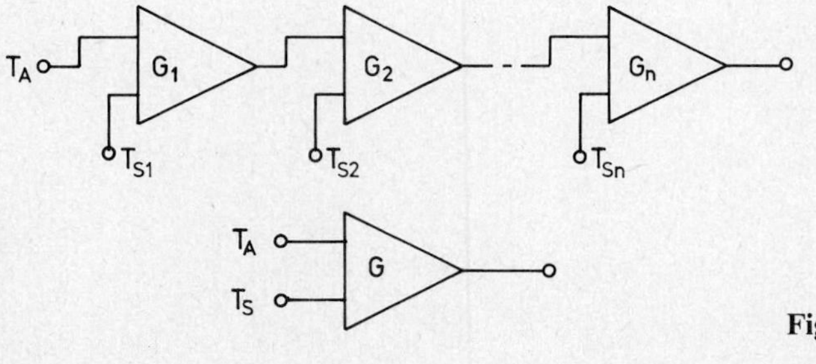

Fig. 7.7. Cascading of amplifiers

The appropriate definition of the total system noise temperature T_S of a system with the total gain $\prod G_i$ is obviously

$$P_n(\nu) = k(T_A + T_S) \prod_{i=1}^{n} G_i(\nu) . \tag{7.52}$$

Substituting (7.50) and (7.51) into (7.52), we obtain (7.53)

$$\boxed{T_S = T_{S1} + \frac{1}{G_1} T_{S2} + \frac{1}{G_1 G_2} T_{S3} + \cdots + \frac{1}{G_1 G_2 \ldots G_{n-1}} T_{Sn}} \quad . \tag{7.53}$$

Therefore, when several amplification stages are necessary, they have to be arranged such that those contributing the least noise temperature are used at the input; for the second and following stage less stringent requirements concerning their noise performance have to be fulfilled.

7.3.3 Mixers

High gain cascaded amplifier chains are quite often troubled by instabilities. If the total gain is of the order of 80–100 dB an exceedingly small amount of power leaking from the output port back to the input port is sufficient to cause the system to oscillate violently. To avoid this, the system has to be shielded heavily, but usually another measure is more effective: the output signal passband is made to differ in frequency from that of the input band by mixing the signal with a monochromatic oscillator signal produced locally, the so-called *local oscillator* (LO). The system then forms a *superheterodyne receiver*.

A *mixer* is the unit that is performing the actual frequency conversion. In principle any device with a nonlinear relation between input voltage and output current can be used for this, but the derivation of its properties are most simple for a pure quadratic characteristic

$$I = \alpha U^2 . \tag{7.54}$$

Let U be the sum of the signal $E \sin(\omega_S t + \delta_S)$ and the local oscillator $V \sin(\omega_L t + \delta_L)$. Then the output is

$$\begin{aligned} I &= \alpha[E \sin(\omega_S t + \delta_S) + V \sin(\omega_L t + \delta_L)]^2 \\ &= \alpha E^2 \sin^2(\omega_S t + \delta_S) + \alpha V^2 \sin^2(\omega_L t + \delta_L) \\ &\quad + 2\alpha E V \sin(\omega_S t + \delta_S) \sin(\omega_L t + \delta_L) \end{aligned} \tag{7.55}$$

or, using the trigonometric addition formulae

$$\begin{aligned} I &= \tfrac{1}{2}\alpha(E^2 + V^2) && \text{(DC component)} \\ &\quad - \tfrac{1}{2}\alpha E^2 \sin(2\omega_S t + 2\delta_S + \tfrac{\pi}{2}) && \text{(2nd harmonic of signal)} \\ &\quad - \tfrac{1}{2}\alpha V^2 \sin(2\omega_L t + 2\delta_L + \tfrac{\pi}{2}) && \text{(2nd harmonic of LO)} \\ &\quad + \alpha V E \sin[(\omega_S - \omega_L)t + (\delta_S - \delta_L + \tfrac{\pi}{2})] && \text{(difference frequency)} \\ &\quad - \alpha V E \sin[(\omega_S + \omega_L)t + (\delta_S + \delta_L + \tfrac{\pi}{2})] && \text{(sum frequency)} . \end{aligned} \tag{7.56}$$

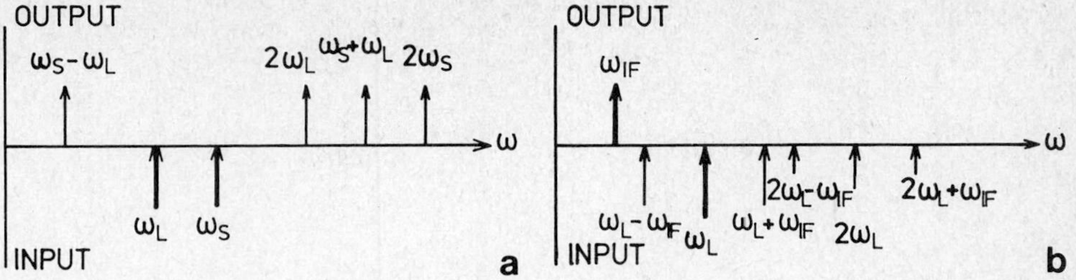

Fig. 7.8 a, b. Input-output frequencies of a mixer (**a**) when two input frequencies are given (**b**) when LO and IF are given

The output then will consist of the superposition of several components at different frequencies (Fig. 7.8 a, b): a DC signal, signals at twice the signal and the local oscillator frequencies, and two components at the difference and the sum of signal and oscillator frequencies. While the amplitude of all other components depends on the second power of signal or local oscillator amplitude, the sum and difference frequency signals depend on the first power only. Their amplitude therefore is a fair reproduction of the input signal amplitude.

By using an appropriate bandfilter, all signals except the wanted one can be suppressed. In this way the mixer can be considered to be a *linear* device producing an output at the frequency $\nu_{\text{if}} = \nu_{\text{S}} - \nu_{\text{L}}$. This is also true for mixers with characteristic curves different from (7.54).

The same difference frequency can be produced in two ways: one in which the local oscillator frequency is below the signal frequency, and another one in which it is above. Therefore a mixer will transform two frequency bands into the intermediate frequency band, and these two are placed symmetrically with respect to the local oscillator frequency ν_{L} – they are called mirror or image frequencies. If measures are taken to eliminate one of the mirrors, the receivers is of the single-sideband type (SSB). Wide-band radiometers usually permit both sidebands. They are double-sideband receivers (DSB) while line spectrometers are most often of the SSB kind.

The output power of a mixer in the IF band is linearly dependent on the local oscillator power. By increasing this power, the gain of the total system will be increased, and usually the LO power is much larger than the signal power. But even more important, any variation in the local oscillator power will appear again as a variation of the total gain of the system. Output power stability for the local oscillator is therefore necessary. One more practical aspect of mixing is that the mixer forms a suitable point in a receiver system for dividing it into a *frontend* section which is specific for the selected frequency range and a more or less standard *backend* which is specific for the intended purpose of the receiver.

7.3.4 Receiver Stability

Because the signal measured at the receiver output is

$$W = k(T_{\text{A}} + T_{\text{sys}})\,G\,\Delta f\,, \tag{7.57}$$

variations of the total system gain ΔG leading to

$$W + \Delta W = k(T_A + T_{sys})(G + \Delta G)\,\Delta f \tag{7.58}$$

are indistinguishable from variations of the antenna temperature

$$W + \Delta W = k(T_A + \Delta T_A + T_{sys})\,G\,\Delta f\ . \tag{7.59}$$

Comparing (7.59) and (7.58) using (7.57) we obtain

$$\boxed{\frac{\Delta T}{T_{sys}} = \frac{\Delta G}{G}} \tag{7.60}$$

indicating that variations of the power gain enter directly into the determination of limiting sensitivity. If such a total power receiver is to be able to measure an increase of the antenna temperature as small as $10^{-4}\,T_{sys}$, its total gain has to be kept constant to within this ratio. This is exceedingly difficult to achieve on a routine basis, and therefore one often resorts to a receiver system based on a differential compensation principle.

The principle generally used was first applied in radio astronomical receivers by Dicke (1946), and it is a straightforward application of the compensation principle as used, for example, in the Wheatstone bridge (Fig. 7.9). A receiver is switched regularly between the antenna producing the signal T_A and a resistive load in thermodynamic equilibrium at the thermodynamic temperature T_R. If both antenna and load are matched equally well to the receiver input, the antenna gives the output

$$W_A = k(T_A + T_{sys})\,G\,\Delta f$$

while the load produces

$$W_R = k(T_R + T_{sys})\,G\,\Delta f\ .$$

At the output of the receiver, the difference of these two signals as measured by a phase-sensitive detector or lock-in amplifier is then

$$W_A - W_R = k(T_A - T_R)\,G\,\Delta f\ .$$

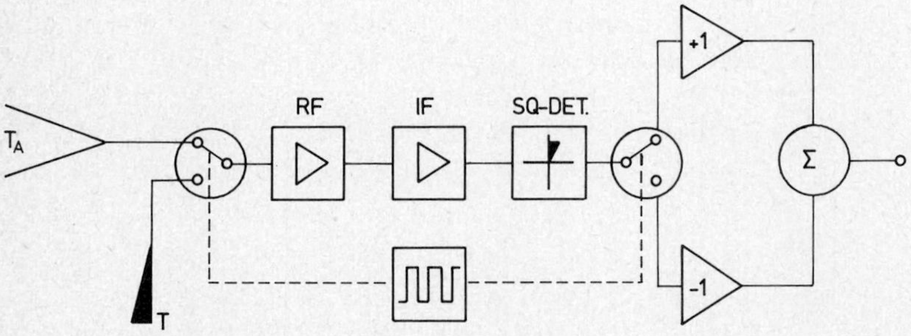

Fig. 7.9. A balanced receiver (Dicke receiver)

Now, if a gain variation ΔG is wrongly interpreted as a variation ΔT of the antenna temperature we have

$$k(T_A - T_R)(G + \Delta G)\Delta f = k(T_A + \Delta T - T_R)G\Delta f$$

or

$$\boxed{\frac{\Delta T}{T_{sys}} = \frac{\Delta G}{G}\frac{T_A - T_R}{T_{sys}}} \quad . \tag{7.61}$$

Depending on the value of $T_A - T_R$ a gain fluctuation will have a much reduced influence; for a *balanced receiver* with $T_A = T_R$ the resolution is completely independent of any gain variations. The receiver then is working as a zero point indicator. But this is true only for weak signals where $T_A \cong T_R$. If strong sources are measured $T_A \neq T_R$ and the receiver is no longer balanced. Gain variations will then influence the signal.

There are different means of producing the comparison load T_R. A straightforward implementation is a resistive load immersed in a thermal bath of the temperature T_R. For a low-noise system the reference temperature has to be fairly low and can be furnished by a liquid nitrogen or liquid helium bath.

Sufficient for many purposes is often a sky-horn giving a T_R that is the mean value over a large part of the visible sky. Another method with the additional advantage that the irregularities introduced by the earth's atmosphere can be largely eliminated is beam-switching, done by alternating the receiver input between two identical feed-horns spaced several beamwidths apart. The difference of the input is then measured at the output of the phase-sensitive detector. Most of the atmospheric irregularities are thereby eliminated since the two beams transverse practically identical volumes of air.

If signals of widely varying strength are to be measured, it is not always possible to maintain a wellbalanced system with $T_A \cong T_R$. Then the system can be stabilized by injecting a constant noise step for some part of the measuring cycle. By comparing the output at appropriate phases of the cycle, both the zero point and the gain of the system can be monitored and kept constant.

If some fraction of the integration time of the receiver is used to control zero point and gain, this part will be missing in (7.48) when the sensitivity of the system is determined. Thus a Dicke system that uses half the integration time to measure T_{ref} will achieve a resolution

$$\frac{\Delta T}{T_{sys}} = \frac{\sqrt{2}}{\sqrt{\Delta f\,\tau}} \quad . \tag{7.62}$$

The resolution of a Dicke receiver is often quoted to be less than (7.62) by another factor of $\sqrt{2}$ resulting from the fact that ΔT is computed as the difference $\Delta z = T_A - T_R$, where both the signals T_A and T_R have equal errors. The difference then has $\sqrt{2}$ times the error. But the same procedure applies for a total power receiver because there the signals to be compared are $\Delta T + T_{sys}$ and T_{sys}.

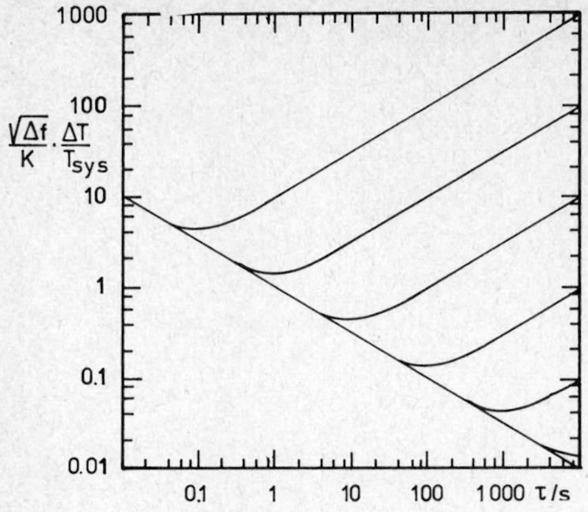

Fig. 7.10. The time dependence of the receiver sensitivity $\Delta T/T_{\rm sys}$ normalized to unit bandwidth Δf. K is a factor comparing the receiver system to a total power system ($K = 1$), and the different curves of the one-parameter field are described by the value τ_m in Eq. (7.64) of the integration time leading to the smallest value for the sensitivity $\Delta T/T_{\rm sys}$

The error of this difference signal has been determined in Sect. 7.2 and is given by (7.48).

If the time variation of G is included in the expression for the sensitivity limit, the obvious generalization of (7.48) for stochastic time variations of $\Delta G/G$ will be (Fig. 7.10 and Table 7.2):

$$\frac{\Delta T}{T_{\rm sys}} = K\sqrt{\frac{1}{\Delta f\,\tau} + \left(\frac{\Delta G}{G}\right)^2} \ . \tag{7.63}$$

For a time dependence

$$\left(\frac{\Delta G}{G}\right)^2 = \gamma_0 + \gamma_1\,\tau \ ,$$

we obtain the smallest value for the resolution $\Delta T/T_{\rm sys}$ at the integration time

$$\tau_m = \frac{1}{\sqrt{\Delta f\,\gamma_1}} \ . \tag{7.64}$$

For well made receivers this time is of the order of hours!

Table 7.2. Noise performance K of different receiver configurations

$$\frac{\Delta T}{T_{\rm sys}} = \frac{K}{\sqrt{\Delta f\,\tau}}$$

Receiver type	K
Total power receiver (7.48)	1
Switched receiver	$\sqrt{2}$
Correlation receiver	$\sqrt{2}$
Adding interferometer	1/2
Correlation interferometer	$1/\sqrt{2}$
1 bit digital correlation spectrometer	3.06
2 bit digital correlation spectrometer	2.42

7.4 Correlation Receivers and Polarimeters

For the comparison of two different stochastic processes $x(t)$ and $y(t)$, the cross-correlation $E\{x(s)\,y(s+\tau)\}$ is of the same importance as the autocorrelation function is in describing the properties of a single process. The definition of the *cross-correlation function* for two ergodic processes is

$$E\{x(s)\,y(s+\tau)\} = \langle x(s)\,y(s+\tau)\rangle = R_{xy}(\tau)$$
$$= \lim_{T\to\infty} \frac{1}{2T}\int_{-T}^{T} x(s)\,y(s+\tau)\,d\tau \ . \qquad (7.65)$$

Examples for such data are the elements of the coherence matrix of a quasi-monochromatic plane wave (3.55) which are the cross-correlation for time $\tau = 0$ of two orthogonal components of the electric field strength.

In order to measure (7.65) a device is needed that multiplies the two processes $x(t)$ and $y(t)$. This is possible by combining two square-law detectors in the proper way (Fig. 7.11), because

$$[x(t)+y(t)]^2 - [x(t)-y(t)]^2 = 4\,x(t)\,y(t) \ .$$

Usually such an analog-multiplication device works properly only for a certain frequency range, outside which phase errors destroy the designed-for purpose. However, by a slight extension of the arguments used to derive the properties of mixers in (7.55) we can see that mixing two partially coherent signals with a local oscillator conserves the correlation between the two signals even if the intermediate frequencies are considered.

Two monochromatic waves

$$x(t) = E_x \sin(\omega t + \delta_x)$$
$$y(t) = E_y \sin(\omega t + \delta_y)$$

have the cross-correlation

$$\langle x(t)\,y(t)\rangle = E_x E_y \lim_{T\to\infty}\frac{1}{2T}\int_{-T}^{T} \sin(\omega t+\delta_x)\sin(\omega t+\delta_y)\,dt$$
$$= E_x E_y\left[\frac{1}{2}\cos(\delta_x-\delta_y) - \lim_{T\to\infty}\frac{1}{2T}\int_{-T}^{T}\frac{1}{2}\cos(2\,\omega t+\delta_x+\delta_y)\,dt\right]$$
$$= \tfrac{1}{2}E_x E_y \cos(\delta_x-\delta_y) \ . \qquad (7.66)$$

If now these two signals are mixed with the same local oscillator signal and the difference frequency is extracted,

$$x_{\rm if}(t) = \alpha\,V E_x \sin[(\omega-\omega_{\rm L})t + (\delta_x - \delta_{\rm L} + \tfrac{\pi}{2})]$$
$$y_{\rm if}(t) = \alpha\,V E_y \sin[(\omega-\omega_{\rm L})t + (\delta_y - \delta_{\rm L} + \tfrac{\pi}{2})] \ ,$$

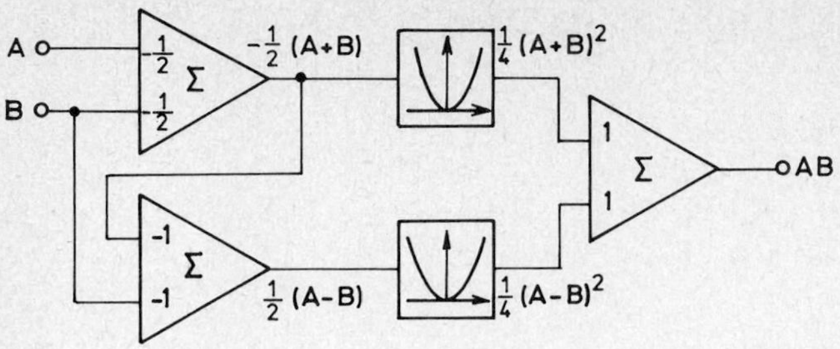

Fig. 7.11. A quarter-square multiplier

so that the cross-correlation is

$$\langle x_{\text{if}}(t)\, y_{\text{if}}(t)\rangle = \tfrac{1}{2}\alpha^2 V^2 E_x E_y \cos(\delta_x - \delta_y) \ . \tag{7.67}$$

Except for the "gain" factor $\alpha^2 V^2/2$ the intermediate frequency cross-correlation (7.67) is identical to the original (7.66). The phase δ_{L} of the local oscillator and any jumps of this phase have cancelled in (7.67). Correlation receivers therefore can be built with the correlation being actually performed in the IF section.

The block diagram for a correlation receiver is shown in Fig. 7.12. The signals from the antenna and from the reference are input to a "3 dB hybrid", a four-port device with two input and two output ports. If the signals at the inputs are $x(t)$ and $y(t)$, the outputs are $\frac{1}{2}[x(t) + y(t)]$ and $\frac{1}{2}[x(t) - y(t)]$. Such hybrids can be built using various techniques, from coaxial to stripline and waveguide, and in general the loss imposed on the signal by such a device is less than that for a switch as is needed in a Dicke receiver.

The two outputs of the hybrid are amplified by two independent radiometer receivers which share a common local oscillator, and the IF signals then are correlated (Fig. 7.13). If the input voltages to the correlator are

$$U_1 = \sqrt{G_1}\,[(U_{\text{A}} + U_{\text{ref}})/\sqrt{2} + U_{\text{N1}}]$$
$$U_2 = \sqrt{G_2}\,[(U_{\text{A}} - U_{\text{ref}})/\sqrt{2} + U_{\text{N2}}] \ ,$$

then the instantaneous output voltage is

$$U = \sqrt{G_1 G_2}\,[(U_{\text{A}}^2 - U_{\text{ref}}^2)/2 + U_{\text{N1}}(U_{\text{A}} - U_{\text{ref}})/\sqrt{2} + U_{\text{N2}}(U_{\text{A}} + U_{\text{ref}})/\sqrt{2} + U_{\text{N1}} U_{\text{N2}}] \ .$$

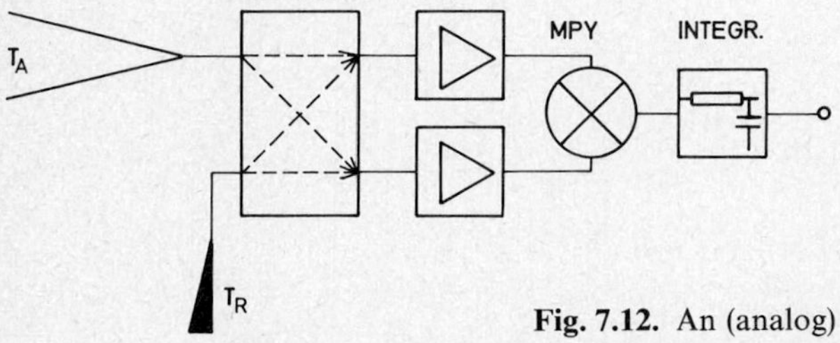

Fig. 7.12. An (analog) correlation receiver

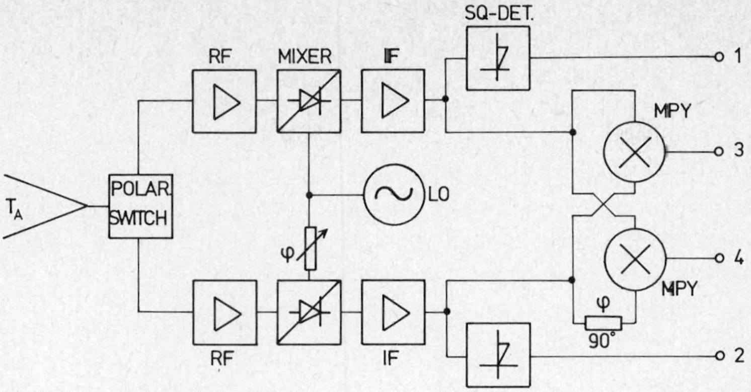

Fig. 7.13. Polarization receiver

But since the stochastic signals U_A, U_{ref}, U_{N1} and U_{N2} are all uncorrelated, the time average of all mixed products will average zero and only

$$\langle U \rangle = \tfrac{1}{2}\sqrt{G_1\,G_2}\,[\langle U_A^2 \rangle - \langle U_{ref}^2 \rangle] \tag{7.68}$$

remains. Gain fluctuations therefore affect only this difference signal; the stability of the correlation receiver is therefore the same as that of a Dicke receiver. For the limiting sensitivity we obtain

$$\frac{\Delta T}{T_{sys}} = \frac{\sqrt{2}}{\sqrt{\Delta f\,\tau}} \ . \tag{7.69}$$

Receivers of a similar kind can be used to measure the polarization of a wavefield as defined by the Stokes parameters (3.67).

The orthogonal linear polarization modes of a partially polarized wavefield as collected by a circular horn are coupled by orthogonal probes into the two input ports of a correlation receiver. A fairly complex cross-correlation device then processes four output signals from the two IF signals:

output z_1 is IF 1 detected by a square-law detector
output z_2 is IF 2 detected by a square-law detector
output z_3 is the correlation of IF 1 and IF 2,
output z_4 is the correlation of IF 1 and IF 2 with a phase delay of $\frac{\pi}{2}$ in one of the channels .

Comparing these outputs with the definition of the Stokes parameters (3.67) we have

$$\begin{aligned} I &= \mathrm{const}\,(z_1 + z_2) \\ Q &= \mathrm{const}\,(z_1 - z_2) \\ U &= 2\ \mathrm{const}\ z_3 \\ V &= 2\ \mathrm{const}\ z_4\ . \end{aligned} \tag{7.70}$$

The output signals z_3 and z_4 come from the cross-correlation of both IF channels; they will therefore be fairly immune to amplification fluctuations. A polarimeter of the kind resulting in (7.70) will therefore result in good values for U and V, that is, in wavefields with circular polarization.

But usually the circular polarization of astronomical sources is exceedingly small, and we are more interested in measuring linear polarization. The polarimeter is quite easily converted for this purpose by including a $\lambda/4$ section in the waveguide section of the horn, so that the probes collect the left- and right-hand circular polarization components. If they are fed into the polarimeter, we will now have

$$\begin{aligned} I &= \mathrm{const}\,(z_1 + z_2) \\ V &= \mathrm{const}\,(z_1 - z_2) \\ Q &= 2\ \mathrm{const}\ z_3 \\ U &= 2\ \mathrm{const}\ z_4 \end{aligned} \tag{7.71}$$

so that the linear polarization components Q and U are now derived by correlated outputs with the corresponding immunity to amplification fluctuations.

7.5 Low Noise Frontend Amplifiers

As stated earlier, the first mixer is a convenient dividing line separating radiometers into the frontend and the backend sections. While the frontend must be specifically tailored to its intended wavelength range, the backend can be adapted to any kind of measurement that is intended, largely independent of the wavelength range.

Often some parts of the network connecting the feed with the receiver input are included in the frontend. This network can consist of many different things such as Dicke-switches, hybrids, directional couplers permitting noise injection for the purpose of calibration, and the like.

As shown in (7.53) the total system noise temperature is mainly determined by the noise temperature of the first one or two amplification stages of the frontend. In addition all losses that occur to the signal before it enters the receiver is another source of noise [cf. Eq. (7.49)]. In order to decrease the noise contribution of all these parts, the first stages of the frontend are often cooled down to a (thermodynamic) temperature of 70 K or even lower to 15 K. Some critical parts in maser amplifiers may even require cooling down to the boiling point of He or slightly above it in order to avoid the handling of liquid He. Often some parts of the connecting network between feed and receiver are included in the cooled section; there exist receivers where even that section of the feed horn with the coupling probes is cooled. How large the influence of cooling of these sections can be on the system noise temperature is shown in the example given in Sect. 7.3.1: while an uncooled insertion loss of 0.3 dB will contribute $T_{\mathrm{N}} = 21$ K to T_{sys}, cooling down to $T = 20$ K reduces this contribution to $T_{\mathrm{N}} = 1.4$ K.

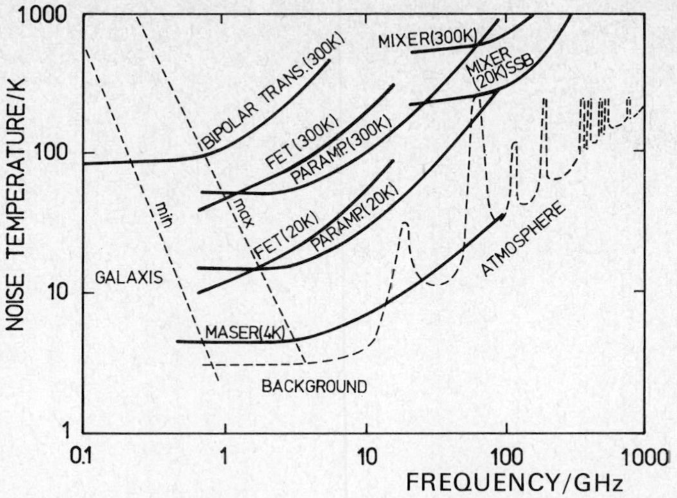

Fig. 7.14. Achievable receiver noise temperatures for different amplifier systems

Low noise amplifiers can be built using a wide variety of techniques; the most suitable and convenient one depends on the frequency range (Fig. 7.14).

7.5.1 Transistor Amplifiers

Advances in the performance of bipolar transistors (pnp or npn silicon transistors) have permitted their application with ordinary lumped-constant-circuits up to frequencies of 1 GHz and even beyond. Such amplifiers are built along the principles used for conventional RF or IF amplifiers. Cooling is seldom required.

A FET transistor, usually of the "metal-semiconductor-FET" (MESFET) kind using a substrate of Gallium-Arsenide (GaAs), can be successfully used for frequencies up to several GHz (6 GHz or higher). While in the lower frequency range lumped-circuits can still be used, both the inductances and the capacitances needed become extremely small when $\nu \gtrsim 2$ GHz. The whole structure then becomes exceedingly sensitive to small mechanical displacements occuring for example when the system is cooled down, and then waveguide or stripline techniques are called for. These systems can be built with fairly large bandwidths, one of the limiting constraints often being the difficulty of avoiding parasitic oscillations (Fig. 7.17 on p. 154).

When transistor amplifiers are built carefully with stable mechanical support, proper thermal insulation or cooling and well stabilized power supplies, gain instabilities are no great problem and exceedingly well-behaved systems are possible.

7.5.2 Parametric Amplifiers

For frequencies larger than a few GHz and up to about 100 GHz parametric amplifiers are frequently used. For low noise performance these system are often

Fig. 7.15. A 21 cm GaAs FET receiver package (courtesy Max Planck-Institut für Radioastronomie, Bonn)

cooled to 15–20 K and thus achieve noise temperatures which can be as low as 35–40 K for $\nu \cong 1-2$ GHz going up to $\cong 200$ K or more at $\nu \cong 100$ GHz (Fig. 7.15).

The principle of parametric amplification was predicted as early as 1931 by Lord Rayleigh, but it was not until 1957 that the first experimental parametric amplifier was built by Weiss. Since the theory of parametric amplifiers is fairly complicated, we will not give an account of it here. It will suffice to refer to the book by Steiner and Pungs (1965) or the review article by Edrich (1977) where ample references to the original literature is given. We will, however, discuss some of the underlying physics to see how amplification can be achieved at all in this way.

The current in a lossless LC circuit is described by the differential equation

$$\ddot{x}(t) + \omega^2 x(t) = 0 \; ; \qquad \omega^2 = \frac{1}{LC} \; . \tag{7.72}$$

If the capacitance in this system consists of a varactor diode with a voltage dependent capacitance, the resonant frequency of (7.72) can be made time variable by a "pumping oscillator" which causes the capacitance to vary periodically with time.

Let

$$\frac{1}{C(t)} = \frac{1}{C_0}(1 + p\cos\gamma t) \ ; \qquad \text{then}$$

$$\ddot{x}(t) + \omega_0^2(1 + p\cos\gamma t)\,x(t) = 0 \ , \qquad \omega_0^2 = \frac{1}{LC_0} \tag{7.73}$$

describes the current in a system with time-variable parameters. This is Mathieu's differential equation; the theory of its solution is given in Whittaker and Watson (1963), p. 404–428 and also Abramowitz and Stegun (1965), Chap. 20, p. 722.

Some basic properties of the solutions of equations with periodic coefficients is given by Floquet's theorem. Considering the fact that (7.73) is periodic with the period $T = 2\pi/\gamma$, the function $x(t + T)$ must also be a solution of (7.73) if $x(t)$ is one. This does not mean that $x(t)$ will be periodic, but it can be shown that there is a solution of the form

$$F_\nu(t) = e^{\nu t} P(t) \tag{7.74}$$

to the system (7.73) where $P(t)$ is periodic with the period T. Another solution is

$$F_\nu(-t) = e^{-\nu t} P(-t) \tag{7.75}$$

and the general solution of (7.73) can be put into the form

$$x(t) = A\,F_\nu(t) + B\,F_\nu(-t) \tag{7.76}$$

where $F_\nu(t)$ and $F_\nu(-t)$ are linearly independent.

The characteristic exponent ν in (7.74) and (7.75) depends in a complicated way on ω_0, p and γ (Fig. 7.16). It can be shown however, that there are combinations of these so that the resulting ν is real. Regions of the parameter space where this is the case will thus posess solutions of (7.73) consisting of a periodic function $P(t)$ with an exponentially increasing amplitude.

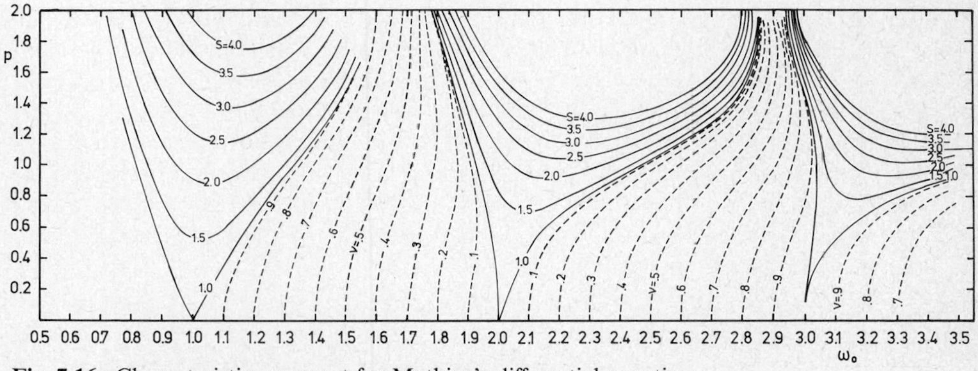

Fig. 7.16. Characteristic exponent for Mathieu's differential equation

$$s = e^{|\nu\pi|}, \quad x'' + \omega_0^2(1 + p\cos 2t)x = 0 \ .$$

Amplification is possible for those ω_0 and p for which $s > 1$

If the LC circuit is not lossless, but has a damping constant λ, the corresponding Floquet solutions are

$$F_\nu(t) = e^{(\nu-\lambda)} P(t)$$
$$F_\nu(-t) = e^{-(\nu-\lambda)t} P(t) \ , \tag{7.77}$$

and the amplification will be correspondingly decreased. From Fig. 7.16 we see that amplification will occur if the pumping frequency γ is chosen so that

$$\gamma = 2\pi\nu_p = 2\omega_0 \pm \varepsilon \quad \text{where} \tag{7.78}$$

$$\varepsilon^2 < \left(\frac{p\,\omega_0}{2}\right)^2 - 4\lambda^2$$

or, for the harmonics of this frequency,

$$\gamma_n = n\gamma = 2\pi n \nu_p \ .$$

The pumping frequency $\gamma_p/2\pi$ must then be close to twice the resonant frequency ω_0 of the circuit or to one of the harmonics thereof.

Thus, in a parametric amplifier, at least two frequencies are of importance: the signal frequency and the pumping frequency causing the variation of the capacitance. For practical designs still another one, the idler frequency, plays a role, and usually this idler frequency is chosen to be equal to the sum or the difference frequency of the first two.

Practical parametric amplifiers are usually built into a waveguide cavity that is simultaneously in resonance for all three frequencies: the signal, the pump and the idler frequency. The varactor diode must be placed at a maximum of the electric field strength for all three frequencies. Another problem with parametric amplifiers is that most of them are only two-port devices: the amplified signal leaves the parametric amplifier through the same port as the input signal enters; the two differ only by their propagation direction and thus have to be separated by circulators.

Well-made circulators present an insertion loss in the propagation direction of 0.1–0.3 dB and thus degrade the noise-performance if they are at the ambient temperature. They are therefore frequently included in the cooled section.

The gain of a single stage is usually kept below 20 dB. Sometimes as little as 10 dB is used so that two cascaded stages of cooled paramps often have to be used in order to obtain a system with an overall low noise temperature. Typical instantaneous bandwidths are 10–20%. Since the gain depends strongly on the pump power this has to be well stabilized, but even then it is sometimes difficult to achieve a satisfactory gain stability.

7.5.3 Maser Amplifiers

Masers – an acronym for Microwave Amplification by Stimulated Emission of Radiation – are amplifiers or oscillators based on the fact first recognized by Einstein in 1917 that a system with discrete internal electronic energy levels will emit electromagnetic radiation in two ways: spontaneous, described by the Einstein coefficients A_{ik}, and stimulated emission, given by $B_{ik}\bar{U}_\nu$, where $\bar{U}_\nu$ is the energy density of the radiation field. While the radiation whose emission is described by A_{ik} is randomly distributed with respect to direction, polarization and phase, the stimulated emission is coherent with the impinging radiation and thus amplifies it. A maser, as the name implies, makes use of this fact.

The interaction of radiation and matter is investigated in Chap. 10 taking into account both emission and absorption. As (10.14) shows, the opacity will become negative, i.e. the radiation intensity is amplified if

$$\frac{N_2}{N_1} > \frac{g_2}{g_1} \; . \tag{7.79}$$

But in a system that is in local thermodynamic equilibrium (LTE), the population of the energy levels is described by the Boltzmann distribution

$$\frac{N_2}{N_1} = \frac{g_2}{g_1} \mathrm{e}^{-\Delta E/kT} \tag{7.80}$$

where ΔE is the energy difference of the two levels, g_1 and g_2 are their statistical weights and T is the (thermodynamic) temperature of the system. Condition (7.79) therefore can never be fulfilled by a system in LTE.

However as first shown by Gordon, Zeiger and Townes in 1954, it is possible to contrive a device that will result in such an inversion of the level populations. Consider a system with three energy levels E_1, E_2 and E_3 $(E_1 < E_2 < E_3)$ and subject it to strong electromagnetic radiation with the frequency

$$\nu_{13} = \frac{1}{h}(E_3 - E_1) \; .$$

Due to absorption and stimulated emission of this pumping radiation, E_1 and E_3 will be equally populated and $N_1 = N_3$ if we consider only non-degenerate systems for the time being. For a stationary system the ratio N_2/N_1 then could be either > 1 or < 1. If the first is the case, then we will have maser action for the transition $2 \to 1$; if the second is true and $N_2/N_1 < 1$, then $N_1/N_2 = N_3/N_2 > 1$, that is, the transition $3 \to 2$ will show maser amplification. Which of these two cases applies for an actual system depends on the transition probabilities of the levels. But in any case we have to make certain that the inversion of the two masering levels is not perturbed by thermal excitations. This will always be the case if $k\,T \leqq \Delta E = h\,\nu$ of the transition in question. For $\nu = 30$ GHz this requires $T \lesssim 1.5$ K, thus the maser material must be cooled.

Usually the cooling temperature condition can be relaxed somewhat from the extreme values quoted because the time needed for this thermal relaxation of the level population is itself strongly temperature-dependent. At low temperatures this relaxation time will be so long that the population inversion can be maintained by strong pumping of the level 3, and therefore masers usually only require $T \cong 4$ K.

Maser amplifiers permit systems with exceedingly low noise temperatures because the spontaneous emission of radiation in the maser material is the only unavoidable noise source. Added to this is the thermal noise of "lossy" parts. In practical systems values as low 6 K for the system noise temperature have been reached. But masers have some fundamental drawbacks, too.

The operating frequency is governed by the transition frequency of the maser material and thus fixed. But for crystals like chromium- or iron-doped ruby (Al_2O_3) or rutile (TiO_3), where the atomic levels of the doting atoms are the masering transitions, the position of these levels can be influenced by external strong magnetic fields. By properly adjusting these magnetic fields, both in strength and in direction, the maser can be tuned. Superconducting electromagnets are usually used for these magnetic fields.

A second, unavoidable drawback is the rather small bandwidth obtainable. It is determined fundamentally by the sharpness of the levels. Although some artificial broadening is possible by permitting the magnetic field to vary inhomogeneously across the maser, the largest band width still remains below 200–250 MHz even for the highest signal frequencies of about 120 GHz that have so far been achieved by masers.

Practical maser-amplifiers are usually built in one of the two following designs. In the cavity maser the maser material is in a cavity which must be in resonance both with the signal and the pumping frequency, and the masering material must be positioned such that it is at the positions of the highest electrical field strength of both signal and pump. The signal enters into the cavity, is then reflected back, amplified, and leaves along the same waveguide along which the signal entered. The two therefore have to be separated by a circulator as in a parametric amplifier.

Another design, the travelling wave maser (TWM), puts the masering material into a waveguide. In order to increase the interaction of the wave and the maser material the waveguide is usually a slow wave structure and in order that only the wave moving in the wanted direction is amplified, ferro-magnetic isolators have to be included.

A maser will give a stable amplification of 20–30 dB; therefore a single stage is usually sufficient, and it provides a rather stable frontend. The low instantaneous bandwidth possible and the large technological effort necessary if successful operation must be guaranteed on a routine basis have prevented most radio astronomical observatories from choosing maser-amplifiers, even if they provide the lowest system noise temperatures.

7.5.4 Frontends for $\nu > 50$ GHz

For frequencies above 50 GHz both low noise parametric amplifiers and masers are difficult to design, and transistor amplifiers are not available neither, because the mobility of the electrons and holes in the valence band of the semiconductor is insufficient for such high frequencies. Therefore mixers are still the best choice for these frequencies. The input signal is mixed with a local oscillator signal and a lower intermediate frequency signal, usually at 0.1 ν_{input} or even less is extracted. All amplification is done in the IF-stages, the mixer itself is causing a loss, which obviously must be kept as small as possible.

One way to achieve this is to cool the first receiver stages to a thermodynamic temperature of $T \cong 15$ K or even less, so that the noise contribution of all ohmic losses will be small. Another reason for the refrigeration is that often the physical effects that cause the nonlinear characteristic curve of the mixer require low temperatures.

In order for the mixer to be applicable for millimeter waves the physical effect causing the nonlinear behaviour must respond to exccedingly fast voltage changes. The *np* transitions of common diodes are not quick enough, but Schottky diodes consisting of a metal and a semiconductor in contact are fast enough for many purposes. When a metal and a semiconductor are brought into contact the Fermi levels in both will attain the same potential, but in the semiconductor close to the contact this level will be deformed. This deformation is called the Schottky effect. If now an external voltage is applied between metal and semiconductor, the Fermi levels of the two will differ, and a current flows. But due to the Schottky effect the resistance depends on the polarity of this voltage, the diode acts as a rectifier. This nonlinearity is responding exceedingly fast to external changes making it possible to use Schottky diodes at very high frequencies.

There is a second reason for this. Any diode in reverse polarity is behaving like a small condensor, which will act as a bypass for the RF signal thus increasing the mixing losses. In order to minimize these, the contact area should be as small as possible, and here Schottky mixers using extremely thin metal whiskers are close to being optimal permitting their use even in the submillimeter range.

Other physical effects leading to a nonlinear characteristic curve with fast response have been used too. Mixers have been built that make use of the Josephson effect as well as socalled SIS-mixers consisting of a sandwich structure of two superconducting leads separated by a thin isolator (Superconductor-Isolator-Superconductor). While Josephson mixers have not been too successful until now, SIS mixers give a good noise performance especially for mm-waves.

The mechanical structure of these mixers can differ widely depending on the physical process used for the mixing as well as on the wavelength range and whether the mixer has to be cooled or not. In the range of long mm-waves waveguide techniques can still be used although the small dimension of the waveguides and cavities and the high surface accuracies required need special technologies.

At higher frequencies, at 1000 GHz and beyond, quasi-optical techniques must be used. Submillimeter waves are not confined in closed waveguide structures that make explicit use of the electromagnetic wave properties, but can be described in many respects by the picture of optical rays, even if diffraction effects still play an important role.

Progress in this field is still very fast so that a general survey of this field will be valid for a short time only. For closer information we have to refer to the lierature. A good accound (in German) of the state of the art at about 1982 is given in Hachenberg and Vowinkel (1982) where references to additional literature can be found too.

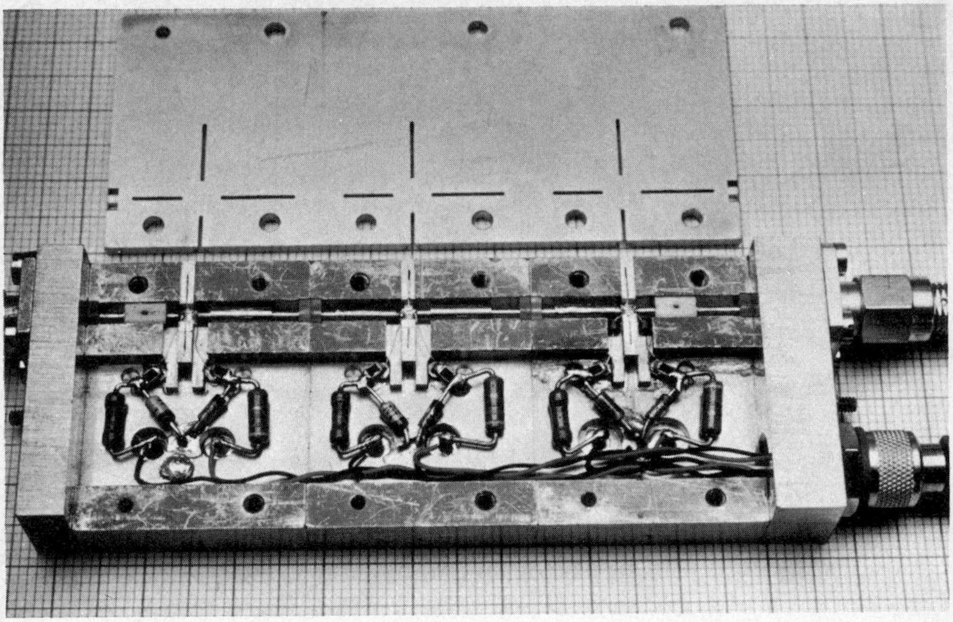

Fig. 7.17. A FET amplifier for $\nu = 10.7$ GHz (courtesy Max Planck-Institut für Radioastronomie, Bonn)

7.6 Spectrometers

Of the many different receiver backends that have been designed for various purposes the most important probably is the spectrometer backend. Fundamentally they are not different from radiometers; they merely put the emphasis on the spectral information contained in the received radiation field. In order to be able to do this they differ in two aspects from general radiometers: the receivers are of the SSB type and the bandwidth Δf is usually much less and in the kHz range.

SSB receivers are necessary because in a DSB type the PSDs contained in both image frequencies are superposed. Even in that case, where we know that line radiation is only contained in one of the sidebands, the other one will contribute to the noise of the measurements and therefore should be excluded.

A bandwidth in the kHz-range is required if narrow spectral features are to be resolved. But since the limiting sensitivity of the spectrometer is also given by (7.48) or its slight generalization

$$\frac{\Delta T}{T_{\mathrm{sys}}} = \frac{K}{\sqrt{\Delta f \, \tau}} \tag{7.81}$$

(a list of values of K for different receiver configurations is given in Table 7.2), the integration time τ needed for a given $\Delta T / T_{\mathrm{sys}}$ can be quite long indeed if Δf is small compared to total width of the spectral feature. This is the reason why specially designed spectrometer backends are essential for an effective operation of a radio-telescope if radio spectroscopy is intended.

If a resolution of Δf is to be achieved for the spectrometer, all those parts of the system that enter critically into the frequency response have to be maintained to within a fraction of Δf, usually at least 0.10 Δf. This applies in particular to the local oscillator, and in this respect the same demand is set on each local oscillator frequency in a double or triple conversion superheterodyne receiver. If possible, therefore the local oscillator signal is derived from a master oscillator such as a rubidium clock or even a hydrogen maser by direct frequency multiplication. If this is not possible for frequencies of the order of 10 GHz and above, frequency stabilization by various lock-in schemes for free-running oscillators have been used. In modern installations these devices are usually computer controlled.

Three types of radio spectrometers have become popular in the last 25 years.

7.6.1 Multichannel Filter Spectrometers

The time needed to measure the power spectrum for a given celestial position can be divided by a factor n if the IF-section with the filters defining the bandwidth Δf, the square-law detectors and the integrators are built not once, but n-times forming n separate channels that simultaneously measure in different parts of the spectrum. The technical details of how such a multichannel spectrometer is built may differ from one instrument to the other, but experience has shown that the following design-aims are essential:

1) The shape of the bandpass $G_i(\nu)$ for the individual channels must be identical. It is not enough that the bandwidths Δf_i are the same.
2) The square-law detectors for the channels must have identical characteristics. This refers both to the mean output power level and to any deviations from an ideal transfer characteristic.
3) Thermal drifts of the channels should be as identical as is technically feasible.

The reason for (1) becomes clear if one considers what becomes of narrow spectral features if Δf is wider than this feature itself. (2) is necessary so that the response of the different channels to a given signal will be identical at any power level (within the appropriate range), while (3) makes certain that the long term behaviour of the different channels is the same. All in all, the stability requirements for the individual channels are very high indeed with optimal stability times according to (7.64) of the order of 0.5–1 hour, and since they are, after all, analogue devices such requirements are not easily met.

Another problem that is complicated to solve with a filter spectrometer is how to change the frequency resolution Δf. This requires the changing of the filters in each of the channels. But, if the channels are arranged such that they form a contiguous comb for one Δf then this will no longer be true if Δf is changed, provided that the central frequency of each channel remains the same. Irrespective how this problem is resolved, it will require rather complicated provisions.

This is the reason why, for many purposes, the following spectrometer type has been used with great success.

7.6.2 Digital Autocorrelation Spectrometer

This spectrometer does not measure the PSD $S(\nu)$ or some convolution of $S(\nu)$ with a filter function $G(\nu)$ directly, but makes use of the Wiener-Khinchin theorem (7.13) relating the autocorrelation function (ACF) $R(\tau)$ to the PSD by a Fourier transform. The ACF $R(\tau)$ is measured by a time-averaging procedure (7.10) from the time series of the signal voltage. An evaluation according to (7.10) makes necessary an analogue-multiplication of the signal voltage with that delayed by the time τ. While this could be done with the help of square-law devices as shown in Sect. 7.4 on correlation receivers, a simple procedure is possible that permits digital techniques to be applied.

We do not determine the ACF $R_x(\tau)$ of the signal $x(t)$ but determine the ACF $R_y(\tau)$ of some transformation $y(t) = \varphi[x(t)]$ chosen such that it can be easily implemented and $R_x(\tau)$ can be computed from $R_y(\tau)$. Let the stochastic process $y(t)$ be defined by (Fig. 7.18)

$$y(t) = \begin{cases} 1 & \text{for } x(t) \geqq 0 \\ -1 & \text{for } x(t) < 0 \end{cases} . \tag{7.82}$$

Then the ACF of y, $R_y(\tau)$ is given by

$$R_y(\tau) = E\{y(t+\tau)\,y(t)\} = P\{x(t+\tau)\,x(t) > 0 - P\{x(t+\tau)\,x(t) < 0\} \tag{7.83}$$

where $P\{x(t+\tau)\,x(t) > 0\}$ is the probability of finding $x(t+\tau)\,x(t) > 0$. This can be computed using the joint probability density function $p(x_1, x_2; \tau)$. For a general stochastic process this is not known, but for a Gaussian process it is given by (7.31) and, since the voltage distribution of the stochastic process describing both the signal and the noise in a receiver are well described by such a distribution, (7.31) should be applicable here.

Obviously

$$\begin{aligned} P\{x(t+\tau)\,x(t) > 0\} &= P\{[x(t+\tau) > 0] \wedge [x(t) > 0]\} \\ &\quad + P\{[x(t+\tau) < 0] \wedge [x(t) < 0]\} = P_{++} + P_{--} \end{aligned} \tag{7.84}$$

and similarly

$$P\{x(t+\tau)\,x(t) < 0\} = P_{+-} + P_{-+} = 2\,P_{+-} \tag{7.85}$$

Fig. 7.18. The arcsin law

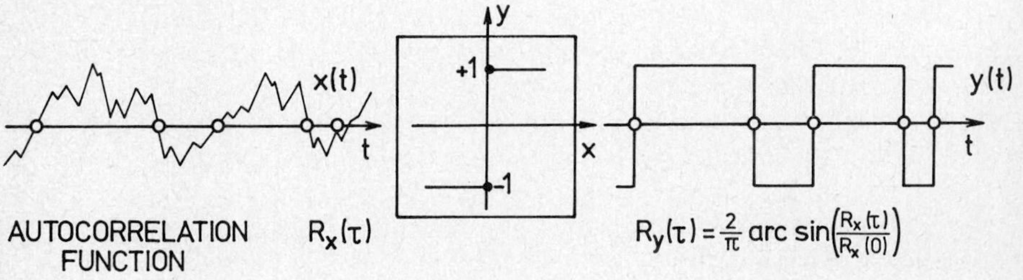

where

$$P_{++} = \int_0^\infty \int_0^\infty p[x(t+\tau), x(t); \tau] dx(t+\tau) dx(t) .$$

Substituting

$$x(t) = z \cos \theta$$

$$x(t+\tau) = z \sin \theta$$

and (7.31) for $p(x_1, x_2; \tau)$ we obtain

$$\begin{aligned} P_{++} &= \frac{1}{2\pi\sigma^2(1-r^2)^{1/2}} \int_0^{\pi/2} \int_0^\infty \exp\left[-\frac{z^2(1-r\sin 2\theta)}{2\sigma^2(1-r^2)}\right] z\, dz\, d\theta \\ &= \frac{1}{4} + \frac{1}{2\pi} \arctan \frac{r}{\sqrt{1-r^2}} \\ &= \frac{1}{4} + \frac{1}{2\pi} \arcsin r . \end{aligned} \tag{7.86}$$

Similarly

$$P_{--} = \frac{1}{4} + \frac{1}{2\pi} \arcsin r \tag{7.87}$$

$$P_{+-} = \frac{1}{4} - \frac{1}{2\pi} \arcsin r . \tag{7.88}$$

Substituting (7.84) to (7.88) into (7.83), we obtain with (7.29)

$$\boxed{R_y(\tau) = \frac{2}{\pi} \arcsin \frac{R_x(\tau)}{R_x(0)}} \quad \text{or} \tag{7.89}$$

$$\boxed{R_x(\tau) = R_x(0) \sin\left[\tfrac{\pi}{2} R_y(\tau)\right]} \; . \tag{7.90}$$

(7.89) is known as the arc sine law, the inverse relation (7.90) is the van Vleck relation. Thus if $R_y(\tau)$ can be measured, $R_x(\tau)$ can be easily computed.

Since the signal $y(t)$ as defined by (7.82) is a random telegraph signal adopting only the discrete values $+1$ and -1, it can be conveniently processed by digital circuits thus avoiding many stability problems.

An autocorrelation spectrometer therefore consists of the building blocks as shown in Fig. 7.19. The clipper transforms the random signal $x(t)$ of the receiver into the signal y, which agrees with x only with regard to its polarity; y is then sampled at equidistant time intervals by the sampler, which is simply an "and" circuit gating y with the clock frequency. The samples then are shifted with the

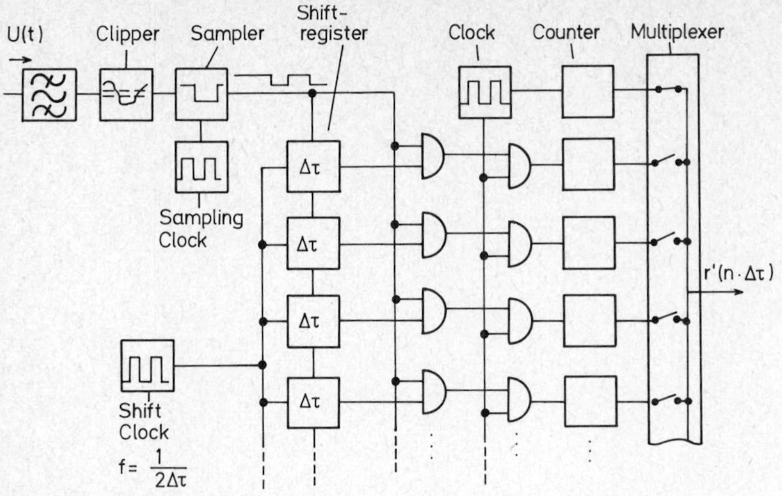

Fig. 7.19. A one-bit autocorrelation spectrometer

clock frequency into a linear shift register which holds 1 bit for each delay. By comparing the shift register content delayed by $\Delta\tau$ steps with the undelayed sample, the contribution to the counters then is proportional to $R_y(\tau)$. By dividing $R_y(\tau)$ through $R_y(0)$, normalized values for this ACF will be obtained. If these values are then transformed using (7.90), the influence of the clipping will be eliminated.

However, by clipping the original signal all information on its total power will be lost, because two signals that differ only by a constant factor will result in identical clipped signals $y(t)$. Therefore the total power is measured independently and used later on to calibrate the spectra.

The PSD of the received signal is then determined according to the Wiener-Khinchin theorem (7.13) by computing the Fourier transform of the measured autocorrelation function. But note that the integral in (7.13) extends to $\pm\infty$; in an actual instrument, however, $R(\tau)$ can be measured only up to a maximum delay τ_m. The measured ACF $R(\tau)$ can thus be considered to be the product of two functions: the true ACF $R(\tau)$ and a function describing the lag window

$$w(\tau) = \begin{cases} 1 & \text{for } |\tau| \leqq \tau_m \\ 0 & \text{else} \end{cases} . \tag{7.91}$$

The convolution theorem of Fourier transforms [see Bracewell (1965)] then states that the measured PSD $\tilde{S}(\nu)$ is the convolution of the true PSD $S(\nu)$ and a filter with the frequency response

$$W(\nu) = 2\,\tau_m\,\mathrm{sinc}(2\,\nu\,\tau_m)\ , \qquad \text{so that} \tag{7.92}$$

$$\tilde{S}(\nu) = S(\nu) \otimes W(\nu)\ . \tag{7.93}$$

The filter frequency response $W(\nu)$ determines the resolution of the autocorrelation spectrometer. If we define the frequency resolution of the spectrometer by the halfwidth of (7.92) we find that

$$\Delta\nu = \frac{0.605}{\tau_m} . \tag{7.94}$$

If the spectral region which is analyzed by the spectrometer has the total bandwidth Δf, the sampling interval for the stochastic time series $x(t)$ must be $\Delta\tau = 1/2\,\Delta f$ according to the sampling theorem. Provided that the autocorrelator has N_0 delay steps and summing counters, then

$$\tau_m = N_0/2\,\Delta f$$

resulting in a frequency resolution

$$\Delta\nu = 1.21\,\frac{\Delta f}{N_0} \quad \text{or} \tag{7.95}$$

$$\Delta\nu = \frac{0.605}{N_0\,\Delta\tau} . \tag{7.96}$$

For a given autocorrelator (N_0 fixed) the frequency resolution can be changed simply by changing the sampling time step $\Delta\tau$, i.e. by changing the clock frequency. But in order to obey the sampling theorem, the total bandwidth covered by the spectrometer has to be simultaneously adjusted too, since $\Delta f = 1/2\,\Delta\tau$.

Using the lag window (7.91) poses another problem: the resulting filter function (7.92) has strong sidelobes which become smaller only proportional to $1/2\,\tau_m\,\Delta\nu$. If narrow, strong features occur in the spectrum $S(\nu)$, $\tilde{S}(\nu)$ will be polluted. This can be strongly diminished by using a lag window different from (7.91). The window first introduced by J. von Hann ("hanning") is given by

$$w_H(\tau) = \begin{cases} \cos^2\left(\dfrac{\pi\,\tau}{2\,\tau_m}\right) & \text{for } |\tau| \leqq \tau_m \\ 0 & \text{else} \end{cases} \tag{7.97}$$

corresponding to the filter frequency response

$$W_H(\nu) = \tau_m\left[\operatorname{sinc}(2\,\nu\,\tau_m) + \frac{2\,\nu\,\tau_m}{\pi\,[1-(2\,\nu\,\tau_m)^2]}\sin(2\,\pi\,\nu\,\tau_m)\right] . \tag{7.98}$$

The frequency resolution corresponding to this lag window is

$$\Delta\nu = \frac{1}{\tau_m} = \frac{2\,\Delta f}{N_0} = \frac{1}{N_0\,\Delta\tau} ; \tag{7.99}$$

that is, the frequency resolution is 40% less than using the window (7.91). The first sidelobe, however, is now at only 2.6% of the peak, while for (7.91) it is at 22%.

Since multiplying the time series $x(t)$ with the lag window $w(\tau)$ is equivalent to convolving $S(\nu)$ with $W(\nu)$, introducing the lag window (7.97) can be done even after performing the Fourier transform of $R(\tau)$ by convolving with (7.98). For a spectrum $S(\nu)$ given at equidistant frequencies with $\Delta\nu = 1/2\,\tau_m$, this amounts to forming the running average with the weights 1/4, 1/2, 1/4. This is called "hanning", and it is good practice to smooth spectra obtained by an autocorrelation spectrometer in this way.

In order to obtain the PSD $S(\nu)$ from the measured $\tilde{R}(\tau)$ a numerical Fourier transform has to be performed. Since $\tilde{R}(\tau)$ is quite naturally obtained for a series of equidistant τ_i this transformation is best done by the fast Fourier transform (FFT) algorithm of Cooley and Tukey. To simplify and to speed up these computations the number of lags in the autocorrelator is usually taken equal to a power of 2, and 256, 512 or 1024 are frequently used, but spectrometers with a different number of channels have been built too.

Clipping of the stochastic time signal according to (7.82) corresponds to a one-bit digitization, and this quite naturally means a loss of information. From (7.90) we see that for small values of R_x we lose by this a factor of $\frac{\pi}{2}$ in the resolution of $R_x(\tau)$ so that the resulting limiting sensitivity turns out to be

$$\frac{\Delta T}{T_{\text{sys}}} = \frac{\frac{\pi}{2}}{\sqrt{\Delta\nu\, t_{\text{int}}}} \tag{7.100}$$

where $\Delta\nu$ is the frequency resolution according to (7.95), (7.96) or (7.99), and t_{int} is the effective total integration time, that is, the time the system has actually looked at the source.

The resolution $\Delta T/T_{\text{sys}}$ can be somewhat improved if the digitization of the time signal is done better than with just one bit. For a two-bit correlator the factor $\frac{\pi}{2}$ in (7.100) would be replaced by 1.14. This is such a small improvement and it is obtained for the price of such a large complication in the autocorrelator that this is very seldom done.

While in a filter spectrometer a high frequency resolution means narrow selection filters with problems of thermal instability and the like, with an autocorrelation spectrometer it is difficult to achieve a large bandwidth and a low frequency resolution. This is so because, according to (7.96) or (7.99) this means a small sampling time interval $\Delta\tau$ or a high clock frequency. The clipper, the sampler and the shift register, and the "XOR" circuits all have to work at this high clock rate. Using the TTL-technique, clock rates above 20 MHz are difficult to achieve, restricting the total bandwidth Δf to 10 MHz. Using modern digital technology it has in recent years been possible to push up the clock frequency so that now 1000 channel autocorrelation spectrometers covering a bandwidth of 100 MHz have been successfully built. But even this is often not wide enough for work on molecular lines in the mm-range, and therefore other kinds of spectrometers are used for this kind of work.

This drawback of autocorrelation spectrometers is regrettable because according to (7.64) they usually permit maximum integration times of several hours. They show no trace of the drift problem or the problem of how to make certain that the different channels give identical responses as encountered with filter spectrometers.

The digital autocorrelation spectrometer was invented by S. Weinreb (1963) and a description of the instrument and its theory was given in his thesis. Since then such instruments have become more or less standard for spectrometry in the cm-wave range.

7.6.3 Acousto-Optical Spectrometers

Since it was discovered that molecular line radiation could be observed in mm-wave radio astronomy there has been a need for spectrometers with bandwidths of several hundred MHz. At 100 GHz a velocity range of 300 km s^{-1} corresponds to 100 MHz, while the narrowest linewidths observed correspond to 30 kHz. A correlation spectrometer is not applicable for such bandwidths and so recourse had often to be taken to multichannel filter spectrometers. But as stated in Sect. 7.6.1 these are rather inflexible and often have differential drift and calibration problems, and therefore the need was felt for a wideband system with good long-term stability that could easily be adjusted to different needs. It now seems that acousto-optical spectrometers (AOS) can meet most of these demands.

This kind of instrument has been developed over the years rather consistently in Australia (Cole 1973) and more recently by many other groups, particularly and with great success in Japan. Today most of the parts needed for such an instrument can be bought from industry.

The AOS makes use of the diffraction of light by ultrasonic waves, an effect already predicted in 1921 by Brillouin. A soundwave causes periodic density variations in the medium through which it passes. These density variations in turn cause variations in the bulk constants ε and n of the medium, so that a plane electromagnetic wave passing through this medium will be affected. In the book "Principles of Optics" by Born and Wolf (1965) on p. 596 and following it is shown using Maxwell's equations how such a medium will cause a plane, monochromatic electromagnetic wave (wave number k and frequency ω) to be dispersed so that the emerging field can be described by the superposition of a sequence of waves with the wave number $k \sin\theta + lK$ and the frequency $\omega + l\Omega$ where K and Ω are wave number and frequency of the sound wave, θ the angle between electric and acoustic wave, and l an index $l = 0, \pm 1, \ldots$. The amplitude of the different emerging waves can then be determined by a recurrence relation.

For a proper understanding of this mechanism the expansion of Maxwell's equation or the equivalent integral equation method [Born and Wolf (1965), § 12.2] must be studied, but often a more intuitive understanding resulting from a physical analogy will suffice. The plane periodic variations of the index of refraction n can be considered to form a 3-dimensional grating that causes diffraction of the electromagnetic wave.

Let a monochromatic lightwave of angular frequency ω and wavelength λ make an angle θ with the y-axis, and let the angle of the diffracted ray be ϕ (Fig. 7.20). Since the velocity v of the compression wave is always much smaller than the velocity of light we can consider the periodic structure in the matter to be stationary. The permitted angles ϕ are then determined by the condition that the optical path difference from neighbouring acoustic wave planes should be integral multiples of λ. With a spacing Λ between adjacent acoustic wave crests, thus (see Fig. 7.20)

$$\Lambda(\sin\phi - \sin\theta) = l\lambda\ , \quad l = 0, \pm 1, \pm 2 \ldots\ . \tag{7.101}$$

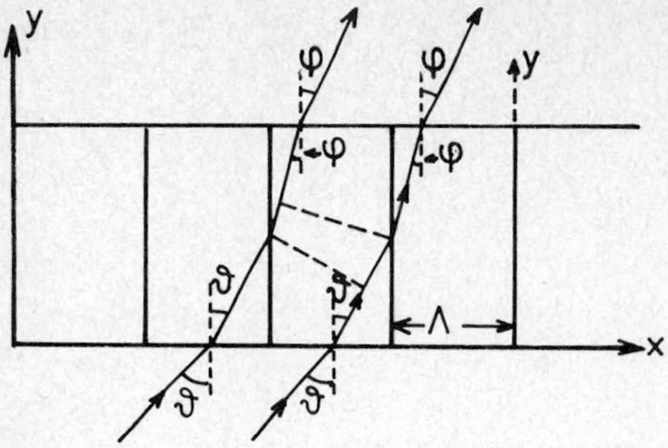

Fig. 7.20. Diffraction by an acoustic wave

This is the Bragg condition met in the diffraction of X-rays in crystals. Because of this the device, in which an acoustic wave is made to affect the index of refraction producing a travelling periodic wave which can then be detected by a monochromatic light beam is called a Bragg cell.

The practical problem is to find a transparent material with a low velocity of sound so that the acoustic wavelength Λ is small for a given sound frequency ν_s. A second problem is how to couple the transducer that converts the electric signal ν_s into an acoustic wave with a reasonably constant conversion factor over a wide bandwidth. And finally, a sump for the acoustic wave has to be provided so that no standing wave pattern develops. Such a pattern will always have resonances and therefore is not suitable for broadband applications.

For the light beam a monochromatic laser is used and if the acoustic wave intensity is small enough not to run into problems with saturation, the intensity of the diffracted light is proportional to the acoustic power. As long as everything is in the linear range, different acoustic frequencies ν_s can be superposed resulting in different diffracted angles ϕ_s.

Differentiating (7.101) and substituting

$$\Lambda_s \nu_s = v_c$$

we obtain

$$\boxed{\cos\phi\,\delta\theta = \frac{l\lambda}{v_c}\,\Delta\nu_s} \quad . \tag{7.102}$$

The construction principle of an AOS is shown in Fig. 7.21, the device itself in Fig. 7.22. The light source is a laser whose beam is expanded to match the aperture of the Bragg cell and the distribution of the light intensity in the focal plane is detected by a photodiode array or a CCD. After integration for some milliseconds, the photodiodes are sampled and read out into a computer where the final integration is done. The maximum number of channels that can be resolved by such an instrument can be determined by considering how well the wavefront for light that is emerging from a monochromatic grating can be determined. This uncertainty is

$$\Delta\theta \cong \lambda/L \tag{7.103}$$

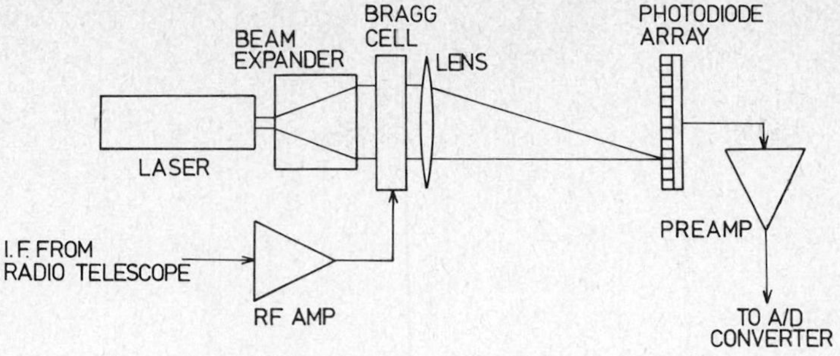

Fig. 7.21. Block diagram of an acousto-optical spectrometer

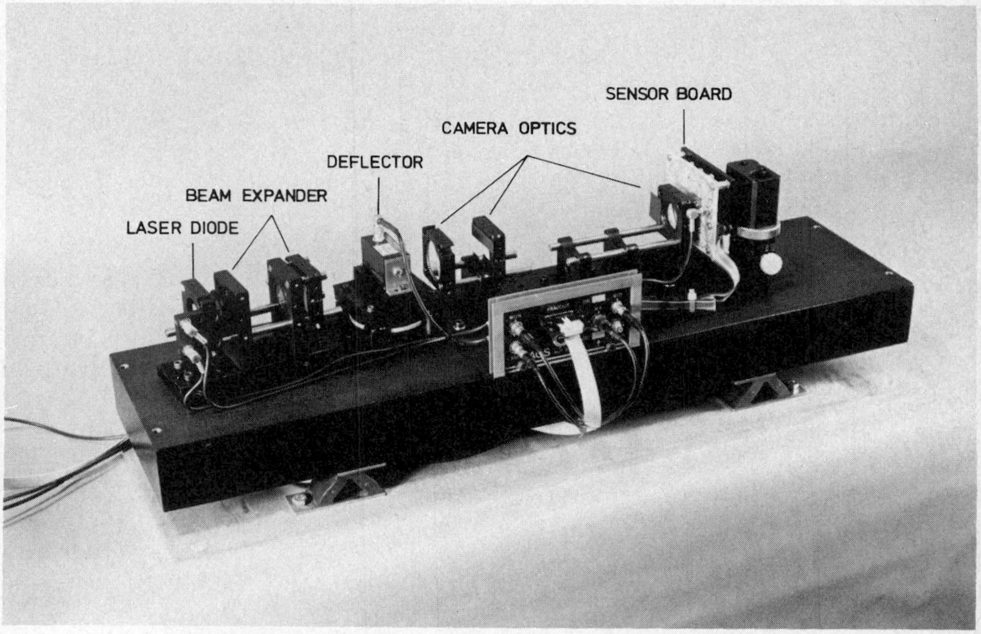

Fig. 7.22. The AOS spectrometer of the University of Cologne

if L is the aperture of the Bragg cell. Therefore a total bandwidth $\Delta\nu$ can at most be resolved into

$$N_0 = \frac{\delta\theta}{\Delta\theta} = \frac{L}{v_c \cos\phi}\Delta\nu = \frac{\tau_c}{\cos\phi}\Delta\nu \tag{7.104}$$

channels. $\Delta\nu$ is limited by the condition that adjacent orders of the diffraction ($\Delta l = \pm 1$) should not overlap. τ_c is the time it takes the acoustic wave to pass through the Bragg cell: $\tau_c = L/v_c$.

The total bandwidth possible for a single Bragg cell is today of the order of 300 MHz but, by placing several such devices in parallel, much larger bandwidths can be achieved.

Another limiting instrumental parameter is the dynamic range that can be covered. It gives the ratio between the largest signal and the smallest. The one is determined by nonlinearities of the response, the other by the dark current of the photodiodes, by internal reflections both of the acoustic signal and the laser light, and by other unwanted parasitic effects. Ranges of about 30 dB have been achieved on a routine basis.

Thermal drift is another problem that has troubled AOS, and this concerns the stability of the frequency scale even more than the intensity. Both temperature compensation and active servo regulation have been employed to overcome this, obviously with success.

7.7 Receiver Calibration

In radio astronomy the noise power is usually measured in terms of the noise temperature, and thus to calibrate a receiver means to determine which noise temperature increment ΔT at the receiver input corresponds to a given measured receiver output increment Δz.

Even if no external noise source is connected to the receiver input the system noise temperature T_{sys} is present and produces an output. In principle this could be computed from the output signal z provided that the powergain transfer functions $G(\nu)$ and $W(\nu)$ as well as the detector characteristics were well known, but in practice this is never done: the receiver is calibrated by connecting the input with known power sources. Usually matched resistive loads at the known (thermodynamic) temperatures T_L and T_H are used. The receiver outputs are then

$$z_L = (T_L + T_{sys})\, G$$

$$z_H = (T_H + T_{sys})\, G$$

from which

$$\boxed{T_{sys} = \frac{T_H - T_L\, y}{y - 1}} \quad \text{where} \tag{7.105}$$

$$y = z_H / z_L \ . \tag{7.106}$$

The noise temperatures T_H and T_L are usually produced by matched resistive loads at the ambient temperature ($T_H \cong 293$ K or 20 °C) and at the temperature of boiling nitrogen ($T_L \cong 78$ K or − 195 °C) or sometimes even of liquid Helium at its boiling point $T_L \cong 4.2$ K. Usually this basic calibrating of the receiver is done rather infrequently, because secondary standards are determined. This could for example be a standard noise source that is fed into the receiver input with directional couplers; the magnitude of this noise step then will be adapted with the help of attenuators. Let a noise source T_N be connected through an attenuator with the loss L; then this noise step is

$$\Delta T = \frac{1}{L} T_N + \left(1 - \frac{1}{L}\right) T_0 = \frac{1}{L}(T_N - T_0) + T_0 \tag{7.107}$$

where T_0 is the (thermodynamic) temperature of the attenuator.

The situation for spectrometers does not differ from that of general radiometers as far as the system noise temperature is concerned but, in addition, the frequency dependence of both the system noise temperature and the gain must also be determined. Usually one is interested only in the frequency-dependent part of the radiation; any wideband contribution should be eliminated. Some of the methods characteristic for spectral line measurements are discussed in Sect. 10.6.

7.8 The Influence of the Earth's Atmosphere

As was stated in Sect. 1.2, the earth's atmosphere is transparent to radio waves if the frequency of this radiation is above a low frequency cut-off given by the critical frequency of free electrons in the earth's ionosphere. This cut-off frequency varies somewhat since it depends on the electron-density in the ionosphere, but usually is in the region of 25 MHz or thereabout. Most radioastronomical work is done at frequencies well above this and therefore the ionospheric effects are of little consequence for radio astronomy and will not be considered further here.

This is different for the absorption in the troposphere of radio waves near the high frequency limit of the frequency band transmitted by the atmosphere. Clouds consisting of water droplets do absorb radio waves even at frequencies as low as 6 GHz – a large raincloud can cause an attenuation as high as 1.5 dB, while the average value for a clear sky is of the order of 0.2 dB. Added to this is the absorption by the lines of water vapour H_2O and O_2, whose center frequencies are at 22 GHz and 60 GHz respectively but their line wings extend into the dm-wave range. In Fig. 7.23 the attenuation of the troposphere is shown for different amounts of H_2O. Obviously it is possible to select sites with a low water content in the atmosphere and to improve in this way the quality of the measurements. But this affects the absorption due to O_2 only in so far as the total atmosphere contribution will be lower for high altitude observatories.

The attenuation of the radiation due to the atmospheric absorption will decrease the measured flux of a radio source. If $S(z)$ is the flux measured at zenith distance z, and S_0 is the flux that would be obtained outside the atmosphere, then

$$S(z) = S_0\, d^{-X(z)} \tag{7.108}$$

where d is the attenuation at the zenith for the airmass 1, and $X(z)$ is the relative airmass in units of the airmass at the zenith. For a plane parallel atmospheric model clearly

$$X(z) = \sec z = \frac{1}{\cos z}\;; \tag{7.109}$$

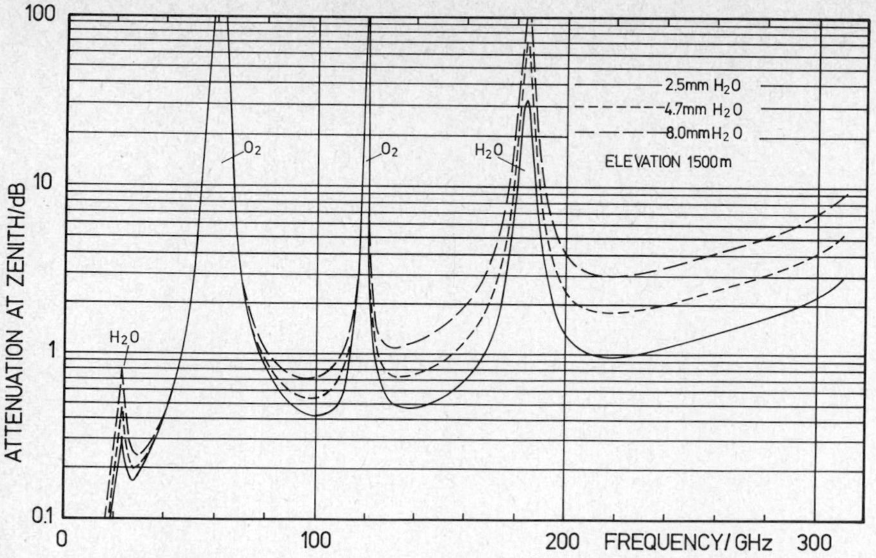

Fig. 7.23. Attenuation of electromagnetic waves in the troposphere (Hachenberg and Vowinkel 1982)

if the curvature of the earth and the troposphere is taken into account we have

$$X(z) = \frac{1}{H} \int_R^{R+H} \frac{\varrho(r)/\varrho(R)}{\sqrt{1 - \left(\frac{R}{r}\frac{n_0}{n}\right)^2 \sin^2 z}}\, dr \tag{7.110}$$

where R is the earth radius, H the height of the atmosphere, $\varrho(r)$ the gas density of the atmosphere at radius r, and $n(r)$ is the index of refraction at r. Bemporad and Schönberg (see Schönberg (1929)) have done extensive investigations of $X(z)$, a Chebyshev-fit to these data up to $X = 5.2$ with an error of less than $6.4 \cdot 10^{-4}$ is given by

$$X(z) = -0.0045 + 1.00672 \sec z - 0.002234 \sec^2 z - 0.0006247 \sec^3 z \ . \tag{7.111}$$

Such a formula should indeed be used in radio astronomy because measurements are often made at zenith distances up to 80° (or even more!).

Due to this attenuation thermal noise is produced by the atmosphere which will increase the system noise temperature. This noise contribution is

$$T_{\text{atm}} = (1 - d^{-X(z)})\, T_{\text{th}} \tag{7.112}$$

Therefore, if the atmospheric absorption is distributed irregularly due to clouds we will observe an irregular distribution of the atmospheric thermal background radiation too.

This can very effectively be eliminated by a beam switching technique in which the receiver input is switched rapidly between two feeds placed several beamwidths apart in the focal plane of the telescope. The noise contribution of the

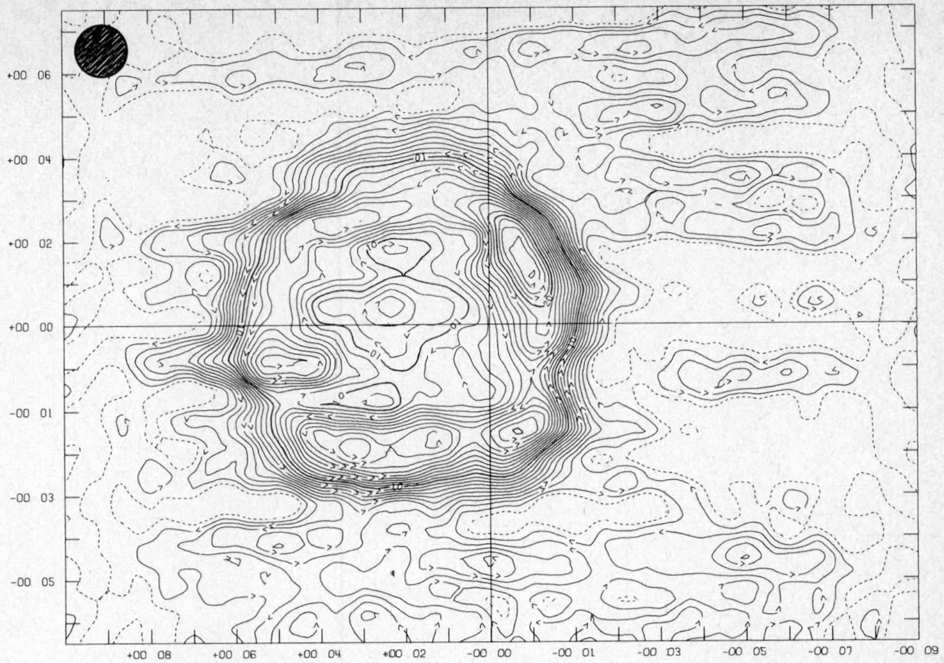

Fig. 7.24. The supernova remnant 3C10 as recorded by a singlebeam receiver at $\nu = 10.7$ GHz ($\lambda = 2.8$ cm) with the Effelsberg 100 m-telescope (beamwidth 70″) on a day with variable atmospheric emission (the map was scanned horizontally) [after Emerson et al. (1979)]

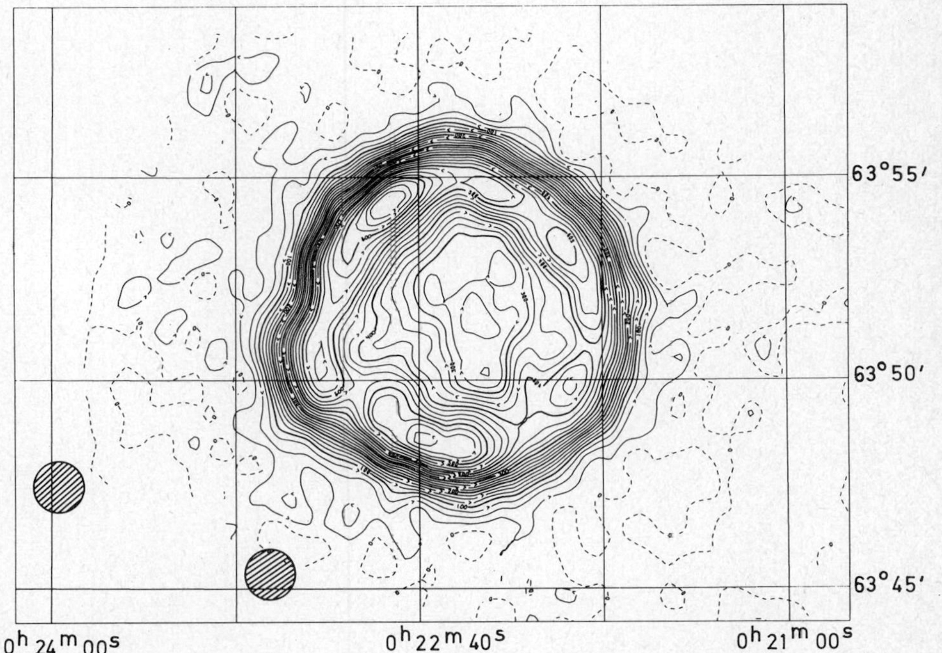

Fig. 7.25. The supernova remnant 3C10 recorded at 2.8 cm by a dual beam system with the Effelsberg 100 m-telescope. The angular separation of the beams was 5′.5, the source has a diameter of $\cong 9'$. The two shaded circles illustrate the beam size and separation. For this figure the same observations have been used as for Fig. 7.24

atmosphere is in the near-field of the telescope, and the air volume in which this is situated is almost identical for both beams, so that practically complete compensation of the additive atmospheric contribution is achieved. It should be remembered, however, that the atmospheric attenuation is not eliminated in this way. But the success of beam switching observation shows (Figs. 7.24 and 7.25) that the additive contributions are the more important [see Emerson et al. (1979)].

Since the index of refraction of the air is different from $n = 1$, radio waves will be refracted like light rays. For frequencies $\nu < 100$ GHz we have

$$(n-1)\cdot 10^6 = 79\left(\frac{p}{\mathrm{hPa}}\right)\left(\frac{T}{\mathrm{K}}\right)^{-1} + 3.79\cdot 10^5\left(\frac{p_\mathrm{w}}{\mathrm{hPa}}\right)\left(\frac{T}{\mathrm{K}}\right)^{-2} \tag{7.113}$$

where p is the air pressure in 10^2 Pa, p_w the partial pressure of water vapour (in 10^2 Pa) and T the gas temperature in K. Therefore the refraction will depend on the humidity of the air and will be time dependent. But if an error of $\cong 15\%$ is acceptable, mean refraction can be used and this closely resembles the optical refraction. Therefore for $z < 80°$

$$\Delta z = \beta \tan z \ , \qquad \beta = 1\overset{'}{.}50 \ . \tag{7.114}$$

The ratio of radio to optical refraction is $\cong 1.56$; thus for larger zenith distances, optical refraction tables can be used, provided the result is correspondingly multiplied.

8. Emission Mechanisms of Continuous Radiation

8.1 The Nature of Radio Sources

In the early days of radio astronomy the resolving power of the then available radio telescopes was low. But even then a considerable part of the received radiation was found to come from localized discrete sources on the celestial sphere. Whether the remainder was due to similar sources which only happened to be unresolved due to insufficient resolving power of the telescopes or whether the source of this radiation was truly continuously distributed in space was largely unclear. Initially only very few of the discrete sources could be identified with objects known from the optical region of the spectrum. But from the investigation of the increase of the number of the sources with decreasing source flux and from the distribution of sources in the sky, two different families of sources could be distinguished: galactic sources concentrated towards the galactic equator and extragalactic sources distributed more or less uniformly in space. The unresolved, spatially continuous radiation belongs to the galactic component, if we neglect for the time the 2.7 K thermal background radiation which was only detected much later and which is cosmological in origin.

It was quite natural to investigate the nature of the discrete sources by making measurements at different frequencies to find out the spectral characteristics. Again two large families of sources appeared. While the flux of one kind of source increases with increasing frequency, the other kind is stronger the lower the frequency (Fig. 8.1). Some of the strongest sources found were of this kind; we will mention only the source Cas A which was later identified as the remnant of a supernova explosion in our Galaxy around the year 1700, and the source Cyg A which is an extragalactic radio source, a radio galaxy, situated very far away with a redshift of $z = 0.057$ so that its absolute radio luminosity must be indeed fantastic.

Some of the other kinds of sources which show an increase of the flux with increasing frequency could be identified with objects well known from the optical range of the spectrum. Both the moon and the sun are radio sources of this kind; for the sun this is true only if restricted to those moments where there are no disturbances, that is, for the quiet sun. But also objects like the Orion nebula and other H II regions like M 17 or the sources W3, W4 and W5 belong to this category. The moon obviously can be taken as an example of a black body and its spectrum is an almost perfect representation of a Rayleigh-Jeans spectrum for a temperature of $T \approx 225$ K. The spectrum of the Orion nebula also indicates a thermal origin. If we restrict ourselves to frequencies below 1 GHz the observa-

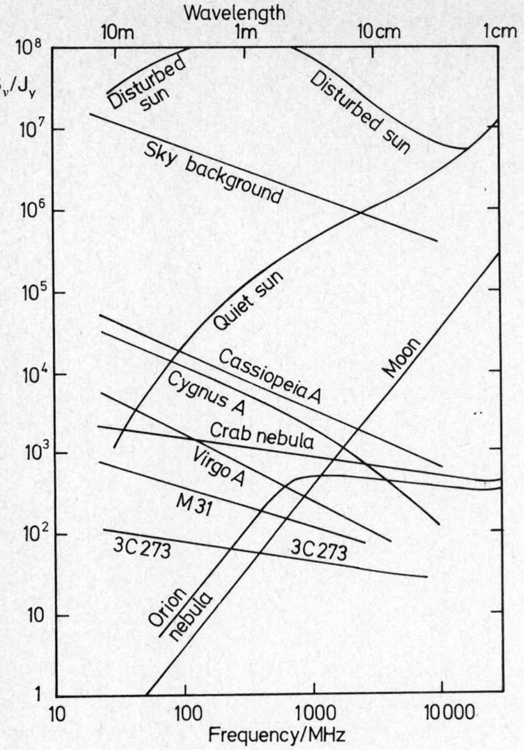

Fig. 8.1. The spectral distribution of various radio sources

tions can be well represented by the Rayleigh-Jeans law, but above this frequency the observed flux is below that extrapolated from lower frequencies.

The explanation is fairly straightforward. If we consider the solution of the equation of radiation transfer (1.18) for an isothermal object

$$I_\nu = B_\nu(T)(1 - \mathrm{e}^{-\tau_\nu}) \ ,$$

we see that $I_\nu < B_\nu$ if $\tau_\nu \lesssim 1$; the actual frequency dependence of I_ν will depend on the run of τ_ν.

The radio sources can thus be classified into two categories: thermal sources which radiate because the medium is hot and all the other, the nonthermal sources. In principle many different radiation mechanisms could be responsible for their emission, but in practice one single mechanism dominates all others: the synchrotron emission or magnetic bremsstrahlung.

The other division of the discrete radio sources into galactic and extragalactic ones, is in principle completely independent of this classification. However, if we find predominantly nonthermal sources amongst the extragalactic sources, this is simply a result of the fact that the strongest sources are nonthermal in origin. Even if thermal sources are abundant in extragalactic objects, they will not be easily perceptible because they are weak compared to the nonthermal sources contained in the same extragalactic system.

With the exception of thermal line emission of atoms and molecules and thermal emission from solid bodies, radio emission always comes from free electrons. And since free electrons can exchange their energy by arbitrary amounts, no definite energy jumps that result in sharp emission or absorption lines will occur: we are dealing with a continuous spectrum.

Now a free electron will only emit radiation if it is accelerated, and therefore such acceleration mechanisms for free electrons will have to be studied if the continuous radio spectrum of discrete sources is to be understood. The survey of all the different acceleration mechanisms for free electrons has only produced two different mechanisms of widespread importance: the bremsstrahlung emitted by thermal electrons moving in the electric field of ions, and the magnetic bremsstrahlung emitted by relativistic electrons gyrating in a cosmic magnetic field. This second mechanism produces the synchrotron radiation that is the source of radiation of the nonthermal radio sources. It will be discussed in the next chapter.

When astrophysicists had to explain the thermal radio emission of gaseous nebulae, it turned out unnecessary to develop from scratch the physics of the processes involved; recourse could be taken to theories that were devised 40 years earlier to explain the continuous X-ray emission spectrum of ordinary X-ray tubes. Both in thermal radio emission and in X-ray tubes the electromagnetic radiation is emitted by free electrons that suffer sudden accelerations in the electric field of ions, and the relative change of the kinetic energy of the electrons is of the same order of magnitude in both cases.

Radio radiation with $\nu = 1$ GHz and a quantum energy $h\nu \cong 4 \cdot 10^{-6}$ eV emitted by the thermal electrons in a hot plasma ($T \approx 10^4$ K, $mv^2/2 = 1$ eV) corresponds to a relative change of the kinetic energy of $\Delta E/E \cong 10^{-5}$, while X-rays produced by a 100 keV beam corresponds to changes $\Delta E/E \cong 10^{-2}$. These values are similar enough so that the same theory can be applied to both situations. We will present this theory not in its most general form but with such simplifications as are acceptable for the radio range.

8.2 Radiation from an Accelerated Electron

For a general theory of the radiation from accelerated charges we must start again with the appropriate electrodynamic potentials, the socalled Liénard-Wiechert potentials. If done properly the results can then be applied to all kinds of accelerated charges, thermal as well as relativistic ones. Such theories are presented in many textbooks on electrodynamics, Jackson (1962), Chaps. 14 and 15; Landau-Lifschitz (1967), Vol. II, §§ 68–71; Panofsky-Phillips (1962), Chaps. 19 and 20; Longair (1981), Chap. 3 just to mention a few.

We will make use here of the formulae derived for the electric dipole in Chap. 4. An oscillating current $I_0 \exp(-\mathrm{i}\,\omega t)$ in an electric dipole is equivalent to moving charges, and we can reformulate the expression such that, instead of the current, the acceleration of an electron is appearing.

Let the moment of the dipole be $d = -e\,\Delta l$, so that $\ddot{d} = -e\,\dot{v}$. Now if the variation of the dipole moment is caused by an oscillation

$$d = -e\,\Delta l\,\mathrm{e}^{-\mathrm{i}\omega t} = d_0\,\mathrm{e}^{-\mathrm{i}\omega t}\ ; \qquad \dot{d} = -\mathrm{i}\,\omega\,d\ ,$$

then the current in the dipole is

$$I = \frac{dq}{dt} = \frac{\dot{d}}{\Delta l} = -\mathrm{i}\,\omega\,\frac{d}{\Delta l}$$

and

$$\ddot{d} = \dot{I}\,\Delta l = -\mathrm{i}\,\omega\,I\,\Delta l \qquad \text{so that}$$

$$I\,\Delta l = \frac{e\,\dot{v}}{\mathrm{i}\,\omega}\ . \tag{8.1}$$

Substituting this into the equation for the electrical field intensity for the far-field or radiation field (4.38), we obtain for the field component directed parallel to the moving charge

$$E_\vartheta = \frac{-1}{4\pi\,\varepsilon_0}\,\frac{e\,\dot{v}\,(t)}{c^2}\,\frac{\sin\vartheta}{r}\,\exp\left[-\mathrm{i}\left(\omega\,t - \frac{2\pi}{\lambda}\,r\right)\right]\ , \tag{8.2}$$

the other components of E being zero. The Poynting flux, that is, the power per surface area and steradian emitted into the direction (ϑ, φ) is then according to (4.40)

$$|\boldsymbol{S}| = \frac{1}{16\,\pi^2\,\varepsilon_0}\,\frac{e^2\,\dot{v}^2}{c^3}\,\frac{\sin^2\vartheta}{r^2}\ . \tag{8.3}$$

Integrating this over the full sphere results in the total amount of power radiated when a charge e is accelerated by $\dot{v}$:

$$P\,(t) = \frac{1}{6\,\pi\,\varepsilon_0}\,\frac{e^2\,\dot{v}^2\,(t)}{c^3}\ . \tag{8.4}$$

$P\,(t)$ is the power emitted at the moment t due to the acceleration $\dot{v}\,(t)$ of the electron. The total amount of energy emitted during the whole encounter is obtained by integrating (8.4); that is,

$$W = \int_{-\infty}^{\infty} P\,(t)\,dt = \frac{1}{6\,\pi\,\varepsilon_0}\,\frac{e^2}{c^3}\int_{-\infty}^{\infty} \dot{v}^2\,(t)\,dt\ . \tag{8.5}$$

In all this we have said nothing about the frequency at which this radiation will be emitted. This frequency is obviously governed by the speed with which E varies, that is, according to (8.2), the speed with which $\dot{v}\,(t)$ is varying. If $\dot{v}$ and E are different from zero only for a very short time interval then we can obtain the frequency dependence of E by a Fourier analysis of the pulse of E. But in order to do this we have to investigate the collision of the electron and the ion in some more detail, This is also necessary in order to be able to compute the integral in (8.5).

8.3 The Frequency Distribution of Bremsstrahlung for an Individual Encounter

The single parameter that governs the speed and strength of the electric field intensity pulse is the ratio p/v of the *collision parameter* p and the *velocity* v which the electron attains when it is at closest approach p to the ion (Fig. 8.2). Now, the required relative change $\Delta E/E \cong 10^{-5}$ of the kinetic energy corresponds to such large values of p that the electron's orbit, which in reality is a hyperbola, can be approximated by a straight line. For the ion-electron distance l we then have

$$l = \frac{p}{\cos\psi} \ . \tag{8.6}$$

The acceleration of the electron is given by Coulomb's law

$$m\dot{\boldsymbol{v}} = -\frac{1}{4\pi\varepsilon_0}\frac{Ze^2}{l^3}\boldsymbol{l} \ .$$

For orbits with large p this acceleration is directed perpendicularly to the orbit at those points where $|\dot{\boldsymbol{v}}|$ is large. We will therefore simplify the problem by only considering this component

$$\dot{v} = |\dot{\boldsymbol{v}}|\cos\psi \ , \qquad \text{so that}$$

$$\dot{v} = -\frac{1}{4\pi\varepsilon_0}\frac{Ze^2}{mp^2}\cos^3\psi \quad \text{and} \tag{8.7}$$

$$W = \frac{4}{3}\frac{1}{(4\pi\varepsilon_0)^3}\frac{Z^2e^6}{c^3m^2p^4}\int\limits_0^\infty \cos^6\psi(t)\,dt \ . \tag{8.8}$$

The functional dependence $\psi(t)$ can be obtained from Kepler's law of the areas. If we define

$$dF = \tfrac{1}{2}l^2\,d\psi \ ,$$

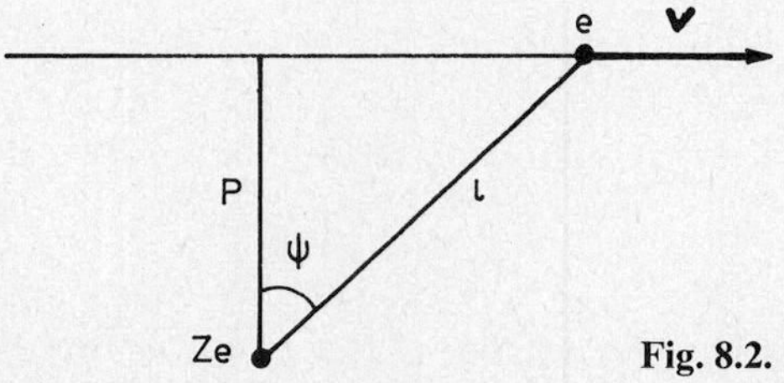

Fig. 8.2. An electron moving past an ion of charge Ze

then in any motion governed by a central force

$$\dot{F} = \frac{1}{2} l^2 \frac{d\psi}{dt} = \text{const} \ .$$

But at the time $t = 0$ the electron attains its closest approach p and has the velocity v, so

$$\dot{F} = \tfrac{1}{2} p v \quad \text{and}$$

$$dt = \frac{l^2}{v p} d\psi = \frac{p}{v \cos^2 \psi} d\psi \ . \tag{8.9}$$

Therefore we find

$$W = \frac{4}{3} \frac{1}{(4 \pi \varepsilon_0)^3} \frac{Z^2 e^6}{c^3 m^2 p^4} \frac{p}{v} \int_0^{\pi/2} \cos^4 \psi \, d\psi$$

or, since

$$\int_0^{\pi/2} \cos^4 x \, dx = \frac{3}{16} \pi \ ,$$

$$\boxed{W = \frac{\pi}{4} \frac{1}{(4 \pi \varepsilon_0)^3} \frac{Z^2 e^6}{c^3 m^2 p^3} \frac{1}{v}} \ . \tag{8.10}$$

This is the total energy radiated by the charge e if it moves in the field of an ion with the charge $Z e$, but this expression is valid only for low energy collisions for which the straight-line approximation is valid. For collisions with small p another approximation would have to be used.

The electrical field intensity induced by the accelerated electron during the encounter with the ion was given by (8.2). This again gives only a pulse $E(t)$ of the electric field intensity. Taking the Fourier-transform of this we can represent this pulse as a wave packet formed by the superposition of harmonic waves of frequency ω; that is,

$$E(t) = \int_{-\infty}^{\infty} A(\omega) \, e^{-i\omega t} \, d\omega \ . \tag{8.11}$$

The wave amplitude $A(\omega)$ in (8.11) is given by

$$A(\omega) = \frac{1}{2\pi} \int_{-\infty}^{\infty} E(t) \, e^{i\omega t} \, dt \tag{8.12}$$

so that $A(\omega)$ is a real quantity if $E(t)$ is symmetric with respect to $t = 0$. According to (8.2) the Fourier analysis of $E(t)$ can be obtained by an analysis of $\dot{v}(t)$. We are therefore interested in

$$C(\omega) = \frac{1}{2\pi} \int_{-\infty}^{\infty} \dot{v}(t) \cos \omega t \, dt \ . \tag{8.13}$$

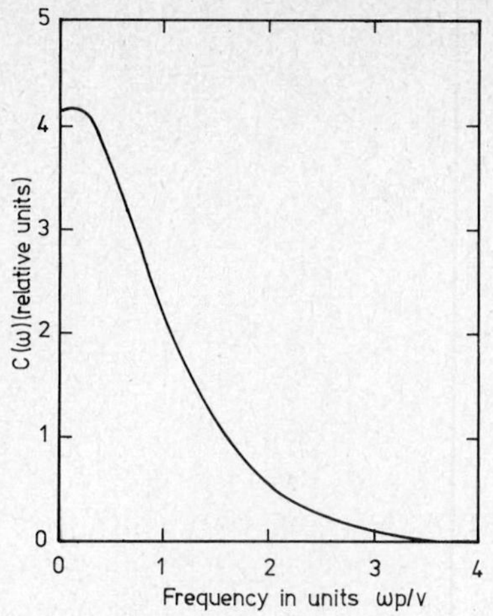

Fig. 8.3. The power spectrum of the radiation of a single encounter (Bekefi 1966)

Substituting $\dot{v}(t)$ from (8.7) this is

$$C(\omega) = -\frac{1}{4\pi\varepsilon_0}\frac{Ze^2}{mp^2}\frac{1}{\pi}\int_0^\infty \cos\omega t \cos^3\psi(t)\,dt \ . \tag{8.14}$$

Using (8.9) this can be written as an integral over ψ. The solution can be written in closed form using modified Bessel functions with an imaginary argument (Hankel-functions), see e.g. Oster (1961), but the precise form of the spectrum $C(\omega)$ is of no great concern here. The only thing that matters is that $C(\omega) > 0$ only for a finite range of ω (Fig. 8.3). We can estimate this limiting ω_g in the following way: The total amount of energy radiated in a single encounter as given by W in (8.10) must be equal to the sum of the energies radiated at the different frequencies. This can be stated as

$$\int_0^\infty |C(\omega)|^2\,d\omega = \frac{1}{2\pi}\int_{-\infty}^\infty |\dot{v}(t)|^2\,dt \ . \tag{8.15}$$

This is nothing but the Parseval theorem from Fourier transformation theory.

The left-hand side of (8.15) can be approximated by $C^2(0)\,\omega_g$, while (8.5) substituted into the right-hand side of (8.15) results in

$$C^2(0)\,\omega_g = \frac{1}{2\pi}\frac{6\pi\varepsilon_0 c^3}{e^2}W \ .$$

From (8.14) and using (8.9) we obtain

$$C(0) = -\frac{1}{4\pi\varepsilon_0}\frac{Ze^2}{\pi m p v} \ , \tag{8.16}$$

so that finally, with (8.10), we find

$$\omega_g = \frac{3\pi^2}{16}\frac{v}{p} = 1.851\frac{v}{p} \ . \tag{8.17}$$

This is the limiting frequency below which the spectral density of the bremsstrahlung of an electron colliding with an ion can be considered to be flat.

8.4 The Radiation of an Ionized Gas Cloud

The radiation emitted by a single encounter of an electron and an ion thus depends in all its characteristic features, be this the total radiated energy W, its average spectral density or the limiting frequency ν_g, on the collision parameters p and v. In a cloud of ionized gas these occur with a wide distribution of values, and thus the appropriate average will be the total radiation emitted by this cloud. In addition, the radiation of each collision will be polarized differently, such that the resulting emission will be randomly polarized. If, therefore, a single polarization component is measured, only 1/2 of the total emitted power will be detected.

The consideration of the last section had the result: a collision with the parameters p and v will emit bremsstrahlung with a flat spectrum which, using (8.10) and (8.17), is

$$P_\nu(p,v) = \begin{cases} \frac{4}{3}\frac{1}{(4\pi\varepsilon_0)^3}\frac{Z^2 e^6}{c^3 m^2}\frac{1}{p^2 v^2} & \text{for } \nu < \nu_g = \frac{3\pi v}{32 p} \\ 0 & \nu \geqq \nu_g \end{cases} . \tag{8.18}$$

Since only collisions with small relative energy changes $\Delta E/E$ for the electrons are considered, their velocity is changed only very slightly during the collision, so that a Maxwell distribution can be adopted for the distribution function of v:

$$f(v) = \frac{4 v^2}{\sqrt{\pi}}\left(\frac{m}{2kT}\right)^{3/2} \exp\left\{-\frac{m v^2}{2kT}\right\} . \tag{8.19}$$

The number of electrons in a unit volume of space with a given velocity $v = \sqrt{\boldsymbol{v}^2}$ that will meet a given ion with a collision parameter between p and $p + dp$ is (Fig. 8.4)

$$2\pi p\, dp\, v\, N_e\, f(v)\, dv \ .$$

But since there are all in all N_i ions per unit volume a total of

$$dN(v,p) = 2\pi N_i N_e v p f(v)\, dv\, dp \tag{8.20}$$

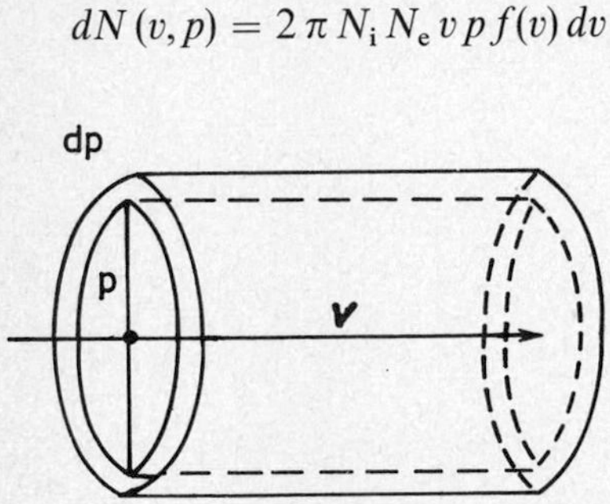

Fig. 8.4. On the probability of collisions with the collision parameter p

collisions with collision parameters between p and $p + dp$ and velocities between v and $v + dv$ will occur per second. Due to these collisions a total power of $4\pi\,\varepsilon_\nu\,d\nu$ will be radiated in the frequency interval ν and $\nu + d\nu$ given by

$$4\pi\,\varepsilon_\nu\,d\nu = P_\nu(v,p)\,dN(v,p)\,d\nu \ .$$

Substituting here (8.18) and (8.20) and integrating p from p_1 to p_2 and v from 0 to ∞ we find

$$4\pi\,\varepsilon_\nu = \int\limits_{p_1}^{p_2}\int\limits_0^\infty \frac{1}{(4\pi\,\varepsilon_0)^3}\frac{4}{3}\frac{Z^2 e^6}{c^3 m^2}\frac{1}{p^2 v^2} N_\mathrm{i} N_\mathrm{e}\,f(v)\,2\pi\,pv\,dp\,dv$$

$$= \frac{1}{(4\pi\,\varepsilon_0)^3}\frac{8\pi}{3}\frac{Z^2 e^6}{c^3}\frac{N_\mathrm{i} N_\mathrm{e}}{m^2}\int\limits_0^\infty \frac{1}{v} f(v)\,dv \int\limits_{p_1}^{p_2}\frac{dp}{p} \ .$$

Using (8.19) the first integral becomes

$$\int\limits_0^\infty \frac{1}{v} f(v)\,dv = \sqrt{\frac{2m}{\pi k T}}$$

so that finally

$$\boxed{\varepsilon_\nu = \frac{1}{(4\pi\,\varepsilon_0)^3}\frac{2}{3}\frac{Z^2 e^6}{c^3}\frac{N_\mathrm{i} N_\mathrm{e}}{m^2}\sqrt{\frac{2m}{\pi k T}}\ln\frac{p_2}{p_1}} \ . \qquad (8.21)$$

This is the coefficient for thermal emission of an ionized gas cloud. But in it the appropriate limits for the collision parameter p_1 and p_2 have still been left open. But both if $p_1 \to 0$ or if $p_2 \to \infty$, ε_ν diverges logarithmically; therefore appropriate values for these limits have to be estimated. For p_2 upper limits should be the mean distance between the ions or the Debye length in the plasma, but for the limits of p_1 we need quantum mechanical considerations. These are traditionally collected into the *Gaunt* factor.

Oster (1961) arrives at

$$\frac{p_\mathrm{max}}{p_\mathrm{min}} = \frac{p_2}{p_1} = \left(\frac{2kT}{\gamma m}\right)^{3/2} \frac{m}{\pi\,\gamma\,Z e^2 \nu} \qquad (8.22)$$

where $\gamma = e^\mathrm{C} = 1.781$ and $\mathrm{C} = 0.577$, Euler's constant, which is valid as long as $T > 20$ K and $\nu_\mathrm{max} > 30$ GHz.

If the emission coefficient is known the absorption coefficient κ_ν can be computed using Kirchhoff's law (1.14):

$$\kappa_\nu = \frac{\varepsilon_\nu}{B_\nu(T)}$$

where $B_\nu(T)$ is the Planck function. Using the Rayleigh-Jeans approximation we obtain

$$\boxed{\kappa_\nu = \frac{2}{(4\pi\varepsilon_0)^3}\frac{Z^2 e^6}{3c}\frac{N_\mathrm{i} N_\mathrm{e}}{\nu^2}\frac{1}{\sqrt{2\pi(m k T)^3}}\ln\frac{p_2}{p_1}} \quad . \tag{8.23}$$

For use in radio astronomy numeric values have to be substituted into (8.21) and (8.23). If in addition the assumption is made that the plasma is macroscopically neutral and that the chemical composition is given approximately by $N_\mathrm{H} : N_\mathrm{He} : N_\mathrm{other} \cong 10 : 1 : 10^{-3}$, then quite closely $N_\mathrm{i} = N_\mathrm{e}$. If further $T_\mathrm{e} = \mathrm{const}$ along the line of sight in an emission nebula, then it is useful to separate in the formula for the optical depth

$$\tau_\nu = -\int_0^s \kappa_\nu\, ds \; . \tag{8.24}$$

The *emission measure* EM is given by

$$\boxed{\frac{\mathrm{EM}}{\mathrm{pc\,cm}^{-6}} = \int_0^{s/\mathrm{pc}} \left(\frac{N_\mathrm{e}}{\mathrm{cm}^{-3}}\right)^2 d\left(\frac{s}{\mathrm{pc}}\right)} \tag{8.25}$$

so that

$$\boxed{\tau_\nu = 3.014\cdot 10^{-2}\left(\frac{T_\mathrm{e}}{\mathrm{K}}\right)^{-3/2}\left(\frac{\nu}{\mathrm{GHz}}\right)^{-2}\left(\frac{\mathrm{EM}}{\mathrm{pc\,cm}^{-6}}\right)\langle g_\mathrm{ff}\rangle} \tag{8.26}$$

where the Gaunt factor for free-free transitions is given by

$$\boxed{\langle g_\mathrm{ff}\rangle = \begin{cases} \ln\left[4.955\cdot 10^{-2}\left(\dfrac{\nu}{\mathrm{GHz}}\right)^{-1}\right] + 1.5\ln\left(\dfrac{T_\mathrm{e}}{\mathrm{K}}\right) \\ 1 \quad \text{for} \quad \dfrac{\nu}{\mathrm{MHz}} \gg \left(\dfrac{T_\mathrm{e}}{\mathrm{K}}\right)^{3/2} \end{cases}} \quad . \tag{8.27}$$

Approximating $\langle g_\mathrm{ff}\rangle$ by $\alpha\, T^\beta \nu^\gamma$ and substituting this into (8.26), the simpler expression (Mezger and Henderson 1967)

$$\boxed{\tau_\nu = 8.235\cdot 10^{-2}\left(\frac{T_\mathrm{e}}{\mathrm{K}}\right)^{-1.35}\left(\frac{\nu}{\mathrm{GHz}}\right)^{-2.1}\left(\frac{\mathrm{EM}}{\mathrm{pc\,cm}^{-6}}\right)a(\nu, T)} \tag{8.28}$$

can be derived. The correction $a(\nu, T)$ is usually $\cong 1$; more precise values are given in Fig. 8.5.

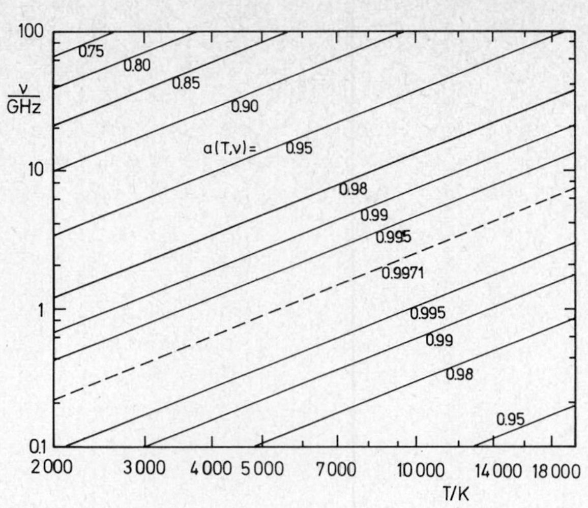

Fig. 8.5. The correction factor $a(\nu, T)$

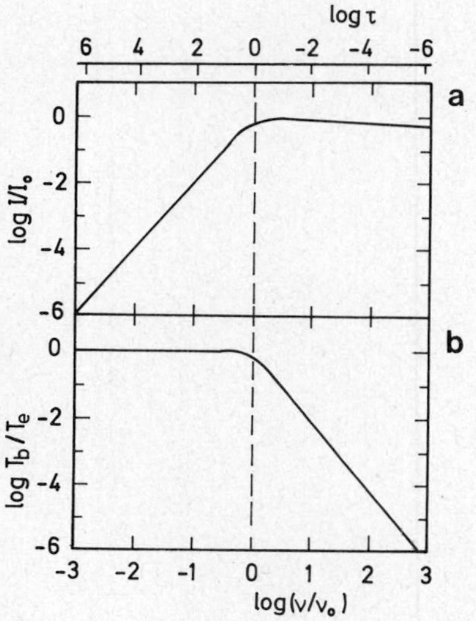

Fig. 8.6 a, b. Thermal radiation of a gas cloud. (**a**) Spectral distribution of the intensity. (**b**) Spectral distribution of the brightness temperature

Substituting (8.28) into (1.18) we find for the brightness of a gas cloud the result given in Fig. 8.6. In this diagram we assumed a background brightness $I_\nu(0) = 0$. In the same diagram the frequency distribution $T_b(\nu)$ according to (1.31) is also given. The frequency is given in units of ν_0, where ν_0 is that frequency at which the optical depth is unity. According to (8.28) this is

$$\boxed{\frac{\nu_0}{\mathrm{GHz}} = 0.3045\left(\frac{T_e}{\mathrm{K}}\right)^{-0.643}\left(\frac{a(\nu, T)\,\mathrm{EM}}{\mathrm{pc\,cm^{-3}}}\right)^{0.476}} \ . \tag{8.29}$$

For typical bright H II regions like M42 ν_0 is of the order of $\simeq 1$ GHz.

8.5 Nonthermal Radiation Mechanisms

Even when it had become clear that there are two kinds of radio sources, thermal and nonthermal, the radiation mechanism of the second type of source remained an enigma since the proposed mechanisms were all unsatisfactory in one way or another. The solution to this problem was presented in two papers in 1950 one by Alfvén and Herlofson, the other by Kiepenheuer.

Alfvén and Herlofson proposed that the nonthermal radiation is emitted by relativistic electrons moving in intricately "tangled" magnetic fields in the radio sources. Kiepenheuer made a similar proposal. He showed that the intensity of the nonthermal galactic radio emission can be understood as the radiation from relativistic cosmic ray electrons that move in the general interstellar magnetic field. He found that a field of 10^{-6} Gauss $\cong 10^{-10}$ Tesla and relativistic electrons of an energy of 10^9 eV would give about the observed intensity if their concentration is a small percentage of that of the more heavy particles in the cosmic rays.

This solution to the nonthermal radiation enigma proved to be tenable and beginning in 1951 Ginzburg and collaborators developed this idea further in a series of papers. The radiation was given the name "synchrotron radiation" since it was observed to be emitted by electrons in synchrotrons (Blewett 1946; Elder et al. 1947, 1948), and its theory was developed by Ivanenko and Sokolov (1948), Vladimirsky (1948) and Schwinger (1949). But much of the relevant theory had already been developed by Schott (1907, 1912) who used the name "Magneto-Bremsstrahlung".

A fairly complete exposition of this theory which includes the polarization properties of this radiation is presented in the books of Pacholczyk (1970, 1977) where extensive bibliographical notes are also given. Very readable accounts of the theory are given by Rybicki and Lightman (1979) or by Longair (1981).

In radio astronomy we usually observe only the radiation from an ensemble of electrons with an energy distribution function that spans a wide range of energies. These distribution functions can quite often be expressed as a power law, and many of the details of the characteristics of synchrotron radiation are thus smoothed out. Therefore we will present here only a more qualitative discussion emphasizing, however, the underlying basic principles. In this we will make heavy use of the text by Rybicki and Lightman.

Since the electrons that emit the synchrotron radiation are highly relativistic, Lorentz transformations have to be applied describing their motion. In the full theory starting with the Liénard-Wiechert potentials of moving charges, all this is automatically taken into account, since Maxwell's equations are invariant against Lorentz transformations. But here we will start with a short recapitulation of these transformations.

8.6 Review of the Lorentz Transformation

The special theory of relativity is based on two postulations:

1) The laws of nature are the same in any two frames of reference that are in uniform relative motion.
2) The speed of light, c, is constant in all such frames.

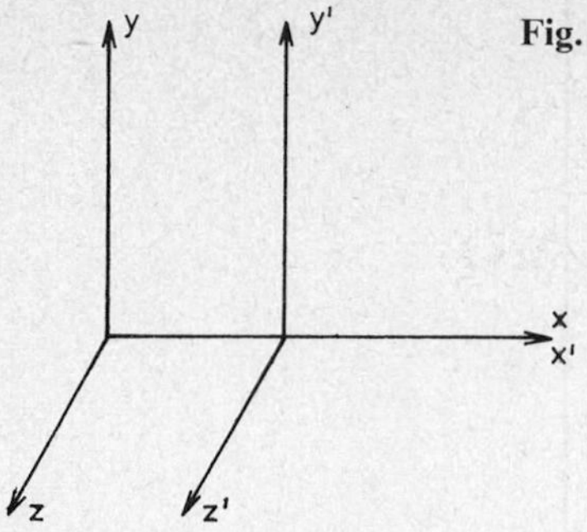

Fig. 8.7. Coordinate systems in relative motion

If we now consider two frames K and K' that move with relative uniform velocity v along the x-axis (Fig. 8.7), and if we assume space to be homogeneous and isotropic, then the coordinates in the two frames are related by the Lorentz transformations

$$\boxed{\begin{aligned} x &= \gamma(x' + v t') \\ y &= y' \\ z &= z' \\ t &= \gamma\left(t' + \beta\frac{x'}{c}\right) \end{aligned}} \quad \text{where} \tag{8.30}$$

$$\beta = \frac{v}{c} \quad \text{and} \tag{8.31}$$

$$\gamma = (1 - \beta^2)^{-1/2} \ . \tag{8.32}$$

Defining the (three) velocity $\boldsymbol{u}$ by

$$\begin{aligned} u_x &= \frac{dx}{dt} = \frac{dx}{dt'}\frac{dt'}{dt} = \gamma(u'_x + v)\frac{dt'}{dt} \\ \frac{dt}{dt'} &= \gamma\left(1 + \beta\frac{u'_x}{c}\right) = \gamma\sigma \\ \sigma &= 1 + \beta u'_x/c \end{aligned} \tag{8.33}$$

so that

$$u_x = \frac{u'_x + v}{1 + \beta\dfrac{u'_x}{c}} = \frac{1}{\sigma}(u'_x + v) \tag{8.34}$$

$$u_y = \frac{u'_y}{\gamma\left(1 + \beta\dfrac{u'_x}{c}\right)} = \frac{1}{\gamma\sigma}u'_y \tag{8.35}$$

$$u_z = \frac{u'_z}{\gamma\left(1 + \beta \frac{u'_x}{c}\right)} = \frac{1}{\gamma \sigma} u'_z \ . \tag{8.36}$$

For an arbitrary relative velocity $\boldsymbol{v}$ that is not parallel to any of the coordinate axis the transformation can be stated in terms of the velocity components of u parallel and perpendicular to $\boldsymbol{v}$

$$\boxed{u_{\parallel} = \frac{u'_{\parallel} + v}{1 + \beta \frac{u'_{\parallel}}{c}}} \tag{8.37}$$

$$\boxed{u_{\perp} = \frac{u'_{\perp}}{\gamma\left(1 + \beta \frac{u'_{\parallel}}{c}\right)}} \ . \tag{8.38}$$

Although the theory of special relativity is concerned with the transformation of physical quantities if subjected to an uniform velocity, it is quite possible to compute how an acceleration is transformed. Differenciating (8.34) to form

$$a_x = \frac{du_x}{dt} = \frac{du_x}{dt'} \frac{dt'}{dt}$$

and using (8.33) so that

$$\frac{d\sigma}{dt'} = \frac{\beta}{c} a'_x \ , \tag{8.39}$$

we obtain

$$a_x = \dot{u}_x = \gamma^{-3} \sigma^{-3} a'_x \quad \text{and} \tag{8.40}$$

$$a_y = \dot{u}_y = \gamma^{-2} \sigma^{-3} \left(\sigma a'_y - \beta \frac{u'_y}{c} a'_x\right) \ . \tag{8.41}$$

For the special case that the particle is at rest in system K', that is, that initially $u'_x = u'_y = u'_z = 0$, we have $\sigma = 1$ and

$$\begin{aligned} a'_{\parallel} &= \gamma^3 a_{\parallel} \\ a'_{\perp} &= \gamma^2 a_{\perp} \ . \end{aligned} \tag{8.42}$$

The transformation equations (8.34)–(8.42) are seen to be fairly complicated and they are different from the Lorentz transformation. This is so because neither the velocity u_x, u_y, u_z nor the accelerations a_x, a_y are components of four-vectors, and only they obey the simple transformation laws of the coordinates.

Equation (8.33) describes the time dilatation, giving the different rate at which a clock at rest in the system K' is seen in the system K:

$$\Delta t = \gamma\,\Delta t' \; . \tag{8.33 a}$$

This should not be confused with the apparent time dilatation (or contraction) known as the *Doppler effect*. Let the distance between clock and observer change with the speed v_r. During the intrinsic time interval $\Delta t'$ the clock has moved, so that the next pulse reaches the observer at

$$\Delta t = \left(1 + \frac{v_r}{c}\right)\gamma\,\Delta t' \; . \tag{8.43}$$

The factor γ is a relativistic effect, whereas $(1 + v_r/c)$ appears even in nonrelativistic physics.

8.7 The Synchrotron Radiation of a Single Electron

The relativistic Einstein-Planck equations of motion for a particle with the charge e and the mass m that moves with a (three) velocity $\boldsymbol{v}$ in a magnetic field with the flux density $\boldsymbol{B}$ is given by

$$\frac{d}{dt}(\gamma\, m\, \boldsymbol{v}) = e\,(\boldsymbol{v} \times \boldsymbol{B}) \; . \tag{8.44}$$

If there is no electric field $\boldsymbol{E}$, then energy conservation results in the equation

$$\frac{d}{dt}(\gamma\, m\, c^2) = 0 \; . \tag{8.45}$$

But this implies that γ is a constant and therefore that $|\boldsymbol{v}|$ is a constant. Projecting $\boldsymbol{v}$ into two components $\boldsymbol{v}_\parallel$ parallel to $\boldsymbol{B}$ and $\boldsymbol{v}_\perp$ perpendicular to it, we find from (8.44) that

$$\frac{d\boldsymbol{v}_\parallel}{dt} = 0 \quad \text{and} \tag{8.46}$$

$$\frac{d\boldsymbol{v}_\perp}{dt} = \frac{e}{\gamma\, m}(\boldsymbol{v}_\perp \times \boldsymbol{B}) \; . \tag{8.47}$$

(8.46) has the solution $\boldsymbol{v}_\parallel$ = constant so that, since $|\boldsymbol{v}|$ is constant, $|\boldsymbol{v}_\perp|$ must also be constant. The solution to (8.47) therefore must obviously be a uniform circular motion with a constant orbital velocity $v_\perp = |\boldsymbol{v}_\perp|$. The frequency of the gyration is

$$\omega_B = \frac{e\,B}{\gamma\, m} \, , \qquad \omega_G = \frac{e\,B}{m} \; . \tag{8.48}$$

Since superimposed on this circular motion is the constant velocity $\boldsymbol{v}_{\parallel}$ the path followed by the electron is a helix winding around $\boldsymbol{B}$.

Inserting numerical values for e and m, we find

$$\frac{\omega_{\mathrm{G}}}{\mathrm{MHz}} = 17.6\left(\frac{B}{\mathrm{Gauss}}\right) \quad \text{or} \quad \frac{\omega_{\mathrm{G}}}{\mathrm{GHz}} = 176.2\left(\frac{B}{\mathrm{Tesla}}\right)$$

so that, for $B \cong 10^{-6}$ Gauss in interstellar space, $\omega_{\mathrm{G}} = 18$ Hz. From (8.47) we see that the electron is accelerated in its orbit, this acceleration is directed $\perp$ to $\boldsymbol{B}$ and its magnitude is

$$a_{\perp} = \omega_B v_{\perp} \ . \tag{8.49}$$

Since the electron is accelerated, it will radiate. We will investigate this radiation in a two-step procedure:

1) We select a inertial frame K' that moves with respect to the laboratory rest frame K such that the electron is at rest at a certain time. This particle will not remain so for long since its acceleration is not zero, but for an infinitesimal time interval we can adopt this. In this frame K', the electron is nonrelativistic and radiates according to the Larmor formula (8.4)

$$P' = \frac{1}{6\pi\varepsilon_0}\frac{e^2}{c^3}a_{\perp}'^2 \tag{8.50}$$

where $a'_{\perp}$ is the acceleration of the electron in the rest frame K'.

2) We will now transform the emitted power P' into the laboratory rest frame K. K moves relative to K' with the velocity $\boldsymbol{v}_{\perp}$. The energy is one component of the four-component vector (momentum, energy), and it is transformed accordingly by

$$W = \gamma W' \ .$$

For a time interval (at constant space position) we have similarly

$$dt = \gamma\, dt'$$

so that for the power emitted

$$P = \frac{dW}{dt} \ , \qquad P' = \frac{dW'}{dt'}$$

we find that

$$P = P' \ . \tag{8.51}$$

Considering in addition the transformation of the acceleration (8.42) we find

$$P = \frac{1}{6\pi\varepsilon_0}\frac{e^2}{c^3}\gamma^4 a_{\perp}^2 = \frac{1}{6\pi\varepsilon_0}\frac{e^2}{c^3}a_{\perp}'^2 = P' \ . \tag{8.52}$$

Introducing here the total energy E of the electrons through

$$\gamma = \frac{E}{m c^2} \tag{8.53}$$

and using (8.49) we obtain as the power emitted by a relativistic ($\beta \cong 1$) electron

$$\boxed{P = \frac{1}{6 \pi \varepsilon_0} \frac{e^4 B^2}{m^2 c} \left(\frac{E}{m c^2} \right)^2} \ . \tag{8.54}$$

Please note that the energy E introduced in (8.53) and (8.54) is a quantity entirely different from the electric field intensity E of (2) or of Chaps. 2–4. This unfortunate double meaning for the same letter E cannot be avoided if we want to conform to the general usage. In the subsequent sections E will always be used with the meaning of (8.53).

P in (8.54) is the total power radiated by the gyrating electron at any moment in the laboratory reference frame K as can be seen by retracing the arguments back to (8.50) which gives this power in the moving frame K' of the electron. This power P' has an angular distribution given by

$$\frac{dP'(\vartheta, \varphi)}{d\Omega} = \frac{1}{16 \pi^2 \varepsilon_0} \frac{e^2}{c^3} a_\perp'^2 \sin^2 \vartheta \tag{8.55}$$

and this distribution has to be transformed into the laboratory reference frame K. This can be done using the addition theorem of the velocities (8.37) and (8.38).

Consider the directions $\vartheta = 0$ and π (Fig. 8.8). According to (8.55) no radiation is emitted into these directions, which have an angle of $\pm \frac{\pi}{2}$ with respect to the relative velocity of the systems K' and K. Since we are considering electromagnetic radiation, obviously $u'_\perp = c$ and $u'_\parallel = 0$ so that

$$\tan \theta = \frac{u_\perp}{u_\parallel} = \frac{c}{\gamma} \frac{1}{v} \quad \text{and} \quad \sin \theta = \frac{1}{\gamma} \tag{8.56}$$

where θ is that angle in the system K into which the angle $\frac{\pi}{2}$ of system K' is transformed. The power P of (8.54) thus is confined to a cone with the angle θ and we obtain a strong beaming effect (Fig. 8.9).

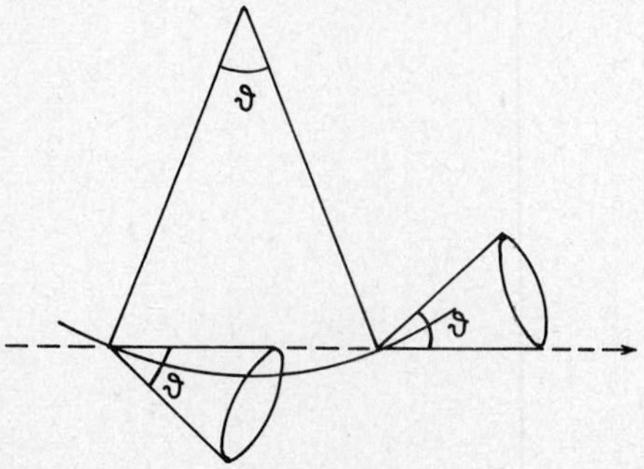

Fig. 8.8. Geometry of the emission cones at various points in an accelerated particle orbit

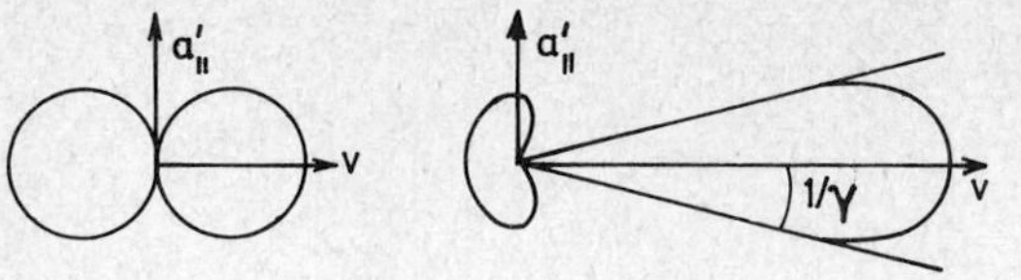

Fig. 8.9. Relativistic transformation of the polar power patterns of accelerated radiating sources

This strong beaming effect for highly relativistic electrons with $\gamma \gg 1$ is the reason why such electrons emit radiation with fairly high frequencies although the cyclotron frequency of these particles according to (8.48) is very low indeed. An external observer will be able to "see" the radiation from the electron only for such a time interval as the cone of emission of angular width $2/\gamma$ includes the direction of observations.

In the frame K' of the electron the time for one gyration is

$$\Delta T' = \frac{2\pi}{\omega_G} .$$

In the relevant part of the orbit the electron moves towards the observer, so that $v_r = -v$, and thus the time for one gyration of the electron is in the frame K of the observer, using the Doppler formula (8.43) and (8.48)

$$\Delta T = \frac{2\pi}{\omega_G}\gamma\left(1 - \frac{v}{c}\right) = \frac{2\pi}{\omega_B}\left(1 - \frac{v}{c}\right) .$$

Since the pulse of radiation is visible only for $2/(2\pi\gamma)$ of one gyration, the effective width of the pulse received is

$$\Delta t^A = \frac{2}{\gamma}\frac{1 - \frac{v}{c}}{\omega_B} .$$

Now

$$\left(1 + \frac{v}{c}\right)\left(1 - \frac{v}{c}\right) = 1 - \frac{v^2}{c^2} = 1 - \beta^2 = \frac{1}{\gamma^2}$$

and

$$1 + \frac{v}{c} \cong 2 \quad \text{for } \gamma \gg 1 , \qquad \text{so that}$$

$$1 - \frac{v}{c} \cong \frac{1}{2\gamma^2} \quad \text{and}$$

$$\Delta t^A = \frac{1}{\gamma^3 \omega_B} .$$

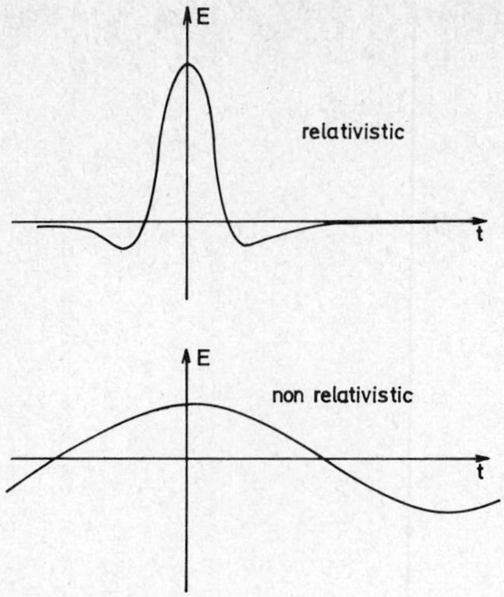

Fig. 8.10

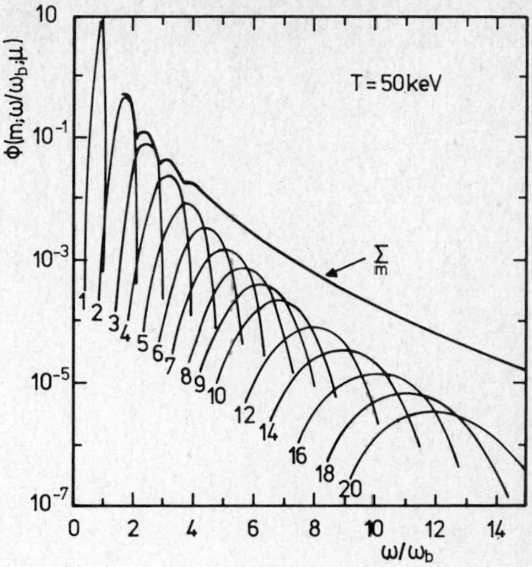

Fig. 8.11

Fig. 8.10. The narrowing of the emitted pulse of radiation and the conversion
Fig. 8.11. Spectrum of the first twenty harmonics of 50 KeV electrons [after Hirshfield et al. (1961)]

If now the electron has some velocity component $\boldsymbol{v}_{\parallel} \neq 0$ parallel to the magnetic field, it will move in a helix with a pitch angle φ given by

$$\tan\varphi = |\boldsymbol{v}_{\perp}|/|\boldsymbol{v}_{\parallel}|$$

and we obtain approximately

$$\Delta T^A = \frac{1}{\gamma^3\,\omega_B}\,\frac{1}{\sin\varphi}\,. \tag{8.57}$$

Radiation from the electron is therefore only seen for a time interval $\sim 1/\gamma^3$; the radiation pulse becomes very short indeed if $\gamma \gg 1$ (Fig. 8.10). If now the pulse is Fourier-analyzed to derive the frequency distribution of the radiation we see that very high harmonics of ω_B will be present (Fig. 8.11).

The result of such a Fourier-analysis [for the details see Pacholczyk (1970) or Bekefi (1966)] is: the total emissivity of an electron of energy E with $E/mc^2 = \gamma \gg 1$ which has a pitch angle φ with respect to the magnetic field is

$$P(\nu, E; \varphi)\,d\nu = \sqrt{3}\,\frac{e^3 B\sin\varphi}{m_e c^2}\,\frac{\nu}{\nu_c}\,d\nu \int\limits_{\nu/\nu_c}^{\infty} K_{5/3}(\eta)\,d\eta\,. \tag{8.58}$$

The critical frequency ν_c is defined by

$$\nu_c = \tfrac{3}{2}\,\gamma^2\,\nu_G\sin\varphi \tag{8.59}$$

where $\nu_G = \omega_G/2\pi$ is the nonrelativistic gyro-frequency according to (8.48).

8.8 The Spectral Distribution of an Ensemble of Electrons

Equation (8.58) gives the spectral density of the emission of a single relativistic electron, obviously the emission of N electrons all with identical velocity and pitch angle relative to the magnetic field is the N-fold of (8.58). But in nature the situation is rarely that simple: the relativistic electrons move with velocities that vary over a wide range into different directions, and in addition the magnetic field is frequently tangled, so that the pitch angle φ will vary widely. The resulting emissivity of the relativistic electrons then is given by

$$\varepsilon(\nu) = \int_{\Omega, E} P(\nu, E; \varphi) N(E, \varphi) \, dE \, d\Omega \ . \tag{8.60}$$

Very little is usually known on the pitch-angle distribution, and therefore it is often assumed to be isotropic, while empirical evidence of the cosmic rays close to the earth (Fig. 8.12) shows that $N(E)$ is well described by a power-law spectrum

$$\boxed{N(E) \, dE = K E^{-\delta} dE \quad \text{for } E_1 < E < E_2} \ . \tag{8.61}$$

According to (8.58), γ appears in (8.60) only through ν_c. This fact alone is sufficient to derive an important result concerning synchrotron spectra. Assuming an isotropic pitch angle distribution and a power law spectrum for the electron energy we have

$$\varepsilon(\nu) \propto \int_{E_1}^{E_2} P(\nu/\nu_c) E^{-\delta} dE = \int_{E_1}^{E_2} P\left(\frac{\nu}{\mu E^2}\right) E^{-\delta} dE$$

since according to (8.53) and (8.59) and μ containing all constants

$$\nu_c = \mu E^2 \ . \tag{8.62}$$

Substituting

$$x = \nu/\mu E^2 \ , \qquad dE = -\frac{1}{2\sqrt{\mu}} \nu^{1/2} x^{-3/2} dx$$

and omitting constant factors, we obtain

$$\varepsilon(\nu) \propto \nu^{(1-\delta)/2} \int_{x_2}^{x_1} P(x) \, x^{(\delta-3)/2} dx \ . \tag{8.63}$$

The limits x_1 and x_2 correspond to the limits E_1 and E_2. If they are sufficiently wide we can approximate $x_1 \cong 0$ and $x_2 \cong \infty$ so that the integral becomes a constant and then we find

$$\varepsilon(\nu) \propto \nu^{(1-\delta)/2} \ . \tag{8.64}$$

Thus the synchrotron spectrum of relativistic electrons will also be a power spectrum with an exponent that is related to the exponent of the distribution

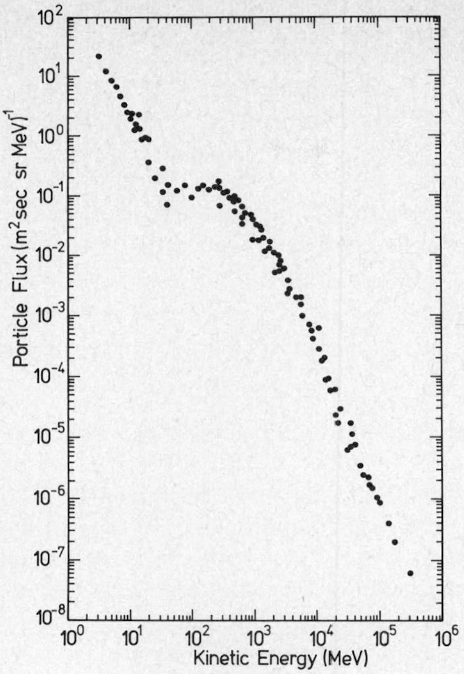

Fig. 8.12. The energy spectrum of cosmic ray electrons [from Meyer (1969)]

function (8.61) of the electrons. If

$$\varepsilon(\nu) \propto \nu^{-\alpha} \quad \text{then} \tag{8.65}$$

$$\boxed{\delta = 2\alpha + 1} \; . \tag{8.66}$$

If this analysis is done with full precision we obtain for the flux of a synchrotron source with a volume V the following expression

$$\boxed{S(\nu) = 1.7 \cdot 10^{-28}\, a(\delta)\, V\, B^{(\delta+1)/2} \left(\frac{6.26 \cdot 10^{18}}{\nu/\mathrm{Hz}}\right)^{(\delta-1)/2} \quad [W\,\mathrm{Hz}^{-1}]} \tag{8.67}$$

Table 8.1. Values of the constant $a(\delta)$ in Eq. (8.67) for different values of δ. The relation of δ and α is given by (8.66)

δ	α	a
1	0	0.283
1.5	−0.25	0.147
2	−0.50	0.103
2.5	−0.75	0.0852
3	−1.00	0.0742
4	−1.50	0.0725
5	−2.00	0.0922

where δ is the exponent of the power-law distribution of the electrons. For the different values of δ, $a(\delta)$ takes on the values given in Table 8.1.

Nonthermal radio sources are observed to have in general a power law emissivity (8.65) with $\alpha \cong 0.7$. If we indeed observe synchrotron emission, then the relativistic electrons responsible for this must have a power-law energy distribution of $N(E) \propto E^{-\delta}$ with $\delta \propto 2.4$.

The discussion of the synchrotron-radiation of nonthermal radio sources as presented here has not dug very deep into the problem; many important problems have not been touched at all. One of these is the question of the state of polarization of synchrotron radiation. Monoenergetic electrons moving in parallel trajectories in a straight magnetic field emit elliptically polarized synchrotron radiation. But if a smooth distribution $N(E)$ for the electrons is assumed, the resulting degree of polarization is very small indeed except for some very special cases. Detailed information on these problems and the application to radio sources can be found in the books by Pacholczyk (1970, 1977).

Another problem that we left out completely arises from the circumstance that the relativistic electrons, that emit the synchrotron radiation do not move in a complete vacuum but in a tenuous plasma. Due to the absorption in this plasma, the low frequency end of the synchrotron spectrum will be affected. For a discussion of these problems and their astrophysical implications we again refer to the books by Pacholczyk. A collection of the relevant formulae are also given in Lang (1974), Chap. 1.

8.9 Energy Requirements of Synchrotron Sources

A question that arises quite naturally when physical models for synchrotron radio sources are considered is that of their total energy content. From the theory of synchrotron radiation we can compute dE/dt, the rate at which this energy is radiated away; the total energy content, however, remains open.

The energy of a source that emits mainly synchrotron radiation can be contained in two forms mainly: as kinetic energy of the relativistic particles W_{part} and as energy stored in the magnetic field W_{mag}. Since thermal radiation is negligible the thermal energy will be small compared to W_{part} and W_{mag}. We know nothing about the gravitational energy as long as no definite model is proposed. Probably then W_{grav} will be the source of the two other forms of energy.

If V is the volume of the source, then

$$W_{\text{tot}} = W_{\text{part}} + W_{\text{mag}} = V\,\varepsilon_{\text{p}} + V\,\frac{B^2}{2\,\mu_0}\,. \tag{8.68}$$

ε_{p} is the energy density of the relativistic particles, that is, of the electrons and protons (and ions with $Z > 1$). Since protons emit only very little synchrotron radiation compared to electrons of the same E, very little is known about the flux of relativistic protons in radio sources and it is customary to assume that

$$\varepsilon_p = \eta\, \varepsilon_e \tag{8.69}$$

where ε_e is the energy density of the electrons and $\eta > 1$ is a factor taking all other particles into account. Now if we again assume a power law (8.61) for the energy spectrum of the electrons, the total energy content of the synchrotron source will be

$$W_{tot} = V\left(\eta \int\limits_{E_{min}}^{E_{max}} K\, E^{-\delta+1}\, dE + \frac{B^2}{2\,\mu_0}\right) . \tag{8.70}$$

In this equation K and B are independent variables and the total energy content of the source can vary within wide limits depending on which values for K and B are chosen. But what can be done is to derive a minimal value for W_{tot} under the constraint that certain observable quantities are preserved.

To arrive at this minimal energy requirement the unknown quantities in (8.70) have to be expressed in terms of observable quantities. Instead of the energy E of an electron we introduce the frequency ν_c of the peak emission like in the computations leading to (8.63). Thus we obtain

$$W_{part} = \frac{\eta\, V K}{\delta - 2}\left[\left(\frac{\nu_{min}}{C\,B}\right)^{(2-\delta)/2} - \left(\frac{\nu_{max}}{C\,B}\right)^{(2-\delta)/2}\right] \quad \text{where} \tag{8.71}$$

$$C = \frac{3}{4\,\pi}\,\frac{e}{m^3\,c^4} . \tag{8.72}$$

The same electrons emit synchrotron radiation, and a certain fraction of this is emitted in the direction of the observer. Therefore the measured flux of the source is given by

$$S_\nu = A\, V\, \varepsilon(\nu)$$

where $\varepsilon(\nu)$ is given by (8.64) and A is a constant that contains the distance to the source as well as a factor that depends on the angular distribution of the relativistic electrons and gives the fraction of the total emitted radiation that is emitted in the direction of the observer. Collecting all the numerical coefficients in (8.67) as well as the value for the integral into a new numerical factor a, we can eventually write

$$S_\nu = a\, V\, K\, B^{(\delta+1)/2}\, \nu^{(1-\delta)/2} . \tag{8.73}$$

Substituting K from this equation into (8.71) we obtain

$$W_{part} = \frac{\eta\, V}{\delta - 2}\,\frac{S_\nu}{a\, V}\, B^{-(\delta+1)/2}\, \nu^{-(1-\delta)/2}\, (C\,B)^{(\delta-2)/2}\, (\nu_{min}^{(2-\delta)/2} - \nu_{max}^{(2-\delta)/2})$$

or

$$W_{part} = G\, \eta\, S_\nu\, B^{-3/2} \tag{8.74}$$

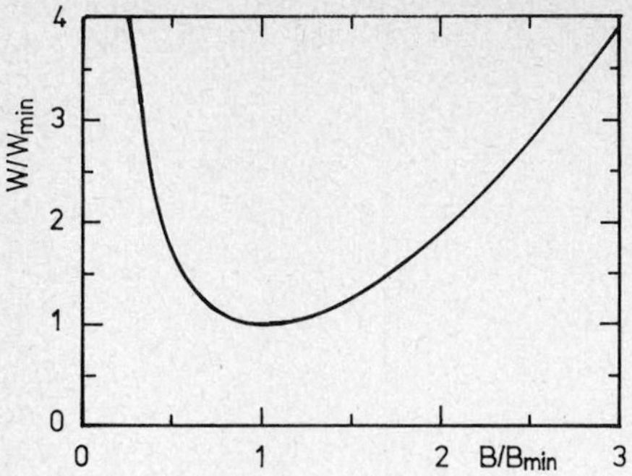

Fig. 8.13. The total energy of a synchrotron radio source

where G is a constant that contains a, C, ν_{min} and ν_{max}. Therefore from (8.74) and (8.68) we find

$$\frac{W_{part}}{W_{mag}} = 2\,\mu_0\,\eta\,\frac{G}{V}\,S_\nu\,B^{-7/2} \tag{8.75}$$

and (Fig. 8.13)

$$W_{tot} = \eta\,G\,S_\nu\,B^{-3/2} + \frac{V}{2\,\mu_0}\,B^2 \ . \tag{8.76}$$

In this, B is the only unknown since S_ν is an observable quantity. W_{tot} reaches its minimum value for

$$B_{min} = \left(\frac{3\,\mu_0\,\eta\,G}{2\,V}\,S_\nu\right)^{2/7} \tag{8.77}$$

for which

$$\frac{W_{part}}{W_{mag}} = \frac{4}{3} \ . \tag{8.78}$$

Thus the minimum total energy to produce the observed radio emission corresponds quite closely to an equipartition of energy between relativistic particles and the magnetic field.

If the constants in (8.76) are to be evaluated, the full theory of synchrotron radiation has to be used. This has been done by various authors. Scheuer (1967) gives the formula

$$\boxed{\left(\frac{W_{min}}{\mathrm{J}}\right) \approx 4.3 \cdot 10^{26}\,\eta^{4/7}\left(\frac{\nu}{\mathrm{GHz}}\right)^{2/7}\left(\frac{I_\nu}{\mathrm{Jy\,sr^{-1}}}\right)^{4/7}\left(\frac{R}{\mathrm{kpc}}\right)^{9/7}} \tag{8.79}$$

where ν is the lowest frequency in which the source is observed, I_ν the brightness in Jy/sr and R its radius in kpc. This equation is usually used in high-energy astrophysics to work out the energy requirements of a synchrotron source.

9. Some Examples of Thermal and Nonthermal Radio Sources

As already mentioned in the preceding chapter, radio sources are most conveniently divided into two classes: the thermal and nonthermal ones. Here we will give an example for each class and discuss the physics involved in as much detail as needed.

As an example for a thermal source, we will describe the quiet sun. We preferred this to the other, perhaps more important kind of thermal source exemplified by galactic H II regions, because they will be discussed later in connection with their recombination line radiation.

For a nonthermal source supernova remnants will be considered and we will present a phenomenological theory of their radiation and its time evolution without discussing the energy sources driving the explosion and the mechanisms producing the high-relativistic electrons and (perhaps) the magnetic field.

9.1 The Quiet Sun

First attempts to identify the radio-emission of the sun were made soon after the detection of radio waves by H. Hertz. One of the assistants of Edison tells in a letter about such investigations in 1890, but these attempts proved unsuccessful just like those of Scheiner and Wilsing in Potsdam (1896), Sir Oliver Lodge in England (1900) and of Nordman in France (1902).

There are two reasons why these experiments failed and the realization of this was the reason why no further attempts were made to detect solar radio radiation for a long time. In 1900 Planck discovered the spectral distribution of thermal radiation, so that the expected intensity of solar radiation at radio wavelength could be computed and was found to be much too small for the sensitivity of the then available radio receivers. In 1902 Kennelly and Heavyside concluded from the possibility of transoceanic radio transmission that there exists a conducting layer high in the earth's atmosphere that reflects radio waves and that this layer would prevent the reception of solar radio radiation.

It took therefore more than 40 years before Southworth (1942) detected the thermal radiation of the quiet sun, while in the same year Hey found a very intensive, time variable solar radiation whose emission was somehow connected to the sunspot phenomenon. Observations of the time- and spatial distribution of these microwave bursts and their theoretical explanation in terms of concepts of plasma physics belong to solar radio astronomy. This is an extensive and specialized field which will not be covered here since several monographs mentioned

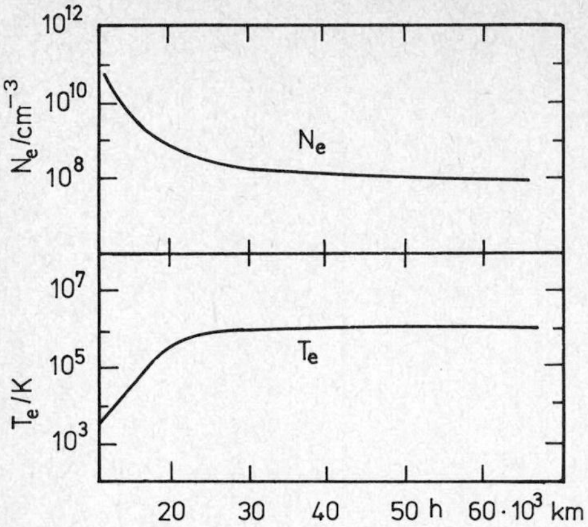

Fig. 9.1. The solar corona after Baumbach and Allen

in the list of general references for this chapter give a good review. We will be concerned only with the radio emission of the quiet sun, that is, with the thermal radiation of the sun's corona.

As was already known from optical measurements there is an extended atmosphere above the solar photosphere in which the gas temperature increases rapidly with height from about 6000 K in the photosphere to a temperature of several million K. The gas density in the corona is quite low compared to that of the photosphere and, because of that, the corona is not conspicuous at all in the visual range; its influence is best seen in observations made at solar eclipses. The opposite is true in the radio range: here the radiation of the hot gas is dominant. The hot corona gas is probably heated by acoustic shock waves that are excited in the convection zone below the photosphere and which then dissipate their energy in the corona. But the details of this mechanism need not interest us here.

Due to the high temperature in the corona the gas is almost fully ionized. The electron density distribution is described by the Baumbach-Allen formula

$$\boxed{\frac{N_e}{\mathrm{m}^{-3}} = \left[1.55\left(\frac{r}{r_0}\right)^{-6} + 2.99\left(\frac{r}{r_0}\right)^{-16}\right] \cdot 10^{14}} \tag{9.1}$$

and for the electron temperature a constant value of $T_e \cong 10^6$ K can be adopted for $h = r - r_0 > 2 \cdot 10^4$ km (Fig. 9.1) where r_0 is the solar radius.

The optical depth τ_ν and the resulting brightness temperature of the corona could now be computed using (8.28) and (1.31) starting with $\tau_\nu = 0$ at $h \to \infty$. But for low frequencies not too far away from the plasma frequency (2.82) the index of refraction (2.86) is significantly less than unity ($n < 1$), so that refraction effects must be taken into account in the radio range. The radio waves will propagate along curved paths. We will now determine the differential equation governing the shape of these rays.

If the corona is considered to consist of concentric shells of constant index of refraction n (Fig. 9.2), then the law of refraction gives

$$n' \sin \varphi' = n \sin \psi \tag{9.2}$$

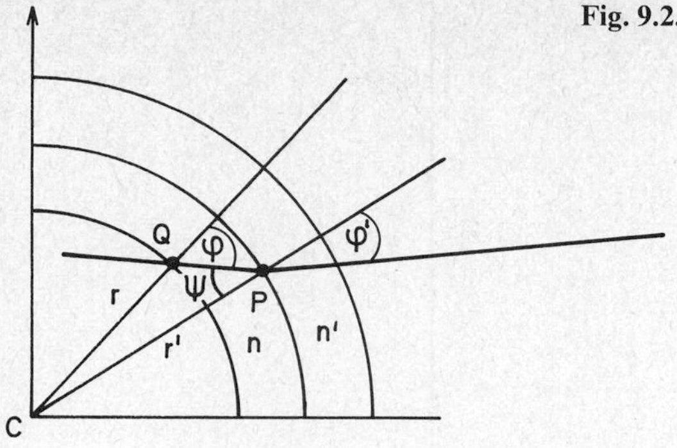

Fig. 9.2. Refraction in the solar corona

while in the triangle CPQ

$$\frac{\sin\psi}{r} = \frac{\sin(180^\circ - \varphi)}{r'} = \frac{\sin\varphi}{r'} \; . \tag{9.3}$$

From this

$$n\,r\sin\varphi = n'\,r'\sin\varphi' = \varrho = \text{const} \tag{9.4}$$

where the parameter ϱ of the ray is equal to the minimum distance from the solar center that a straight line tangent to the exterior part of the ray would have attained. The angle $\varphi(r)$ between the ray and the solar radius is closely related to the slope of the ray, so that (9.4) or the equivalent

$$\boxed{n(r)\,r\sin\varphi(r) = \varrho} \tag{9.5}$$

is a differential equation for the rays in the solar corona.

If the brightness distribution across the solar disk is now to be computed, the equation of transfer (1.16) has to be solved along a representative sample of rays as shown in Fig. 9.3. Each ray has a point of closest approach and it is symmetric around this point. The radius of this point of closest approach is found from (9.5) by putting $\varphi = \frac{\pi}{2}$.

The brightness distribution across the solar disk now depends on how the optical depth is distributed along the ray in relation to this point of closest approach (Fig. 9.4).

1) For low frequencies $\nu < 0.1$ GHz ($\lambda > 3$ m), the refraction effects in the outer parts of the solar corona are so strong that the points of closest approach are situated well above those layers where τ_ν approaches unity. The solar disk is brightest at the center and the brightness decreases smoothly with r reaching zero only after several solar radii. Because $\tau_\nu \ll 1$, the brightness temperature is less than the corona temperature.

2) For 0.1 GHz $< \nu <$ 3 GHz, the optical depth reaches unity $\tau_\nu \cong 1$ near the point of closest approach. Refraction effects can be neglected. The situation can

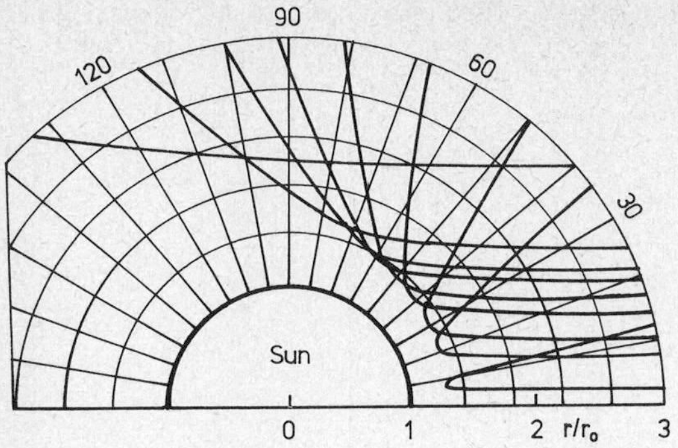

Fig. 9.3. Ray geometry in the solar atmosphere [after Jaeger and Westfold (1949)]

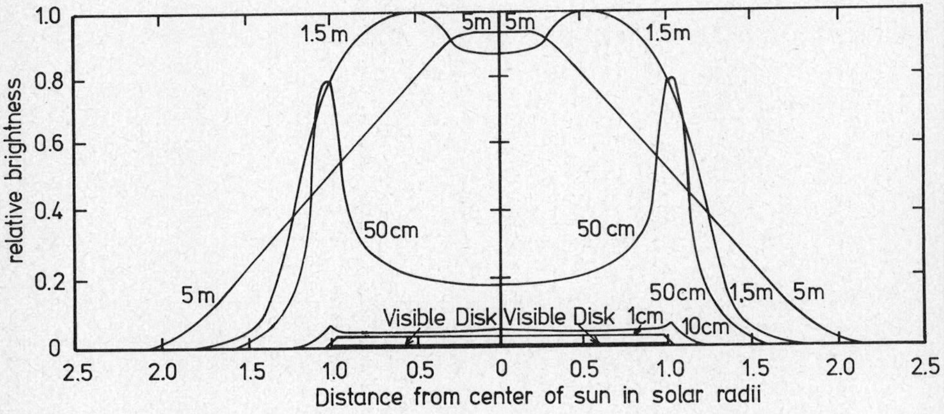

Fig. 9.4. Brightness distribution across the solar disk at different wavelengths [after Smerd (1950)]

be modeled by a luminous atmosphere above a nonluminous stellar body resulting in a bright rim.

3) For $\nu > 3$ GHz conditions approach those met in the optical range. The optical thickness produced by the corona is so small that very little radiation is emitted by it and there is no bright rim.

The thermal radiation of the quiet sun is quite often completely negligible compared to the slowly varying component and, even more so, to the solar bursts. But as both the instruments and observational techniques needed to observe these effects and the physics needed for an explanation differ considerably from those used in the other parts of radio astronomy, they are the subject of the separate discipline of solar radio astronomy.

9.2 Supernovae and Supernova Remnants

Many of the brightest radio sources at low frequencies ($\nu < 1$ GHz) turn out to be nonthermal. The identification of the source Tau A by Bolton and Stanley

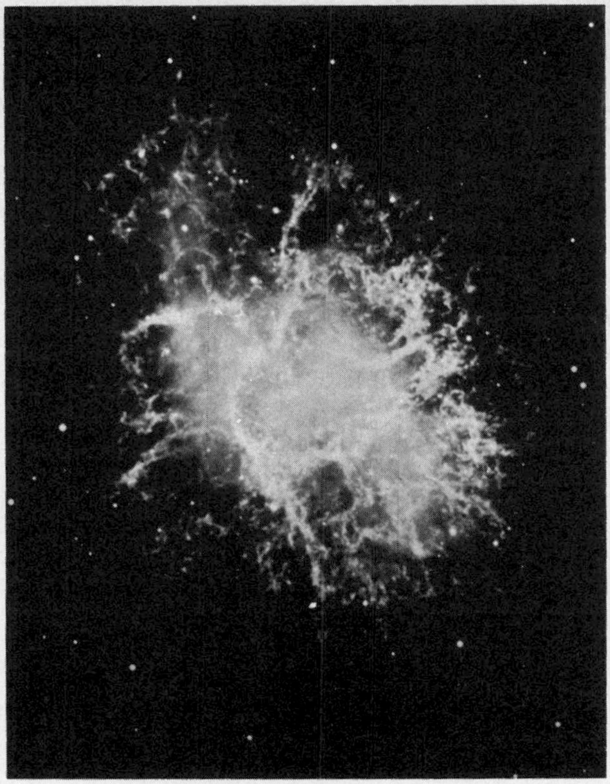

Fig. 9.5. The Crab nebula, the remnant of the supernova of AD 1054

with the Crab Nebula (Fig. 9.5), the remnant of the supernova explosion observed by chinese astronomers in 1054, and that of the source Cas A with the remnant of a supernova that exploded around 1700 but was not observed because of heavy interstellar absorption (Fig. 9.6), makes it plausible that supernova remnants might be sources of nonthermal radio emission.

While the Crab Nebula radio source Tau A is an elliptical region completely filled with radio emission, Cas A is a spherical shell source where the radio emission is confined almost exclusively to a thin shell and observations made a different epochs show this shell to be expanding. Many other nonthermal radio sources were also found to be of a similar shell type and are quite often incomplete. In some cases, additional evidence was found showing that these sources are indeed remnants of supernova explosions. This evidence can consist of wisps of H_α filaments with exceedingly high radial velocities or of X-ray emission. Indeed, X-ray emission is quite commonly observed in supernova remnants, and such measurements are now about as important in their study as radio data.

The number of known supernova remnants has increased considerably in the last ten years due to the increasing resolving power and sensitivity of the surveys. Usually a procedure consisting of several steps is needed before a source is considered to be a bona-fide supernova remnant.

As a first step, one tries to sift out from a catalogue of radio source flux and spectral index those objects that have a negative spectral index and are probably

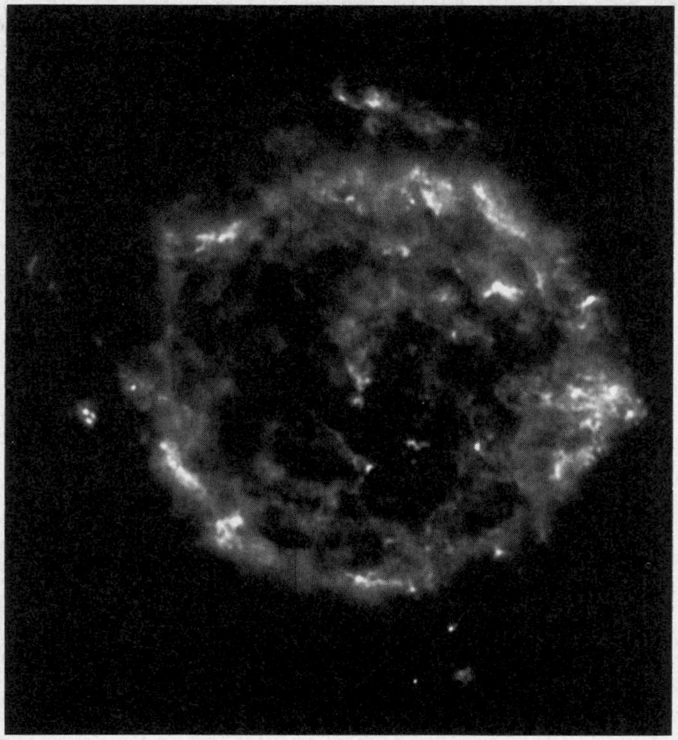

Fig. 9.6. The supernova remnant Cas A from observations with the Cambridge 5 km-radio synthesis telescope (courtesy Mullard Radio Astronomy Observatory)

extragalactic. If the photographic identifications show a galaxy at the source position the situation is clear; in other cases, where interstellar dust extinction prevents such a direct identification, a small source diameter of 1′ or less can be taken to indicate an extragalactic origin for the source. The remaining sources with a negative spectral index are tentatively supernova remnants but, before a definite identification is possible, additional evidence is required. This might be high-resolution measurements showing a shell structure, a distance estimation from 21-cm absorption measurements, or expanding gas filaments visible in the optical range. Or eventually X-ray observations may confirm the identification.

Until now about 200 supernova remnants have been identified in this way, but eventually the total list of such sources in our Galaxy will be much larger because each high-sensitivity survey results in many more low surface brightness objects. It is therefore rather difficult to estimate the completeness of the present surveys.

9.3 The Hydrodynamic Evolution of Supernova Remnants

When a star implodes as a supernova, a part of its gravitational binding energy is set free ejecting some fraction of the stellar mass as a shell expanding with a high velocity. Initial expansion velocities of 20 000 km s^{-1} or thereabout are observed. This ejected shell can be considered initially to expand into a vacuum, because its density is so much larger than that of the ambient density of interstellar gas. But if the analogy with terrestrial explosions is correct, we expect a shock to form when the shell has swept over a distance about equal to the mean free path in the surroundings.

However, what is the mean free path for the protons ejected by the supernova? Protons with $v = 20\,000$ km s^{-1} have a kinetic energy of 2 MeV and a range of about 500 pc if the mean density of interstellar neutral hydrogen is 1.2 cm^{-3} and if all possible loss mechanisms are taken into account. Obviously, no shock would form under such conditions. But protons from a supernova are charged particles and therefore they will gyrate around interstellar magnetic field lines. A magnetic flux density of $B = 3 \cdot 10^{-6}$ Gauss would result in a Larmor radius of only $R_L \cong 10^9$ m $\cong 10^{-8}$ pc and, even though the energy density of such a field is much too small to affect directly the outward motion of these protons, it will serve as a massless barrier between the supernova gas and the ionized interstellar material so that a hydromagnetic shock will form. The magnetic field is swept up by this process and is collected along with the gas by the outward moving shock front. Since the motion of the charged particles along the field lines is not affected, part of the kinetic energy contained in the relativistic particles will be able to seep through to greater distances. However, the details of the scattering processes for these particles are complicated and as yet are incompletely understood, so that the number of relativistic particles produced by a supernova and made available to the general interstellar field is still rather uncertain.

Three different stages may be distinguished in the evolution of a supernova remnant. In the *first stage* the mass of the gas swept up by the expanding shell is still less than the initial mass M_s:

$$\tfrac{4}{3}\pi r_s^3 \varrho_1 \leqq M_s \; . \tag{9.6}$$

The details of this phase are governed by the *SN* explosion. For $\varrho_1 \cong 2 \cdot 10^{-21}$ kg m^{-3}, $M_s = 0.25\, M_0 \cong 0.5 \cdot 10^{30}$ kg, $r_s \cong 1.3$ pc and this stage will last for about 60 years.

In the *second phase* the remnant is dominated by swept-up material, but radiative losses are still negligible compared to the total amount of energy that was produced by the supernova. This is therefore expansion under energy conservation and the evolution of the remnant is governed by the interaction between the high-energy particles in the remnant that put pressure on the shell and the surrounding material.

For a realistic treatment, numerical models are needed but many features of the time evolution can be described by the similarity solution of Sedov. Most of these can even be obtained by using still simpler arguments and first principles if some general properties of the similarity solutions are adopted. If the supernova explosion deposits the total energy E in the remnant, the similarity solution shows that the fraction $K_1 E$ is in the form of heat energy and the remainder in kinetic energy. This factor $K_1 = 0.72$ is a constant in the similarity solution and independent of time.

A second result taken from this solution is the ratio $K_2 = 2.13$ of the pressure immediately behind the shock to the mean pressure of the heated gas within the spherical volume enclosed by the shock.

These two factors permit us now to connect this pressure p_2 behind the shock with the total energy of the supernova explosion. For an ideal gas, the specific internal energy e is related to the gas pressure p and the specific volume v by

$$e = \tfrac{3}{2} p v \ .$$

Solving this for the mean pressure of the gas enclosed by the shock with the volume

$$V = \tfrac{4}{3} \pi r_s^3$$

we find

$$\boxed{\begin{aligned} p_2 &= K_2 \frac{2}{3} \frac{3 K_1 E}{4 \pi r_s^3} = \frac{K E}{2 \pi r_s^3} \\ K &= K_1 K_2 = 1.53 \end{aligned}} \ . \tag{9.7, 9.8}$$

The velocity U_s with which the shock is expanding into the surrounding gas of density ϱ_1 and temperature T_1 can be obtained from the "jump conditions" that relate physical quantities on each side of the front. If u_1 and u_2 are the streaming velocities in the gas, conservation of mass requires

$$\varrho_1 u_1 = \varrho_2 u_2 \tag{9.9}$$

while the conservation of momentum gives

$$p_1 + \varrho_1 u_1^2 = p_2 + \varrho_2 u_2^2 \ . \tag{9.10}$$

Since we assumed radiation losses to be negligible in this phase, the total energy E in the gas must be conserved during the expansion and the shock must be treated as an adiabatic transition. The third shock condition is then

$$\frac{1}{2} u_1^2 + \frac{p_1}{\varrho_1} + \frac{e_1}{\varrho_1} = \frac{1}{2} u_2^2 + \frac{p_2}{\varrho_2} + \frac{e_2}{\varrho_2}$$

which, for an ideal gas, becomes

$$u_1^2 + \frac{2\gamma}{\gamma - 1} \frac{p_1}{\varrho_1} = u_2^2 + \frac{2\gamma}{\gamma - 1} \frac{p_2}{\varrho_2} \tag{9.11}$$

where $\gamma = c_p/c_v = 5/3$ for a monoatomic gas. This γ here should not be confused with $\gamma = (1 - v^2/c^2)^{-1/2}$ of the special theory of relativity in Chap. 8.

From (9.9)–(9.11) we obtain after some algebra [see Landau-Lifschitz (1967)]

$$\frac{p_2}{p_1} = \frac{2}{\gamma + 1} \frac{\varrho_1}{p_1} u_1^2 - \frac{\gamma - 1}{\gamma + 1} \quad \text{and} \tag{9.12}$$

$$\frac{\varrho_1}{\varrho_2} = \frac{\gamma - 1}{\gamma + 1} + \frac{2\gamma}{\gamma + 1} \frac{p_1}{\varrho_1 u_1^2} . \tag{9.13}$$

For strong shocks with

$$p_2 \gg \frac{\gamma - 1}{\gamma + 1} p_1 \tag{9.14}$$

we find asymptotically

$$p_2 = \frac{2\varrho_1}{\gamma + 1} u_1^2 \quad \text{and} \tag{9.15}$$

$$\frac{\varrho_1}{\varrho_2} = \frac{\gamma - 1}{\gamma + 1} . \tag{9.16}$$

(9.16) is the largest possible jump of the densities for an adiabatic shock, that is, for $\gamma = 5/3$ we obtain $\varrho_2/\varrho_1 \lesssim 4$, even if the ratio of the pressures $p_2/p_1 \to \infty$.

For the expansion velocity of the shock we obviously must put $U_s = u_1$. Substituting (9.7) into (9.15), we find that

$$\boxed{U_s^2 = \frac{2KE}{3\pi \varrho_1 r_s^3}} . \tag{9.17}$$

The temperature behind the shock is obtained using the equation of state as well as (9.15) and (9.16):

$$T_2 = \frac{\mu m_H}{k} \frac{p_2}{\varrho_2} = \frac{\mu m_H}{k} \frac{2}{\gamma + 1} \frac{\varrho_1}{\varrho_2} u_1^2 = \frac{\mu m_H}{k} \frac{2}{\gamma + 1} \frac{\gamma - 1}{\gamma + 1} u_1^2$$

or

$$\boxed{T_2 = \frac{3}{16} \frac{\mu m_H}{k} U_s^2 = 0.061 \frac{\mu m_H}{k} \frac{E}{\varrho_1 r_s^3}} . \tag{9.18}$$

Here μ is the mean molecular weight per particle for fully ionized gas with a cosmic abundance ($N_H/N_{He} \cong 10$) of $\mu = 0.61$. Since $U_s = dr_s/dt$ Eq. (9.17) can be integrated resulting in

$$r_s = \left(\frac{5}{2}\right)^{2/5} \left(\frac{2KE}{3\pi \varrho_1}\right)^{1/5} t^{2/5} \quad \text{or}$$

$$\boxed{\frac{r_s}{\text{pc}} = 0.26 \left(\frac{N_H}{\text{cm}^{-3}}\right)^{-1/5} \left(\frac{t}{a}\right)^{2/5}} \tag{9.19}$$

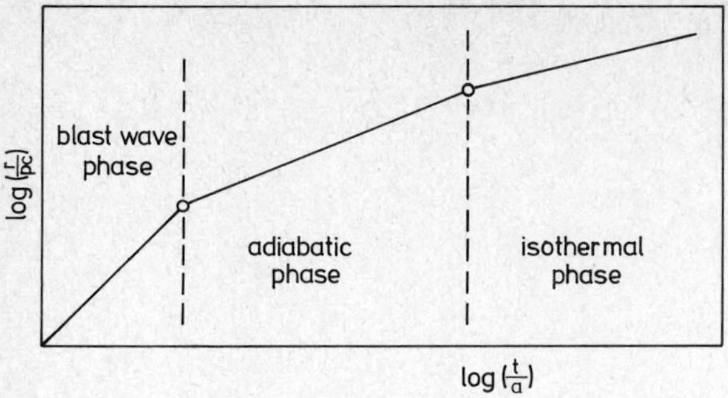

Fig. 9.7. The diameter of a supernova remnant as function of the time elapsed since the supernova explosion

for $E \cong 4 \cdot 10^{43}$ J $= 4 \cdot 10^{50}$ erg as a representative mean value (see also Fig. 9.7). Substituting this into (9.18) we obtain

$$\boxed{\frac{T_2}{\mathrm{K}} = 1.5 \cdot 10^{11} \left(\frac{N_\mathrm{H}}{\mathrm{cm}^{-3}}\right)^{-2/5} \left(\frac{t}{a}\right)^{-6/5}} \quad . \tag{9.20}$$

T_2 is the temperature right behind the shock; for smaller values of r/r_s, T/T_2 decreases as more detailed model computations show. The same applies to the density ϱ/ϱ_2. Consequently, most of the gas of the supernova remnant is collected in a thin layer in which practically all of the radiation, the radio emission as well as the X-rays, is emitted.

According to (9.20), T_2 is decreasing with time and, when it falls below 10^6 K, the abundant ions C, N and O are able to acquire bound electrons so that they become efficient cooling agents. The thermal energy of the supernova remnant will be radiated away in a short time so that the shock can no longer be treated as adiabatic.

We then enter into the *third phase.* Let t_rad be the time at which the remnant has radiated away half of the initial energy E ejected by the supernova. The cooling time then is so short that the matter behind the shock cools quickly and there are no longer any pressure forces to drive the shock. The shell will move at a constant radial momentum piling the swept-up interstellar gas like a snowplow. Constant radial momentum implies that

$$\tfrac{4}{3}\pi r_s^3 \varrho_1 U_s = \mathrm{const}$$

which can be integrated to

$$\boxed{r_s = r_\mathrm{rad} \left(\frac{8}{5}\frac{t}{t_\mathrm{rad}} - \frac{3}{5}\right)^{1/4}} \quad . \tag{9.21}$$

Here r_rad is the radius of the shell at the time t_rad when the adiabatic relation (9.19) ceases to be applicable. The supernova remnant will eventually be lost in the fluctuating density distribution of the interstellar gas when the expansion velocity of the shell approaches $U_s \cong 10$ km s^{-1}, a value close to the rms velocity dispersion of the interstellar gas.

The scenario of the evolution of a supernova remnant outlined here is very schematic only, and many of its features are not beyond dispute. In this the least of the problems are those posed by the semi-analytic approach. Precise numerical models like those by Chevalier (1974, 1975) give a picture not too different from that outlined here. More serious is that there are still doubts about which of the three phases the observed supernova remnants belong to: whether to the freely expanding first phase, the adiabatic Sedov phase or the isothermal phase. This is of great practical importance, especially if the important surface brightness-diameter relation which will be derived in the next section, is to be understood.

9.4 The Radio Evolution of Supernova Remnants

Supernova remnants emit radiation in the radio range and thus the question of their radio evolution is of importance. In numerical models the radio emission is computed along with the hydrodynamic evolution; here we will estimate this semi-analytically using ideas first presented by Shklovsky (1960).

As outlined in Sect. 9.2 we will assume that the radio emission of the remnant is synchrotron radiation so that the total flux of a source assumed to be optically thin is, according to (8.73)

$$\boxed{S_\nu = \frac{G}{d^2} V K B^{(1+\delta)/2} \nu^{-(\delta-1)/2}} \tag{9.22}$$

where V is the volume of the source, d its distance to the observer, B the average magnetic flux density, and ν the observing frequency. K is a constant appearing in the distribution function of the differential number density of the relativistic electrons

$$\boxed{N(E)\,dE = K E^{-\delta}\,dE} \tag{9.23}$$

per unit volume with energies between E and $E + dE$, and G is a constant depending on the spatial distribution of both the relativistic electrons and the magnetic field throughout the emitting volume. The constant K must somehow depend on the total power output of the supernova: the larger its power output the larger K probably is. But no specific relation between these two quantities is known until now.

The problem stated here is: How will S_ν evolve with time if the remnant expands according to (9.19) or (9.21)?

Following (9.20) the temperature T in the supernova remnant remains $> 10^4$ K throughout the whole evolution. Therefore the gas is strongly ionized and of great conductivity, so that any magnetic field remains "frozen" into the gas. If the gas expands, the magnetic field will as well and B will decrease. It is difficult to estimate the precise rate, but since expansion along the field lines should be without effect, a dependence

$$B(r) = B_0 \left(\frac{r_0}{r}\right)^2 \tag{9.24}$$

seems reasonable. The density of the high-energy electrons will be affected by this expansion too, and in two ways. If we assume that the energy output of the supernova has ceased, all relativistic electrons in the source have been formed, and we are concerned only with the fate of these during the evolution of the remnant.

This assumption of an initial formation of all relativistic electrons is an approximation that cannot be valid for some of the remnants, especially with regard to their X-ray radiation, because the loss rate $\dot{E}/E$ turns out to be so large that the mean life expectancy of the high-energy electrons must be much shorter than the observed life of, say, the Crab nebula. In this case, as well as for other objects, there obviously must exist some continuously flowing source of relativistic electrons. In the case of the Crab Nebulae this is probably connected somehow with the pulsar, but some other electron source is needed in objects where no pulsar is known.

But let us consider now the fate of existing high-energy electrons. These electrons are confined in a volume V whose expansion is caused by their pressure. If the radiation losses are negligible, this expansion is adiabatic, and the work done by the electrons is

$$dW = -p\,dV \qquad \text{where}$$

$$W = V \int E\,N(E)\,dE \ .$$

Now, for a relativistic gas

$$p = \frac{1}{3} e = \frac{1}{3}\frac{W}{V} = \frac{1}{3} \int E\,N(E)\,dE$$

and

$$\frac{dW}{W} = -\frac{1}{3}\frac{dV}{V} \ .$$

But, since for a relativistic gas we have

$$\frac{dW}{W} = \frac{dE}{E} \ ,$$

spherical expansion

$$\frac{V}{V_0} = \left(\frac{r}{r_0}\right)^3 \tag{9.25}$$

then results in

$$\frac{dE}{E} = -\frac{dr}{r} \ . \tag{9.26}$$

If now the number of relativistic electrons remains constant in $V(r) = V_0\,(r/r_0)^3$ although the energy of each electron decreases as

$$E(r) = E_0\,\frac{r_0}{r}\,, \tag{9.26 a}$$

we obviously must have

$$V_0 \int_{E_1}^{E_2} K_0\,E^{-\delta}\,dE = V_0\left(\frac{r}{r_0}\right)^3 \int_{E_1 r_0/r}^{E_2 r_0/r} K(r)\,E^{-\delta}\,dE$$

or

$$\boxed{\frac{K(r)}{K_0} = \left(\frac{r}{r_0}\right)^{-(2+\delta)}}\;. \tag{9.27}$$

The exponent δ remains constant during adiabatic expansion

$$\delta = \delta_0\;. \tag{9.28}$$

Substituting (9.24) to (9.28) into (9.22) we obtain

$$\boxed{S_\nu(r) = S_\nu(r_0)\left(\frac{r}{r_0}\right)^{-2\delta}} \tag{9.29}$$

or, if the time dependence of r in the adiabatic or Sedov phase according to (9.19) is substituted,

$$\boxed{S_\nu(t) = S_\nu(t_0)\left(\frac{t}{t_0}\right)^{-4\delta/5}}\;. \tag{9.30}$$

This then is the time evolution of a supernova remnant. In differential form (9.30) becomes

$$\boxed{\frac{\dot{S}_\nu}{S_\nu} = -\frac{4}{5}\,\frac{\delta}{t}} \tag{9.31}$$

and it relates the spectral index $\alpha = (\delta - 1)/2$ to the time-dependence of the source flux.

This relation can be tested. For Cas A with $\alpha = 0.77$, $\delta = 2.54$, and $t \cong 300$ years an annual decrease $S_\nu/S_\nu = -0.7\%$ is predicted, while the observed $\dot{S}_\nu/S_\nu = -(1.3 \pm 0.1)\%$ is in reasonable agreement.

Both (9.29) and (9.30) compare the fluxes that a single supernova remnant has at different times (9.30) or at different radii (9.29). If we want to use (9.29) to compare different supernova remnants $S_\nu^{(a)}$ and $S_\nu^{(b)}$ we have to form

$$\frac{S_\nu^{(a)}(r_1)}{S_\nu^{(b)}(r_2)} = \frac{S_\nu^{(a)}(r_0)}{S_\nu^{(b)}(r_0)}\left(\frac{r_1}{r_2}\right)^{-2\delta} \tag{9.32}$$

and $S_\nu^{(a)}(r_0)/S_\nu^{(b)}(r_0)$, the ratio of the fluxes of these two sources at identical radii, would have to be known. But as stated earlier, the total radio flux of a source depends in a yet unknown way on the total energy output of the supernova, and, amongst other things, on the density of the interstellar medium into which the source expands. Thus the ratio is a priori unknown. Experience shows however that, for many purposes, sufficient precision is achieved if we put $S_{\nu 0}^{(a)}/S_{\nu 0}^{(b)} = 1$, and thus interprete (9.29) as the relation between different supernova remnants. In principle both S_ν and r of a SNR are observable, but both depend on the distance d to the SNR, which is usually not a well-determined quantity. Thus errors in d can introduce a spurious correlation between S_ν and r.

To avoid this the surface brightness is introduced by

$$\Sigma_\nu = \frac{S_\nu}{\pi r^2}$$

and this is independent of the distance. (9.29) then is modified to

$$\boxed{\Sigma_\nu(r) = \Sigma_\nu(r_0)\left(\frac{r}{r_0}\right)^{-2(\delta+1)}} \;. \tag{9.33}$$

This relation will be compared with the data for SNR with well-determined distances following a compilation by Caswell and Lerche (1979 a). They obtain

$$\frac{\Sigma_\nu}{\mathrm{W\,m^{-2}\,Hz^{-1}\,sr^{-1}}} = 10^{-15}\left(\frac{D}{\mathrm{pc}}\right)^{-3.0} \exp\left\{-\left(\frac{z}{\mathrm{pc}}\right)\Big/175\right\} \tag{9.34}$$

$$\frac{D}{\mathrm{pc}} = 0.93\left(\frac{t}{a}\right)^{2/5} \exp\left\{\frac{z}{\mathrm{pc}}\Big/900\right\} \quad \text{and} \tag{9.35}$$

$$\frac{\Sigma_\nu}{\mathrm{W\,m^{-2}\,Hz^{-1}\,sr^{-1}}} = 1.25\cdot 10^{-15}\left(\frac{t}{a}\right)^{-6/5} \exp\left\{-\left(\frac{z}{\mathrm{pc}}\right)\Big/110\right\} \tag{9.36}$$

where $D = 2r$ is the linear diameter of the remnant, z is the height of the SNR above the galactic plane, and the factor $\exp\{-z/z_0\}$ takes into account the decrease of the gas density in the interstellar medium with z.

These empirical relations should be compared with the theoretical relations (9.19) and (9.33). The mean spectral index of the SNR is $\bar{\alpha} = 0.46 \pm 0.13$, which according to (8.66) and (9.33) would require an exponent n in the $\Sigma_\nu \sim D^n$ relation of $n = -5.8 \pm 0.52$; the observed value is in (9.34) $n = -3$, however. On the other hand, D (or r) depends on the ambient gas density by only $N_\mathrm{H}^{-1/5}$ (9.19) and, since the factor $\exp\{-z/175\}$ is roughly proportional to $N_\mathrm{H}(z)$, the $\exp\{z/900\} \sim N_\mathrm{H}(z)^{-1/5}$ is of the right magnitude.

It is, however, not at all difficult to devise models that result in a different power n because many assumptions enter into the derivation of (9.29). Caswell and Lerche (1979) showed that, if we change the variation of the magnetic field with r from (9.24) to r^{-a}, change the variation of the energy of a single electron with r from (9.26) to r^{-b}, let the total number of electrons be conserved in a volume that varies as r^c, and let the emitting volume of the source vary as r^d instead of (9.25) then

$$\boxed{S_\nu(r) = S_\nu(r_0)\left(\frac{r}{r_0}\right)^n} \quad , \tag{9.37}$$

$$\boxed{n = d - c - b(\delta - 1) - \frac{a}{2}(\delta + 1)} \quad , \quad \text{and} \tag{9.38}$$

$$\boxed{\Sigma_\nu(r) = \Sigma_\nu(r_0)\left(\frac{r}{r_0}\right)^{n-2}} \quad . \tag{9.39}$$

The observed value $n = -3.0$ could then be reconciled with large variety of different configurations, but no definite choice is possible yet. In addition, modern observations seem even to cast doubts on the existence of a unique $\Sigma - D$ relation.

10. Line Radiation Fundamentals

10.1 The Einstein Coefficients

If local thermodynamic equilibrium prevails, the intensity of emitted and absorbed radiation is not independent but related to each other by Kirchhoff's law (1.14). This not only applies for continuous radiation but is just as valid for line radiation. The Einstein coefficients give the convenient means to describe the interaction of radiation with an atomic system by emission and absorption of photons.

Consider a cavity containing atoms with discrete energy levels E_i. According to Einstein (1916) an atom in the excited level E_2 will return spontaneously to the lower level E_1 with a certain probability A_{21} such that $N_2 A_{21}$ is the number of such spontaneous transitions per second in a unit volume element if N_2 is the density of the atoms in the state E_2 (Fig. 10.1). Now both the states E_2 and E_1 are not infinitesimally sharp but have a certain finite energy width so that $E_2 - E_1$ can have a certain energy distribution. Converting this into frequencies using $E_2 - E_1 = h\nu$, the absorption line will be described by a line profile function $\varphi(\nu)$ which is sharply peaked at ν_0 and usually normalized so that

$$\int_0^\infty \varphi(\nu)\,d\nu = 1 \ . \tag{10.1}$$

If the intensity of the radiation field is I_ν (for a definition see Sect. 1.3) we can define an average intensity by

$$\bar{I} = \int_0^\infty I_\nu \varphi(\nu)\,d\nu \tag{10.2}$$

and the probability of an absorption of a photon then is $B_{12}\bar{U}$ such that the number of absorbed photons is $N_1 B_{12}\bar{U}$ where $\bar{U} = 4\pi\bar{I}/c$ is the average energy density of the radiation field. Einstein found that to derive Planck's law another

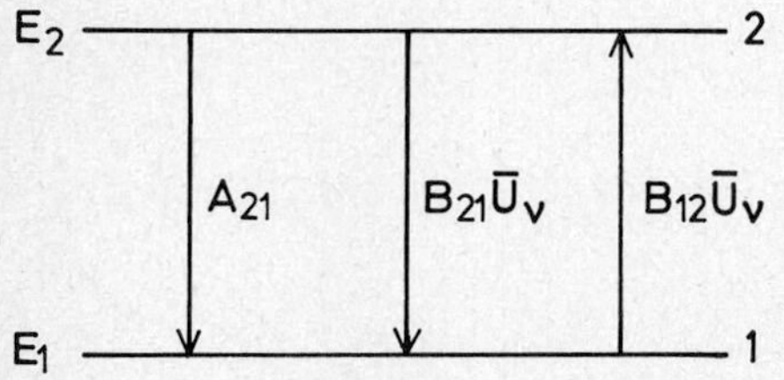

Fig. 10.1. Transitions between the states 1 and 2 and the Einstein probabilities

emission process proportional to $\bar{U}$ was needed, and so he put $N_2 B_{21} \bar{U}$ equal to the number of photons emitted by stimulated emission.

Assuming now the system to be in a stationary state, the number of absorbed and emitted photons must be equal, so that

$$\boxed{N_2 A_{21} + N_2 B_{21} \bar{U} = N_1 B_{12} \bar{U}} \quad . \tag{10.3}$$

These Einstein coefficients A_{21}, B_{21} and B_{12} are not independent of each other as can be seen if we consider a system in full thermodynamic equilibrium. The atoms in the cavity then are not all in one state but are distributed over different states so that the different atomic levels are populated according to the Boltzmann distribution

$$\frac{N_2}{N_1} = \frac{g_2}{g_1} \exp\left(-\frac{h\nu_0}{kT}\right) \tag{10.4}$$

where g_1 and g_2 are the statistical weights of the states and T is the absolute temperature of the cavity. Solving (10.3) for $\bar{U}$

$$\bar{U} = \frac{A_{21}}{\frac{N_1}{N_2} B_{12} - B_{21}} , \tag{10.5}$$

and we obtain with (10.4) the expression

$$\bar{U} = \frac{A_{21}}{\frac{g_1}{g_2} \exp\left\{\frac{h\nu_0}{kT}\right\} B_{12} - B_{21}} . \tag{10.6}$$

But in thermodynamic equilibrium we know that $\bar{U}$ must be given by the Planck function (1.13) according to

$$\bar{U} = \frac{4\pi}{c} B_\nu(T) = \frac{8\pi h\nu_0^3}{c^3} \frac{1}{\exp\left\{\frac{h\nu_0}{kT}\right\} - 1} . \tag{10.7}$$

These two expressions (10.6) and (10.7) obviously should be identical, and this is so only if

$$\boxed{g_1 B_{12} = g_2 B_{21}} \quad \text{and} \tag{10.8}$$

$$\boxed{A_{21} = \frac{8\pi h\nu_0^3}{c^3} B_{21}} \quad . \tag{10.9}$$

No reference is made in these relations to any thermodynamic property of the cavity. They must therefore be valid conditions for atomic properties independent of whether these atoms are in an environment with thermodynamic equilibrium or not.

10.2 Radiative Transfer with Einstein Coefficients

When the radiative transfer was considered in Sect. 1.4, the material properties were condensed into the emission coefficient ε_ν and the absorption coefficient κ_ν. Both ε_ν and κ_ν are phenomenological parameters; for a physical theory they have to be related to atomic properties of the matter in the cavity. If now line radiation is considered, the Einstein coefficients as introduced in Sect. 10.1 turn out to be very useful because they can be linked in a direct way to the atomic physics of the transition responsible for the line investigated. But for the radiative transfer ε_ν and κ_ν are needed and so we have to investigate the relation between say κ_ν and the A_{ik} and B_{ik}. This is best done by considering the possible changes of a beam of intensity I_ν in a slab of material of thickness ds just as in Sect. 1.4 but now using A_{ik} and B_{ik}.

According to Einstein there are three different processes contributing to the intensity I_ν. Each atom returning from E_2 to E_1 contributes the energy $h\nu_0$ distributed over the full solid angle 4π so that the total amount of energy emitted *spontanteously* is

$$dE_e(\nu) = h\nu_0 N_2 A_{21} \varphi_e(\nu)\, dV \frac{d\Omega}{4\pi} d\nu\, dt \ . \tag{10.10}$$

For the total energy *absorbed* we similarly obtain

$$dE_a(\nu) = h\nu_0 N_1 B_{12} \frac{4\pi}{c} I_\nu \varphi_a(\nu)\, dV \frac{d\Omega}{4\pi} d\nu\, dt \tag{10.11}$$

and for the *stimulated* emission

$$dE_s(\nu) = h\nu_0 N_2 B_{21} \frac{4\pi}{c} I_\nu \varphi_e(\nu)\, dV \frac{d\Omega}{4\pi} d\nu\, dt \ . \tag{10.12}$$

The line profiles $\varphi_a(\nu)$ and $\varphi_e(\nu)$ for absorbed and emitted radiation could be different, but in astrophysics it is usually permissable to put $\varphi_a(\nu) = \varphi_e(\nu) = \varphi(\nu)$. For the volume element we put $dV = d\sigma\, ds$, and so we find for a stationary situation

$$\begin{aligned} & dE_e(\nu) + dE_s(\nu) - dE_a(\nu) = dI_\nu\, d\Omega\, d\sigma\, d\nu\, dt \\ & = \frac{h\nu_0}{4\pi}\left[N_2 A_{21} + N_2 \mathrm{B}_{21}\frac{4\pi}{c} I_\nu - N_1 B_{12}\frac{4\pi}{c} I_\nu\right]\varphi(\nu)\, d\Omega\, d\sigma\, ds\, d\nu\, dt \end{aligned}$$

resulting in the equation of transfer with Einstein coefficients

$$\boxed{\frac{dI_\nu}{ds} = -\frac{h\,\nu_0}{c}(N_1\,B_{12} - N_2\,B_{21})\,I_\nu\,\varphi(\nu) + \frac{h\,\nu_0}{4\,\pi}\,N_2\,A_{21}\,\varphi(\nu)} \quad . \tag{10.13}$$

Comparing this with (1.9) we obviously obtain agreement only by putting

$$\boxed{\kappa_\nu = \frac{h\,\nu_0}{c}\,N_1\,B_{12}\left(1 - \frac{g_1\,N_2}{g_2\,N_1}\right)\varphi(\nu)} \tag{10.14}$$

and

$$\boxed{\varepsilon_\nu = \frac{h\,\nu_0}{4\,\pi}\,N_2\,A_{21}\,\varphi(\nu)} \tag{10.15}$$

where we used (10.8) to relate B_{12} and B_{21}. The term in brackets in (10.14) is the correction for stimulated emission. In many situations this correction is negligible, but there are circumstances in radio astronomy where the stimulated emission almost completely cancels the effect of the true absorption. How this comes about is best seen if we investigate what becomes of (10.13)–(10.15) if the condition of local thermodynamic equilibrium (LTE) is imposed.

From (10.14) and (10.15) we find by substituting (10.8) and (10.9) that

$$\frac{\varepsilon_\nu}{\kappa_\nu} = \frac{2\,h\nu^3}{c^2}\left(\frac{g_2\,N_1}{g_1\,N_2} - 1\right)^{-1} .$$

But for LTE, according to (1.14), this should be equal to the Planck function (1.13), resulting in

$$\boxed{\frac{N_2}{N_1} = \frac{g_2}{g_1}\exp\left\{-\frac{h\,\nu_0}{k\,T}\right\}} \quad . \tag{10.16}$$

The energy levels of matter in local thermodynamic equilibrium are thus populated according to the same Boltzmann distribution (10.4) for the temperature T that applied to full thermodynamic equilibrium, and the absorption coefficient becomes

$$\boxed{\kappa_\nu = \frac{c^2}{8\,\pi\,\nu_0^2}\,\frac{1}{}\frac{g_2}{g_1}\,N_1\,A_{21}\left(1 - \exp\left\{-\frac{h\,\nu_0}{k\,T}\right\}\right)\varphi(\nu)} \tag{10.17}$$

where we have put

$$B_{12} = \frac{g_2}{g_1}\,A_{21}\,\frac{c^3}{8\,\pi\,h\,\nu^3}$$

Table 10.1. Physical line parameters at different frequencies. Column (2) gives the frequency, column (3) $T_0 = h\nu/k$ and column (4) the temperature at which the correction for stimulated emission is just 10%

Line	$\frac{\nu}{\mathrm{Hz}}$	$\frac{h\nu}{k}/\mathrm{K}$	$T_{10\%}/\mathrm{K}$
Ly cont.	$3.29 \cdot 10^{15}$	$1.58 \cdot 10^5$	$6.9 \cdot 10^4$
H_α	$4.57 \cdot 10^{14}$	$2.19 \cdot 10^4$	$9.5 \cdot 10^3$
21 cm	$1.42 \cdot 10^9$	$6.82 \cdot 10^{-2}$	$3.0 \cdot 10^{-2}$

which is derived from (10.8) and (10.9). In (10.17) the expression in brackets is again the correction for stimulated emission. Now

$$\frac{h}{k} = 4.799\,27\,(15) \cdot 10^{-11}\,\mathrm{K\,Hz^{-1}} \tag{10.18}$$

and so we obtain the values in Table 10.1.

For radiation in the ultraviolet and visual range the correction for stimulated emission $1 - \exp\{-h\nu/kT\} \cong 1$ even for $T \cong 10^4$ K; only in high-temperature environments must stimulated emission be taken into account. This is different in the radio range. Because usually $h\nu/kT \ll 1$ it is sufficient to use the first term of the Taylor series

$$1 - \exp\left\{-\frac{h\nu}{kT}\right\} \cong h\nu/kT - \tfrac{1}{2}(h\nu/kT)^2 + \dots\ . \tag{10.19}$$

Thus stimulated emission here cancels most of the absorption. Only for molecular line radiation in the mm-wave range of low-temperature regions ($T < 10$ K) might some of the higher terms in (10.19) or even the full exponential function be necessary. The last column in Table 10.1 gives the temperature at which the correction for stimulated emission amounts to 10%, that is, at which $\exp\{-h\nu/kT\} = 0.1$.

10.3 Dipole Transition Probabilities

The simplest sources for electromagnetic radiation are oscillating dipoles. Radiating electric dipoles have already been treated classically in Chap. 4, but it should also be possible to express these results in terms of the Einstein coefficients. There are two kinds of dipoles that can be treated by quite similar means: the electric and the magnetic dipole.

Electric Dipole:

Consider an oscillating electric dipole

$$d(t) = e\,x(t) = e\,x_0 \cos\omega t\ . \tag{10.20}$$

According to electrodynamic theory it will radiate and the power emitted by it into the full sphere of 4π steradians is, according to (8.4),

$$P(t) = \frac{e^2}{6\pi\varepsilon_0 c^3}\dot{v}(t)^2 \ . \tag{10.21}$$

Expressing $x = d/e$ and $\dot{v} = \ddot{x}$, we obtain for the mean power emitted over a full oscillation

$$\langle P \rangle = \frac{16\pi^3}{3\varepsilon_0 c^3}\nu_{mn}^4\left(\frac{e x_0}{2}\right)^2 \ . \tag{10.22}$$

But this mean emitted power can also be expressed in terms of the Einstein probabilities

$$\langle P \rangle = h\nu_{mn} A_{mn} \ . \tag{10.23}$$

Equating (10.22) and (10.23) we obtain

$$\boxed{A_{mn} = \frac{16\pi^3}{3\varepsilon_0 h c^3}\nu_{mn}^3 |\mu_{mn}|^2} \tag{10.24}$$

where

$$\mu_{mn} = \frac{e x_0}{2} \tag{10.25}$$

is the mean electric dipole moment of the oscillator for this transition.

Strictly speaking, this expression (10.24) is applicable only to classical electric dipole oscillators but for order-of-magnitude estimates we can use it for atomic systems too. Thus we find a transition probability for atomic hydrogen close to the Lyman-limit of about $10^9\ \mathrm{s}^{-1}$ by putting $x_0 = a_0 = 5.29 \cdot 10^{-11}$ m = Bohr radius, $\nu_{mn} = c R_\infty = 3.29 \cdot 10^{15}$ Hz = frequency at the Lyman limit, and $\mu_{mn} = e a_0/2 = 4.24 \cdot 10^{-30}\ Cm$.

Magnetic Dipole:

For a magnetic dipole

$$m(t) = m_0 \cos\omega t \ , \tag{10.26}$$

the corresponding Einstein coefficient can be computed quite similarly resulting in

$$A_{mn} = \frac{16\pi^3\mu_0}{3hc^3}\nu_{mn}^3 |\mu_{mn}^*|^2 \tag{10.27}$$

where μ^*_{mn} is the mean magnetic dipole moment of the oscillator for this transition. If we again apply this relation to the hydrogen atom and compute

$$|\mu^*_{mn}| = \frac{e\,\hbar}{2\,m_e} \cong 9.3 \cdot 10^{-24}\,\mathrm{A\,m^2} \tag{10.28}$$

in terms of the magnetic moment of the lowest Bohr orbit we obtain

$$A_{mn} \cong 10^4\,\mathrm{s}^{-1}\;. \tag{10.29}$$

The transition probability for a magnetic dipole is thus smaller than that of an electric dipole by a factor of 10^5 provided all other parameters of the two dipoles are identical. The reason is that the typical dipole moment of a magnetic dipole is a factor 692 smaller than that of an equivalent electric dipole.

The Einstein coefficients for transitions in atomic systems in which the electric dipole moment is changing are therefore much larger than those for transitions in which only the magnetic dipole moment or the electric quadrupole moment is changing. Electric dipole transitions therefore are called "allowed", while the others are termed "forbidden".

10.4 The Rate Equation and Some Simple Solutions

In order to compute the absorption or emission coefficients for a selected transition according to (10.14) and (10.15), both the Einstein coefficients and the number densities N_i and N_k must be known. In case of local thermal equilibrium (LTE), the radio of N_i to N_k is given by the Boltzmann-function (10.16) of the local temperature leading thus to (10.17). But if LTE does not apply, the individual processes that lead to the population or depopulation of an energy level have to be considered. Usually such processes involve not only the two levels in question but the whole system of all transitions.

Let R^y_{jk} be the transition probability for the transition $j \to k$ caused by the process y and let N_j be the number density of atoms in the state j. Then obviously

$$\frac{dN_j}{dt} = -\,N_j \sum_k \sum_y R^y_{jk} + \sum_k N_k \sum_y R^y_{kj}\;. \tag{10.30}$$

For a stationary situation obviously $dN_j/dt = 0$. Depending on which processes are causing the transitions, the solution of (10.30) can be rather complicated. We will consider here only two simple cases; a slightly more complicated situation will be met in the case of radio-recombination lines in Chap. 12.

Consider first the case of atoms with two states 0 and 1 where the only way to change state 0 into state 1 and vice-versa is by emission and absorption of radiation. We do not, however, assume that LTE applies. The only transition rates in (10.30) then are given by the Einstein coefficients and for a stationary situation we obviously must have

$$N_0\,B_{01}\,\bar{U} = N_1\,(A_{10} + B_{10}\,\bar{U}) \tag{10.31}$$

where $\bar{U}$ is given by (10.2). $\bar{U}$ is a single number and can be formally expressed by a "brightness temperature T_b" given by

$$\bar{U} = \frac{4\pi}{c}\bar{I} = \frac{8\pi h \nu_0^3}{c^3} \frac{1}{\exp\left\{\frac{h\nu_0}{k T_b}\right\} - 1} \tag{10.32}$$

or

$$T_b = \frac{h\nu_0}{k} \frac{1}{\ln\left(\frac{8\pi h \nu_0^3}{c^3 \bar{U}} + 1\right)} . \tag{10.33}$$

From (10.31) we then obtain using (10.9)

$$\frac{N_1}{N_0} = \frac{B_{01}}{B_{10}} \frac{\bar{I}}{\frac{2h\nu_0^3}{c^2} + \bar{I}} = \frac{g_1}{g_0} \exp\left(-\frac{h\nu_0}{k T_b}\right) . \tag{10.34}$$

The number densities N_1 and N_0 thus will be described by a Boltzmann distribution as in the case of LTE. But the temperature T_b in this distribution describes only the radiation density at the frequency corresponding to the transition $1 \to 0$ and it need have nothing to do with a thermodynamic temperature of the system.

Another simplification of (10.30) is possible if it is used to describe the number densities of the cool parts of the interstellar gas. Let us assume that only the two lowest states are populated to any extent, but it is not radiation that governs the transition rates but mainly collision processes.

Let C_{12} and C_{21} be the collision rates of the transitions $1 \to 2$ and $2 \to 1$, respectively; then the rate equation (10.30) for a stationary situation will be

$$N_1 (C_{12} + B_{12}\bar{U}) = N_2 (A_{21} + B_{21}\bar{U} + C_{21}) . \tag{10.35}$$

The collision rate obviously depends on the number of colliding particles, N:

$$C_{ik} = N\gamma_{ik} = N \int_0^\infty \sigma_{ik}(v)\, v f(v)\, dv \tag{10.36}$$

where γ_{ik} is the collision probability per colliding particle, σ_{ik} the collision cross-section and $f(v)$ the velocity distribution function of the colliding particles. Using arguments similar to those leading to (10.16), the principle of detailed balancing leads to

$$g_m \gamma_{mn} = g_n \gamma_{nm} \exp\left\{-\frac{h\nu_{mn}}{k T_K}\right\} \tag{10.37}$$

where T_K is a formal temperature describing the velocity distribution

$$f(v) = \left(\frac{2}{\pi}\right)^{1/2} v^2 \left(\frac{m_r}{k T_K}\right)^{3/2} \exp\left\{-\frac{m_r v^2}{2 k T_K}\right\} \tag{10.38}$$

and

$$m_r = \frac{m_a\, m_b}{m_a + m_b}$$

is the reduced mass of the colliding particles. T_K is usually called the *kinetic temperature.*

Substituting (10.36) and (10.37) together with (10.32) into (10.35) we obtain

$$\begin{aligned}\frac{N_2\, g_1}{N_1\, g_2} &= \exp\left(-\frac{h\nu}{k T_{ex}}\right)\\ &= \exp\left(-\frac{h\nu}{k T_b}\right)\frac{A_{21} + C_{21}\exp\left(-\frac{h\nu}{k T_K}\right)\left[\exp\left(\frac{h\nu}{k T_b}\right)-1\right]}{A_{21} + C_{21}\left[1-\exp\left(-\frac{h\nu}{k T_b}\right)\right]}\ .\end{aligned} \tag{10.39}$$

If we therefore describe N_2/N_1 by a formal excitation temperature T_{ex} defined as

$$\frac{N_2}{N_1} = \frac{g_2}{g_1}\exp\left(-\frac{h\nu}{k T_{ex}}\right) \tag{10.40}$$

then this excitation temperature is some weighted mean between the radiation temperature T_b and the kinetic temperature T_K. If radiation is dominating the rate equation ($C_{21} \ll A_{21}$), then (10.39) → (10.34) and $T_{ex} \to T_b$. If on the other hand collisions are dominating ($C_{21} \gg A_{21}$) then $T_{ex} \to T_K$. Since C_{ik} increases with increasing N collisions will dominate the distribution in high-density situations and the excitation temperature of the line will be equal to the kinetic temperature. In low density situations $T_{ex} \to T_b$. The density at which this transition from T_b to T_K occurs is called the *thermalization density*. The smaller A_{21}, the lower it is. We will return to this point later in Chap. 12.

10.5 The Radial Velocity

If the observed frequency of a line is compared to the known rest frequency, the relative radial velocity of the line emitting (or absorbing) source and the receiving system can be determined. But this velocity contains the motion of the source as well as that of the receiving system. Both are measured relative to some conveniently chosen standard of rest and usually only the motion of the source is of interest.

The velocity of the receiving system will conveniently be separated into several independent components.

1. Earth Rotation

Due to the rotation of the earth the receiving system moves with a velocity $v = 0.46510 \cos \varphi$ km s^{-1} into the direction straight East in the horizontal coordinate system. Here φ is the geographic latitude of the observing station. If the contribution of this velocity is subtracted the resulting radial velocity is said to refer to the *geocentric* system.

2. The Motion of the Center of the Earth Relative to the Barycenter of the Solar System

If this contribution is eliminated the radial velocity is said to be reduced to the *heliocentric* system. This velocity of the earth could be computed from the anually published Astronomical Ephemeris, but due to the many effects that have to be taken into account, this is a complicated procedure. For low precision purposes appropriate for stellar radial velocity data in the optical spectral range, suitable tables have been published by Herrick (1935) and for early low-precision 21 cm-observations by MacRae and Westerhout (1956).

Today convenient computer algorithms are available in most observatories. One that is rather well documented and for which the theory, error discussion (Stumpff 1979) and the FORTRAN code (Stumpff 1980) have been published is the program BARVEL. In it the perturbed elliptic motion of the Earth-Moon barycenter, the transformation of the Earth-Moon barycenter to the center of the Earth and finally the transition from the sun to the center-of-mass of the Solar System, that is, the perturbed motion of the planets, have all been included. The velocity vector computed by BARVEL deviates at most by 42 cm s^{-1} from such a vector obtained from the best available ephemerides. The motion of the center-of-mass of the Solar System can be considered to be as close to an inertial system as we can hope to come, so there is no physical reason to transform the observed radial velocities any further. Stellar radial velocity observations and practically all extragalactic work is therefore usually published in this system.

In galactic work it is, however, convenient to give the radial velocities in a system such that the gas of the solar neighbourhood shows as little motion as possible. Therefore the motion of the center-of-mass of the Solar System relative to the local gas has to be determined. For this, neutral hydrogen gas as given by the 21 cm-line both in emission and in absorption is best suited, but stellar data can be used too.

The result obtained by many independent investigations [for a summary see Crovisier (1978)] all show that the Solar System moves with a velocity given by the standard solar motion ($v_0 = 20$ km s^{-1} towards $\alpha_{1900} = 18^{\mathrm{h}}, \delta_{1900} = +30°$). This is the solar motion relative to stars most commonly listed in general catalogues of radial velocity and proper motion; these stars are mostly of spectral type A through G, and there is no obvious physical interpretation for this motion.

Observations from which the standard solar motion has been eliminated are said to refer to the "Local Standard of Rest" (LSR). This is a point coinciding with the position of the sun but moving with the local circular velocity around the galactic center. Sometimes it may be advantageous to refer velocities to a system in which the galactic center is at rest. The required correction obviously depends on the adopted galacted circular velocity of the LSR. For many years the value $\Theta_0 = 250 \text{ km s}^{-1}$ as proposed by IAU convention has been used; recently slightly smaller values of $\Theta_0 = 220-230 \text{ km s}^{-1}$ are preferred and even values as low as $\Theta_0 = 185$ have been proposed. The rotational velocity vector is always directed towards $l = 90°$, $b = 0°$.

10.6 Observing the Line

In radio astronomy the line radiation is always only a small fraction of the total power received. It sits on top of a large pedestal of wideband noise signals contributed by different sources: the system noise, spillover from the antenna and in some cases, a true background noise. To avoid the stability problems encountered in total power systems (see Chap. 7) this signal must be compared with another signal that contains the same total power and differs from the first only by the fact that it contains no line radiation. To achieve this aim modern receivers usually permit three different observing modes that differ only in the way in which the comparison signal is produced.

1. Total Power Mode. The received signal "on source" is compared with another signal obtained at a closeby position on the sky. If the receiver is stable so that any gain changes and changes in the band-pass for the signal occur only over time scales which are long compared to the time scale of this position change, and if there are good chances that there is indeed only little line radiation at the comparison region, then this method is efficient and produces excellent line profiles. This method is especially advantageous if baseline ripple is a problem since this ripple will cancel quite well using this method provided that the continuum signal at both positions is similar.

2. Dicke Switch. If the time-scale of the receiver-stability is too short for position switching in the total power mode, the receiver can be connected alternately to the antenna and to a matched resistive load. By noise injection the output power in both switch positions can be equalized, and then the difference of the signals is the line radiation. A good noise balance for both switch positions is essential if good results are to be achieved by this method (see also Sect. 7.3.5 for a discussion of the receiver stability achieved by this).

3. Frequency Switching (Fig. 10.2). Another method to obtain a comparison signal which differs from the measured signal only by the line radiation makes use of the fact that the line radiation is a narrow-band feature while all other signals change very little over large bandwidths. Retuning the receiver over a few MHz therefore produces a comparison signal without line radiation while the other noise power contributions hardly differ. This frequency-switching can be done with almost any chosen speed.

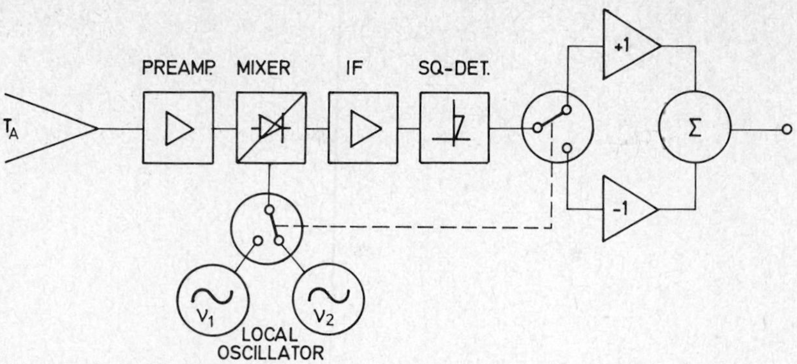

Fig. 10.2. Block diagram of a receiver using frequency switching

4. The Baseline. The result of any of the three observing methods with a radio spectrograph should be a spectrum of the line in which $T_A(\nu) \rightarrow 0$ for ν outside the frequency range of the line. However, quite often this is not so because signal and reference were not balanced well enough, and then a "baseline" must be subtracted from the measurements. Often a linear function of the frequency is sufficient for this, but sometimes some curvature is visible so that polynomials of second or even higher degree have to be subtracted. This should, however, be done with great care because such high-order polynomials can easily "explode" and introduce spurious effects.

On many occasions a sinusoidal baseline ripple with amplitudes of a few 0.1 K is visible. This ripple appears because quite often a small fraction of the signal is reflected off the aperture plane of the feed horn back to the paraboloid and from there back again to the feed. The two waves moving in parallel direction towards the receiver frontend will interfere and produce a standing-wave pattern, the relative phase of which will depend on the pathlength in units of λ of the reflected wave.

A full phase-change of 2π will occur if the frequency of the signal is changed by

$$\Delta\nu = \frac{c}{2f} \tag{10.41}$$

where f is the focal length of the paraboloid.

For the 100 m telescope at Effelsberg with $f \approx 30$ m, $\Delta\nu \approx 5$ MHz is obtained for the "wavelength" of the baseline ripple in Fig. 10.4. If some of the reflected radiation is reflected twice, a ripple with $\Delta\nu = 2.5$ MHz should be present and indeed such a ripple with $\Delta\nu = 5$ MHz and harmonics thereof can be detected in observed line profiles.

There are several possible sources of the reflected radiation. One is the frontend of the receiver that injects into the antenna some noise power, part of which then is reflected back. Another source is strong continuum radiation from cosmic sources. But in both cases the partial reflection of the radiation in the horn aperture are the main cause of the baseline ripple (Figs. 10.3 and 10.4). Experience

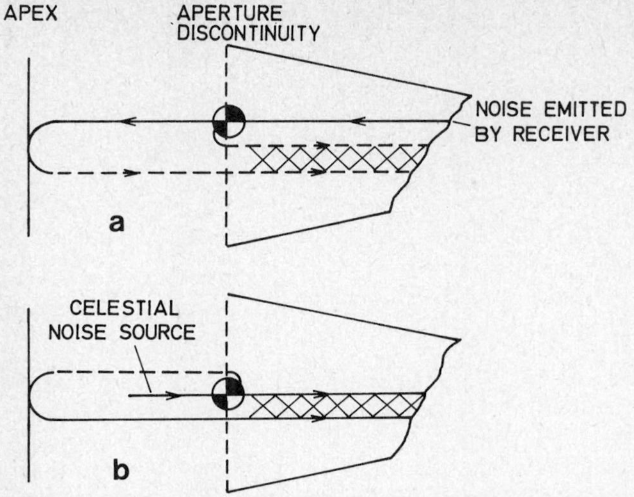

Fig. 10.3 a, b. Formation of baseline ripple by a discontinuity of the waveguide structure in the feed horn: (**a**) noise emitted by the first stages of the receiver and the input structure (**b**) noise from a strong source reflected back

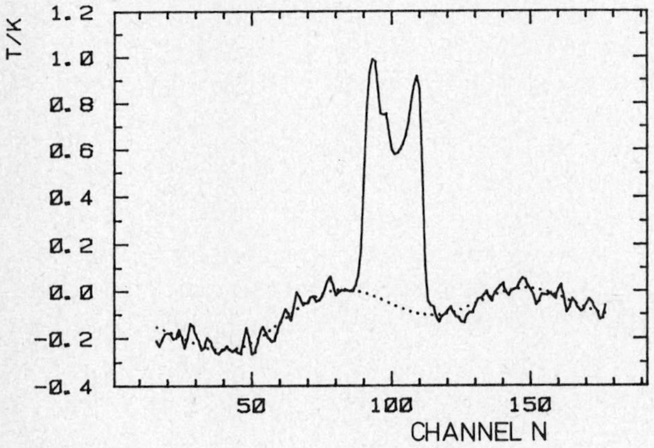

Fig. 10.4. Baseline ripple in 21 cm-emission line profile of the galaxy NGC 1232 (100 m-telescope Effelsberg)

has shown that the amplitude of this ripple is not time independent in a radio telescope. Both changes in the position of the telescope and small changes in the receiving equipment can cause large changes in the amplitude of ripple observed.

The different receiving methods outlined above are susceptible to baseline ripple by quite different amounts. If the background emission is fairly position-independent the method of position switching in the total power mode should result in the least ripple, while frequency switching produces the most.

10.7 Calibration

In principle the calibration of line signals is not different from that of wideband signals: the power available at the receiver input has to be quantified in terms of the chosen (SI) units. For the sake of convenience this power is expressed in terms of the noise power of a matched resistive load

$$W = k\,T_A\,\Delta\nu \tag{10.42}$$

and the (fictitious) thermodynamic temperature of this load is then called the antenna temperature. This antenna temperature can be related to the flux scale S_ν of radio sources with vanishing angular extent as outlined in Chap. 5. The power available at the receiver input is given by

$$W = \tfrac{1}{2}\eta_A\,A_g\,S_\nu\,\Delta\nu \tag{10.43}$$

where A_g is the geometric aperture of the antenna, η_A the *aperture efficiency* and S_ν the flux of the source. (10.42) and (10.43) then result in

$$\boxed{T_A = \Gamma\,S_\nu} \quad \text{where} \tag{10.44}$$

$$\boxed{\Gamma = \frac{\eta_A\,A_g}{2\,k}} \tag{10.45}$$

is the *sensitivity* of the antenna.

These considerations are independent of the frequency and the bandwidth of the signal; therefore they apply to line radiation just as well as to wideband continuum signals. For sources of line radiation with an angular extent which is small compared to the angular resolution of reasonably sized filled aperture antennas, only the total flux in Jy can be determined, and this can be done by comparing the line signal directly with continuum sources. The properties of the antenna cancel completely and neither η_A nor η_{mb} need to be known for this. This applies for most galactic molecular line sources, and it is true also for the 21 cm neutral hydrogen radiation of remote and small diameter galaxies.

The situation is much more complicated, however, if the surface brightness of radiation can be expressed in terms of *brightness temperature* (1.28) and, according to (4.64), this again can be related to the antenna temperature

$$\boxed{T_A(x,y) = \frac{\int P(x-x', y-y')\,T_b(x',y')\,dx'\,dy'}{\int P(x',y')\,dx'\,dy'}}\;. \tag{10.46}$$

Here we have put the radiation efficiency η_R of the antenna equal to 1. For mechanically well-built antennas with low ohmic losses this should be a reasonably good approximation. (10.46) is a two-dimensional convolution equation: the brightness temperature $T_b(x,y)$ is convolved with the power pattern $P(x,y)$ of the antenna and it is written in terms of the cartesian coordinates x, y. In reality spherical coordinates have to be used, and this introduces additional practical and mathematical complications.

What is wanted is the inversion of (10.46), an expression giving T_b in terms of the measured T_A. To derive an approximate expression of this kind the radiation received by the antenna is separated into one part received by the *main beam* (mb) and another one coming from the *stray pattern* (sp). The integral (10.46) can then be separated into

$$T_A(x,y) = \frac{1}{\Omega_A}\left[\int_{(mb)} P(x-x', y-y')\, T_b(x',y')\, dx'\, dy' + \int_{(sp)} P(x-x', y-y')\, T_b(x',y')\, dx'\, dy'\right]. \quad (10.47)$$

If we now suppose that the position dependence of T_b varies very little over angular scales comparable to the beamwidth, T_b can be extracted from underneath the integral for the main beam, yielding for that part of T_A received by the main beam

$$\begin{aligned} T_{AM}(x,y) &= \frac{1}{\Omega_A}\int_{(mb)} P(x-x', y-y')\, T_b(x',y')\, dx'\, dy' \\ &= \frac{\bar{T}_b(x,y)}{\Omega_A}\int_{(mb)} P(x,y)\, dx\, dy \\ T_{AM} &= \eta_M \bar{T}_b\ . \end{aligned} \quad (10.48)$$

there we have introduced the *main beam efficiency*

$$\eta_M = \frac{\Omega_M}{\Omega_A} = \frac{\int_{(mb)} P(x,y)\, dx\, dy}{\int_{(A)} P(x,y)\, dx\, dy} \quad (10.49)$$

as defined in (4.53). Substituting this into (10.47) and solving for $\bar{T}_b$ which we will call the corrected brightness temperature, we obtain

$$\bar{T}_b(x,y) = \frac{1}{\eta_M}\left[T_A(x,y) - \frac{1}{\Omega_A}\int_{(sp)} P(x-x', y-y')\, T_b(x',y')\, dx'\, dx'\right]. \quad (10.50)$$

The integral is over the unknown $T_b(x, y)$, but since it is an average over large angles, it will not cause too big a change if T_b is replaced by $\bar{T}_b$ from (10.50). The computation of T_b thus results in an iterative procedure leading to a Neumann series. As shown by Kalberla et al. (1980) this sequence can be solved by

$$\bar{T}_b(x,y) = \frac{1}{\eta_M}\left[T_A(x,y) - \int_{(sp)} R(x-x', y-y')\, T_A(x',y')\, dx'\, dx'\right]. \quad (10.51)$$

where R is the socalled resolving kernel which can be derived from P by successive approximation. For practical antennas R usually differs very little from P. The correction procedure is very complicated and requires the knowledge of the radiation T_A across the full sky! An example is given in Fig. 10.5.

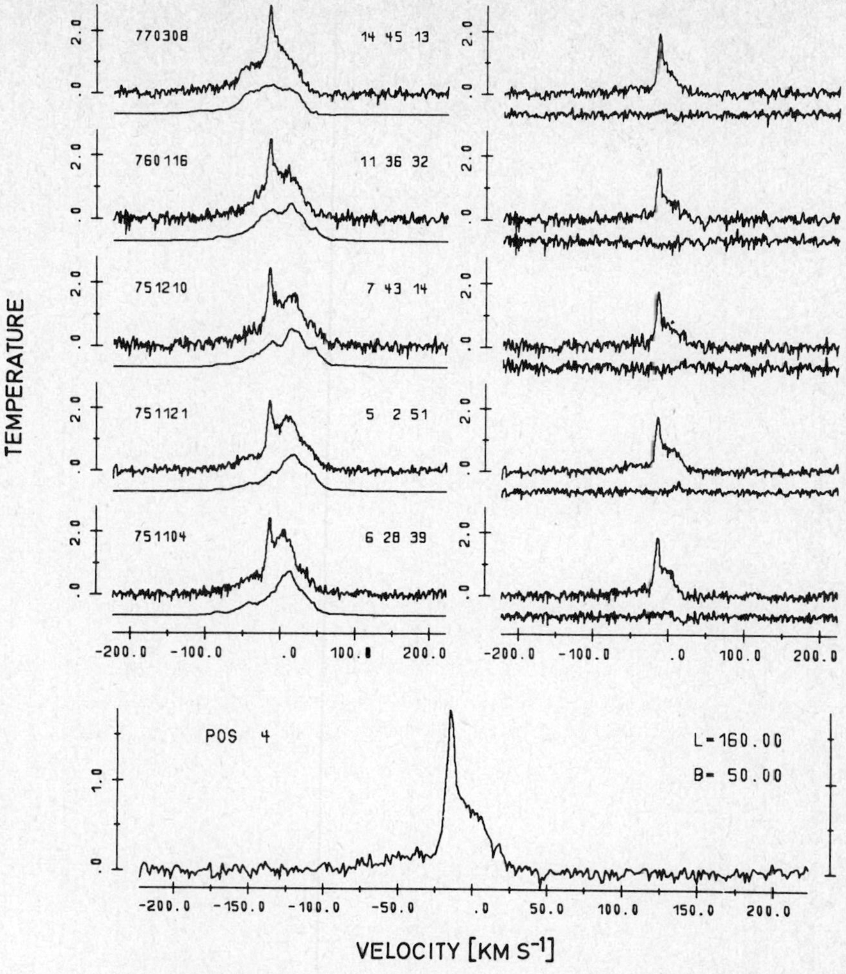

Fig. 10.5. Variations of the measured line profile ($l = 160°$, $b = 50°$) caused by the ground blocking varying parts of the far sidelobe structure. The computed stray radiation profile is given for each profile, at the bottom the corrected profile is plotted [after Kalberla et al. (1980)]

All these arguments apply equally well to wideband continuum radiation and to line radiation. However, for line radiation we meet with another difficulty. In computing the stray pattern of a telescope we will find that quite a large part of this pattern is directed towards the ground. But while the ground does emit thermal continuum, it does not produce line emission. Therefore we have to introduce a time-dependent cut-off into the second integral in (10.47), and this will result in a time variation of T_A. Observations of the 21 cm line emission from a fixed position obtained at different hour angles do indeed show lineshape variations which can be explained by the time-variation of the radiation received through the stray pattern.

Complications of this kind are only to be expected if T_b shows a very wide distribution over the sky. They are present in 21 cm-line measurements where the line radiation is strong in a 10°-wide band along the galactic equator. Probably such effects will also be observed in CO-line emission at $\lambda = 2.6$ mm because CO is widely distributed too. All other molecular and recombination line measurements are probably free from this effect because the lines are emitted by discrete sources that cover only a tiny fraction of the sphere.

This time-variability of the contribution to T_A from the stray pattern puts rather stringent limits on the usefulness of a differential calibration scheme used extensively in 21 cm-line work. The idea is the following: A number of calibration regions have been measured with a well calibrated antenna so that line profiles $T_b(v)$ can be given. If these regions now are measured with some other antenna, scaling factors transforming the measured output of the new system into T_b values can be determined directly. Unfortunately both the output of the new antenna and the published reference profiles are contaminated by time-dependent stray radiation. Kalberla et al. (1980, 1982) have shown how to correct the reference regions but the stray radiation contribution of the new system is still present. It will thus be very difficult to achieve a T_b scale with a precision of better than 10%. This is different and slightly better if the calibration is done in the same way as continuum work. If the aperture efficiency of the antenna is known, the T_A scale can be calibrated properly. Then uncertainties enter only into the transformation of T_A into T_b, and again the different observing modes outlined in Sect. 10.6 are susceptible to quite a different degree.

Rewriting (10.47) for profile areas F_A in units of K km s^{-1} and separating the integration range into the main-beam area reaching to the first nulls of P, the near side-lobes (ns) within a few degrees of the main beam and the remaining far sidelobes (fs) then

$$F_A = \eta_M F_{mb} + \eta_{ns} F_{ns} + \eta'_{fs} F_{fs} \tag{10.52}$$

where η'_{fs} has to be modified to include only those areas of the sky that are above the horizon.

Now if in the total power mode the comparison region can be chosen within a few degrees of the measured position, we can assume with good confidence that $F_{ns\,on} = F_{ns\,off}$, $F_{fs\,on} = F_{fs\,off}$ and so

$$\Delta F = F_{A\,on} - F_{A\,off} = \eta_M (F_{mb\,on} - F_{mb\,off}) \,. \tag{10.53}$$

There should thus be no problem in the transformation of the antenna temperature into brightness temperature. This is different if either load switching or frequency switching is used. Now both F_{ns} and F_{fs} enter the equation. The transformation of T_A into T_b will then depend on the angular distribution of the radiation. Sharp peaks of radiation will transform differently from distributions with angular scale sizes of a few degrees.

If accuracies better than 10% are needed for T_b, a full time-dependent solution of the convolution equation as given by (10.51) is needed.

11. Line Radiation of Neutral Hydrogen

11.1 The 21 cm-Line of Neutral Hydrogen

The 21 cm-line of neutral hydrogen owes its existence to transitions between the hyperfine structure levels $1^2 S_{1/2}$, $F = 0$ and $F = 1$ of H I. The energy of these differs slightly due to the interaction of the spin of the nucleus and that of the electron. Although an explicit expression for the separation of these levels can be given in terms of fundamental atomic constants, the frequency of the resulting line has been measured with high precision in the laboratory; it is, in fact, one of the most precisely measured physical quantities with a mean relative error of only $2 \cdot 10^{-11}$:

$$\nu_{10} = 1.420\,405\,751\,786\,(30) \cdot 10^9 \text{ Hz} \tag{11.1}$$

(Peters et al. 1965). The radiation is that of a magnetic dipole with a dipole matrix element of one Bohr magneton. Substituting (11.1) into (10.27) we then obtain

$$A_{10} = 2.86888\,(7) \cdot 10^{-15}\,\text{s}^{-1} \ . \tag{11.2}$$

This transition probability is about a factor 10^{23} smaller than that of an allowed optical transition, mostly due to the difference in wavelength [the frequency enters as ν^3 in (10.24)]; a factor of $5 \cdot 10^5$ comes from the small magnitude of the magnetic dipole moment.

The mean half-life time of the excited state with regard to spontaneous transitions downwards is

$$t_{1/2} \cong 1/A_{10} = 3.49 \cdot 10^{14}\,\text{s} \cong 1.11 \cdot 10^7\,\text{a} \ .$$

Since a typical interstellar hydrogen atom will change the spin of the electron due to collisions about every 400 years, only a very small fraction of all spin-flips will be connected with the emission or absorption of a photon. It is therefore understandable that in practically all astronomical situations the relative population of the hyperfine structure levels will be controlled by collisions.

Let the relative population of the levels be described by an excitation temperature which in this case is usually called the *spin temperature* T_s

$$\frac{N_1}{N_0} = \frac{g_1}{g_0} \exp\left\{-\frac{h\nu_{10}}{k\,T_s}\right\} \ . \tag{11.3}$$

Now

$$T_0 = \frac{h\nu_{10}}{k} = 0.0682\ \mathrm{K}\ , \tag{11.4}$$

and so

$$\frac{N_1}{N_0} = \frac{g_1}{g_0} = 3 \quad \text{for } T_s \gg T_0\ .$$

For the same reason, the exponential function in (10.17) can be replaced by the first two terms of the Taylor series, so that

$$\boxed{\kappa_\nu = \frac{3c^2}{32\pi}\frac{1}{\nu_{10}} A_{10} N_\mathrm{H} \frac{h\nu_{10}}{k\,T_s}\varphi(\nu)} \tag{11.5}$$

where the total number of neutral hydrogen atoms (per unit volume) has been introduced by $N_\mathrm{H} = N_0 + N_1 = 4N_0$.

Equation (11.5) gives the absorption coefficient per unit frequency interval. Since in radio astronomy the lineshapes are usually given in terms of the corresponding Doppler velocities

$$\frac{\nu_{10} - \nu}{\nu_{10}} = \frac{V}{c}\ , \tag{11.6}$$

Eq. (11.5) can be transformed into

$$\boxed{\begin{aligned} d\tau(V) &= -\kappa_V(s)\, d\left(\frac{s}{\mathrm{cm}}\right) \\ &= -5.4873(10)\cdot 10^{-19}\left(\frac{N_\mathrm{H}}{\mathrm{cm}^{-3}}\right)\left(\frac{T_s(s)}{\mathrm{K}}\right)^{-1}\left(\frac{\varphi(V)}{\mathrm{s\,km}^{-1}}\right) d\left(\frac{s}{\mathrm{cm}}\right)\end{aligned}} \tag{11.7}$$

where V is measured in $\mathrm{km\,s}^{-1}$, see, e.g., Fig. 11.1.

If the spin temperature T_s is independent of s along the line-of-sight, we obtain from (11.7) by integrating both over s and over V

$$\int_{-\infty}^{\infty} \tau(V)\, d\left(\frac{V}{\mathrm{km\,s}^{-1}}\right) = 5.4873(10)\cdot 10^{-19}\left(\frac{T_s}{\mathrm{K}}\right)^{-1}\int_0^{\infty} N_\mathrm{H}(s)\, ds$$

or

$$\boxed{\frac{\mathscr{N}_\mathrm{H}}{\mathrm{cm}^{-2}} = 1.8224(3)\cdot 10^{18}\left(\frac{T_s}{\mathrm{K}}\right)\int_{-\infty}^{\infty}\tau(V)\, d\left(\frac{V}{\mathrm{km\,s}^{-1}}\right)} \tag{11.8}$$

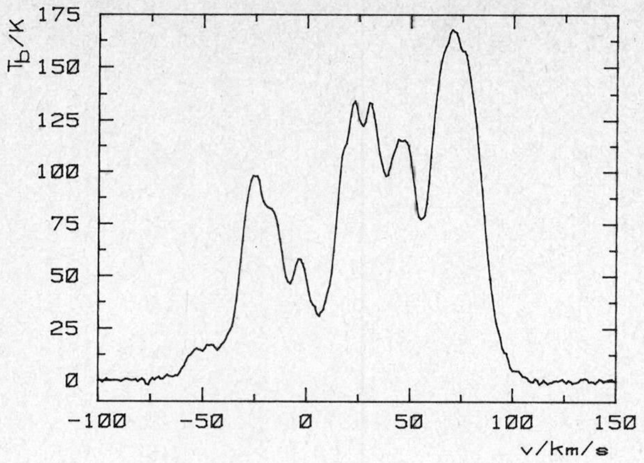

Fig. 11.1. 21 cm-line profile for $l = 41.9$, $b = 0°$ measured with the Effelsberg 100 m-telescope

if we define the *column density* $\mathcal{N}_H$ by

$$\boxed{\frac{\mathcal{N}_H}{\text{cm}^{-2}} = \int_0^\infty \frac{N_H(s)}{\text{cm}^{-3}}\, d\left(\frac{s}{\text{cm}}\right)} \quad . \tag{11.9}$$

In some contexts it is convenient to use the parsec (1 pc = 3.0856 (3) · 10^{16} m) as the unit of distance along the line-of-sight, resulting in the *hydrogen measure* HM

$$\boxed{\frac{\text{HM}}{\text{cm}^{-3}\,\text{pc}} = \int_0^\infty \frac{N_H(s)}{\text{cm}^{-3}}\, d\left(\frac{s}{\text{pc}}\right)} \tag{11.10}$$

formed similarly to the emission measure EM in (8.25). Equation (11.8) then reads

$$\boxed{\frac{\text{HM}}{\text{cm}^{-3}\,\text{pc}} = 0.5906\,(2)\left(\frac{T_s}{\text{K}}\right) \int_{-\infty}^{\infty} \tau(V)\, d\left(\frac{V}{\text{km s}^{-1}}\right)} \quad . \tag{11.11}$$

11.2 The Spin Temperature

The excitation temperature for a given transition in a stationary state will be some average between the brightness temperature describing the ambient radiation field at the wavelength of the transition considered and the kinetic temperature describing the local velocity distribution of the colliding particles. Because $T_s \gg T_0$ is usually the case, the expression can be considerably simplified for most radio lines using

$$e^{-T_0/T} \cong 1 - T_0/T$$

where T_0 is given by (11.4). Equation (10.39) then becomes

$$T_s = T_K \frac{T_b A_{10} + T_0 C_{10}}{T_K A_{10} + T_0 C_{10}} \quad \text{or} \tag{11.12}$$

$$\boxed{T_s = \frac{T_b + y\,T_K}{1 + y}} \quad \text{where} \tag{11.13}$$

$$\boxed{y = \frac{T_0 C_{10}}{T_K A_{10}}} \quad . \tag{11.14}$$

T_s is thus a weighted mean of the kinetic gas temperature and the brightness temperature of the radiation field. The weighting factor y depends on the collision probabilities of the colliding partners HI – HI and HI – e which have to be computed by quantum mechanical methods.

A survey of the methods and the results are given by Purcell and Field (1956), Field (1958) and Elwert (1959); numerical values from Field (1958) are given in Table 11.1. For a gas with the density $N_H > 1\ \text{cm}^{-3}$, the weighting factor y is such that $T_s \cong T_K$ is always true irrespective of whether the gas is mainly neutral or ionized, and this also applies for low-density gas ($N_H < 0.1\ \text{cm}^{-3}$) if it is partly ionized. Only completely neutral, low-density gas with $T_K \gtrsim 10\,400$ K could have $T_s < T_K$, but it is doubtful whether such gas exists in the Galaxy in large amounts. Therefore we can generally adopt $T_s = T_K$ for neutral hydrogen gas.

The spin-temperature T_s will quite often vary with s because the line-of-sight intersects clouds of different kinetic temperature. Then it is possible to determine an average spin temperature which, according to Kahn (1955), will be the harmonic mean value of the temperatures encountered. Let

$$N_V(s) = N_H(s)\,\varphi(V\,|\,s)$$

Table 11.1. Weighting factors for the determination of the spin temperature of neutral hydrogen

$\frac{T_K}{\text{K}}$	$y_H \Big/ \frac{N_H}{\text{cm}^{-3}}$ [a]	$y_e \Big/ \frac{N_e}{\text{cm}^{-3}}$ [b]
1	1200	6700
3	490	3900
10	190	2100
30	85	1200
100	35	650
300	16	350
1000	6.7	130
3000	3.9	66
10 000	1.3	18

[a] computed for collisions with neutral hydrogen atoms
[b] computed for collisions with electrons

be the space density of neutral hydrogen atoms with velocities between V and $V + dV$ at the the position s. Equation (11.7) then can be integrated from 0 to s yielding

$$\tau_V(s) = 5.4873 \cdot 10^{-19} \int_0^s \frac{N_V(s)}{T_s(s)}\, ds = 5.4873 \cdot 10^{-19} \left\langle \frac{1}{T_s(s)} \right\rangle \int_0^s N_V(s)\, ds$$

where $\langle 1/T_s \rangle$ is the appropriate mean value of the inverse spin temperature

$$\boxed{\left\langle \frac{1}{T_s(V)} \right\rangle = \frac{\int_0^s \frac{N_V(s)}{T_K(s)}\, ds}{\int_0^s N_V(s)\, ds}} \quad . \tag{11.15}$$

This average is a weighted harmonic mean value where the neutral hydrogen gas density is the weighting factor and, since N_V depends on the velocity, the harmonic mean spin temperature will depend on V too.

11.3 Emission and Absorption Lines

If we neglect for the time being the problems connected with the lineshape $\varphi(V)$, we see from (11.7) that the strength of the 21 cm-line radiation is mainly governed by two parameters of the interstellar gas: the gas density $N_H(s)$ of the neutral hydrogen and its spin temperature. How these parameters, or at least some appropriate average value, can be determined from line measurements will be investigated in this section.

Consider an isothermal cloud of gas which is traversed by radiation from some background source. The solution of the equation of radiation transfer (1.31) in terms of the brightness temperature is then

$$T_b(V) = T_s[1 - e^{-\tau(V)}] + T_c e^{-\tau(V)} \tag{11.16}$$

where T_s is the spin temperature of the cloud, T_c the brightness temperature of the background source, and $\tau(V)$ the optical depth of the cloud at the radial velocity V. For positions without a background source obviously $T_c = 0$ and we observe a pure *emission line profile*. If $\tau(V) \ll 1$, quadratic and higher terms in the Taylor series $e^{-\tau} = 1 - \tau + \tau^2/2 - \dots$ can be neglected resulting in

$$T_b(V) = T_s \tau(V) \quad \text{for } \tau(V) \ll 1 \ . \tag{11.17}$$

Substituting this into (11.8) or (11.11) we find

$$\boxed{\frac{\mathcal{N}_H}{\mathrm{cm}^{-2}} = 1.8224(3) \cdot 10^{18} \int_{-\infty}^{\infty} \left(\frac{T_b(V)}{\mathrm{K}}\right) d\left(\frac{V}{\mathrm{km\,s^{-1}}}\right)} \tag{11.18}$$

or

$$\boxed{\frac{\mathrm{HM}}{\mathrm{cm}^{-3}\,\mathrm{pc}} = 0.5906(2) \int_{-\infty}^{\infty} \left(\frac{T_{\mathrm{b}}(\mathrm{V})}{\mathrm{K}}\right) d\left(\frac{V}{\mathrm{km\,s}^{-1}}\right)} \quad . \tag{11.19}$$

For optically thin radiation the column density is thus independent of the spin temperature of the gas. It can be determined unambiguously from the integral over the emission line.

While (11.18) and (11.19) are valid only for optically thin radiation, (11.8) and (11.11) are applicable irrespective of the value of τ. But then the column density depends critically on the adopted value for T_s. Solving (11.16) with $T_c = 0$ for τ and substituting this into (11.8) we find that

$$\frac{\mathcal{N}_{\mathrm{H}}}{\mathrm{cm}^{-2}} = -\,1.8224(3)\cdot 10^{18}\left(\frac{T_{\mathrm{s}}}{\mathrm{K}}\right) \int_{-\infty}^{\infty} \ln\left[1 - \frac{T_{\mathrm{b}}(V)}{T_{\mathrm{s}}}\right] d\left(\frac{V}{\mathrm{km\,s}^{-1}}\right) . \tag{11.20}$$

Precise measurements of T_s for an actual cloud of gas are rather difficult, but some indications can be obtained from (11.16) by realizing that $T_b(V) \to T_s$ for $\tau(V) \to \infty$ (Fig. 11.2). This is the basis for the "classical" value $T_s = 125$ K obtained by Dutch radio astronomers. The precise value depends on the calibration of the antennas used but is of little consequence anyway since it is a harmonic mean where the relative amount of low- and high-temperature gas is not known.

The chances of determining a full set of parameters for the interstellar gas are somewhat better if there is a background source whose brightness temperature T_c is of the same order of magnitude as T_s.

Quite often T_c is caused by a background source with a continuous spectral distribution that has an angular diameter which is small compared with the resolving power of the telescope. If S_ν is the flux of this source then from (6.47) the brightness temperature T_c is

$$T_{\mathrm{c}} = \frac{c^2}{2k}\frac{1}{\nu^2}\frac{S_\nu}{\Omega_{\mathrm{M}}}$$

or, using (4.57) and (4.59)

$$T_{\mathrm{c}} = \frac{\eta_{\mathrm{A}}}{\eta_{\mathrm{M}}}\frac{A_{\mathrm{g}}}{2k} S_\nu = \frac{\eta_{\mathrm{A}}}{\eta_{\mathrm{M}}}\frac{\pi}{8k} D^2 S_\nu \tag{11.21}$$

where D is the diameter of the telescope. Therefore, the larger D the larger T_c will be. Absorption effects therefore attain importance for large telescopes only; with small telescopes they can only be detected by using strong background sources.

Let (11.16) be the line radiation of such an "on source" position. Usually a baseline is subtracted from $T(V)$ as described in Chap. 10 so that $T_{\mathrm{on}}(V) = 0$ for V outside the range of the line radiation. Thus

$$\Delta T_{\mathrm{on}}(V) = (T_{\mathrm{s}} - T_{\mathrm{c}})(1 - \mathrm{e}^{-\tau(V)}) \ . \tag{11.22}$$

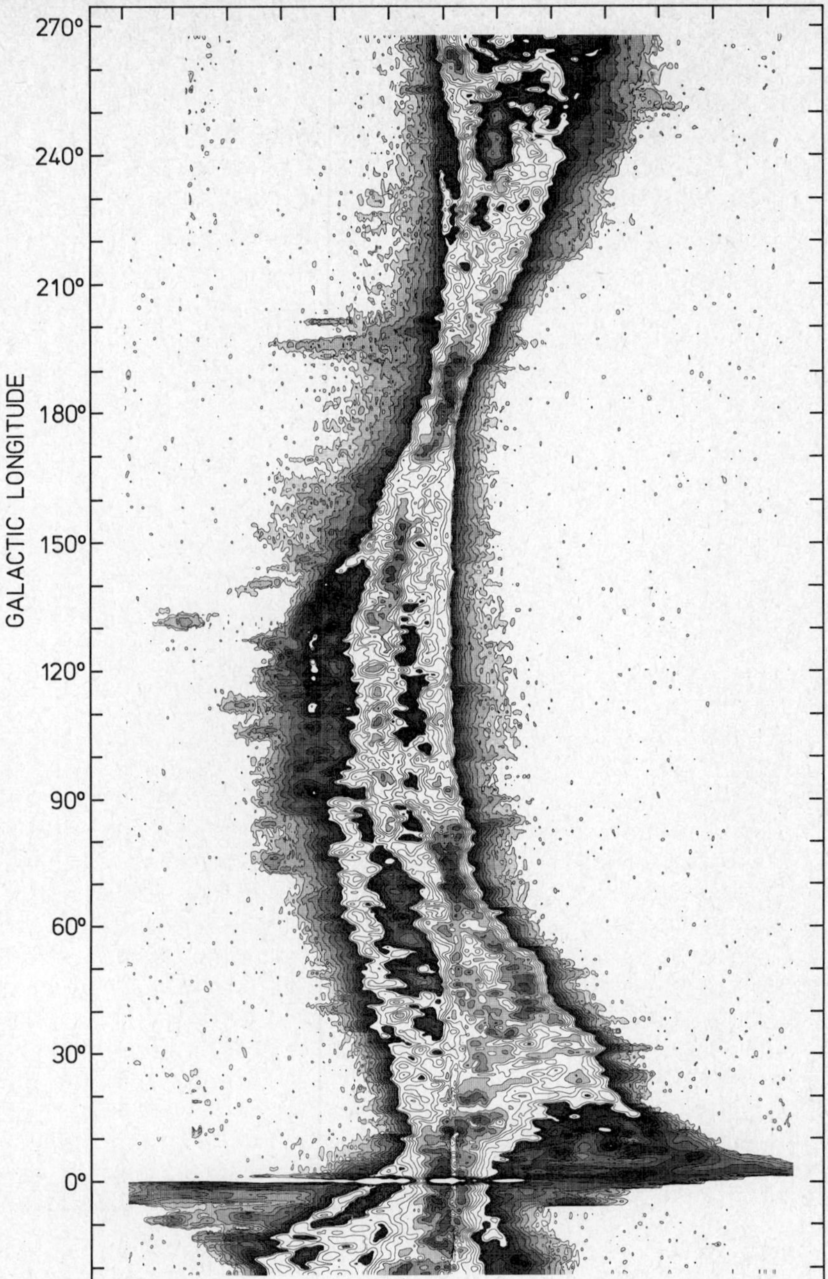

Fig. 11.2. The brightness temperature T_b as function of radial velocity (relative to LSR) along the galactic equator ($b = 0$). Near $l = 0$ and $l = 0\overset{\circ}{.}6$ the effects of absorption are clearly seen (after Burton)

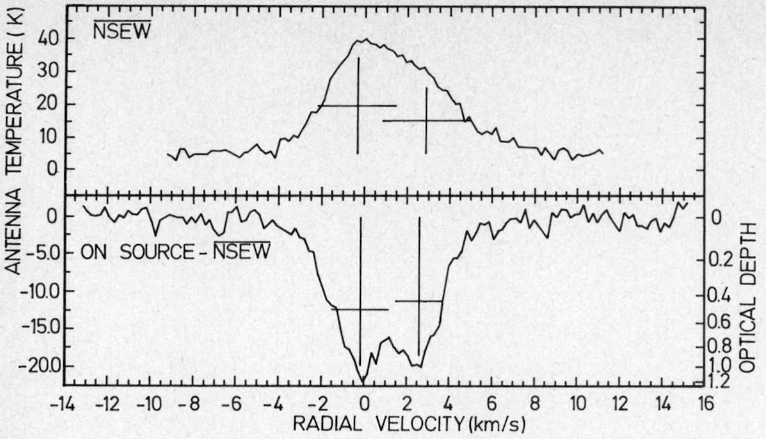

Fig. 11.3. Expected profile and absorption profile for the radio source 3C53 [after Radhakrishnan (1972)]

Next an estimate of the line radiation has to be obtained if $T_c = 0$ but all other gas parameters remain unchanged. This line radiation is usually called the "expected profile" (Fig. 11.3)

$$\Delta T_{\mathrm{ex}}(V) = T_s(1 - \mathrm{e}^{-\tau(V)}) \ . \tag{11.23}$$

If the background source is a pulsar (Fig. 11.4), this expected profile can be measured directly: ΔT_{on} is measured at those time instants when the pulsar is radiating and ΔT_{ex} at those when the pulsar is off.

For a background source with an angular diameter that is small compared to the telescope beam, the variation of T_c with the position on the sky is quite often faster than that of the emission line radiation. Then the expected profile can be obtained with some success by interpolation from neighbouring positions where $T_c = 0$. The uncertainty of this interpolated T_{ex} quite naturally depends on the average variation observed for the line radiation over distances similar to those for which the interpolation was performed. At high galactic latitudes uncertainties as small as < 1 K can be expected, while closely to the galactic plane errors as large as 10 K are possible. This has obviously far-reaching consequences for the optical depth and the other parameters of the interstellar gas derived from these measurements. An example of the problems encountered in the investigation of the absorption of low-latitude extended sources can be found in Rohlfs and Braunsfurth (1982) where the absorption of the gas in front of the extended galactic center source Sgr A is discussed.

Assuming that ΔT_{ex} has been determined by some means, we can obtain from (11.22) and (11.23)

$$T_s = T_c \frac{1}{1 - (\Delta T_{\mathrm{on}}/\Delta T_{\mathrm{ex}})} \ . \tag{11.24}$$

An example is given in Fig. 11.5.

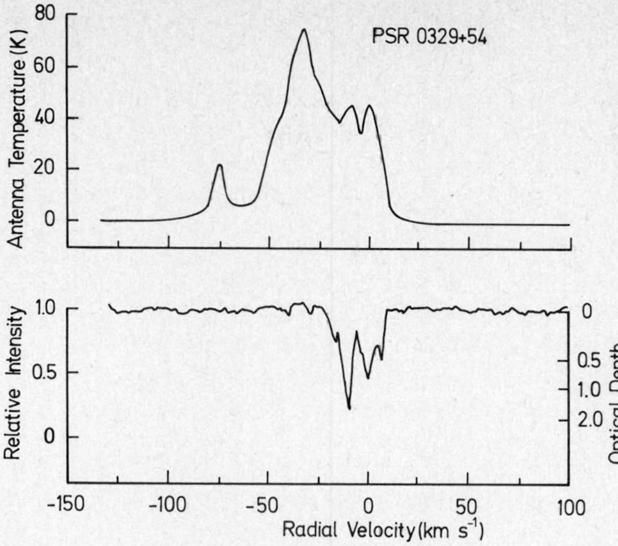

Fig. 11.4. H I $\lambda = 21$ cm-emission and absorption line profiles in the direction of the pulsar PSR 039 + 54. The expected line profile is measured at those moments where the pulsar emission is switched off (Manchester and Taylor 1977)

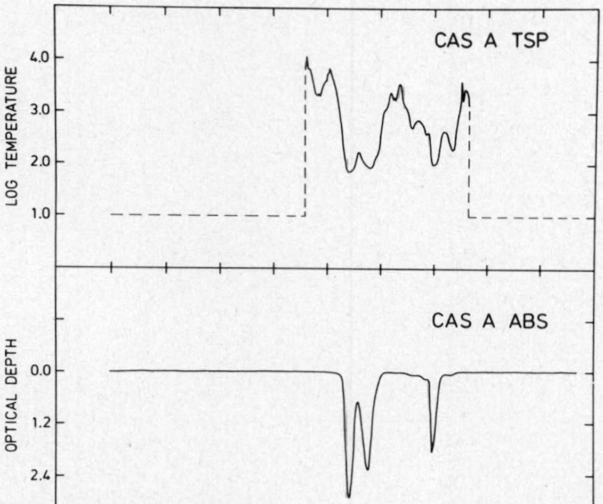

Fig. 11.5. The determination of T_s by (11.24) in the direction of Cas A (after Mebold and Hills)

The spin temperatures thus obtained cover a wide range reaching from several 10^3 K down to about 10 K. It is therefore evident that the assumption of a single value for T_s valid for all neutral hydrogen gas would be a far too great simplification. In the interstellar medium, gas with widely different physical parameters exists, and it is still in dispute whether we are allowed to use a static picture of this or whether a dynamical model has to be used.

11.4 Differential Velocity Fields and the Shape of Spectral Lines

Velocity fields in gas that emit line radiation will affect this radiation in a number of different ways. The bulk velocity of the gas will shift the mean frequency. Random velocities of the gas atoms will influence the line shape. And a change of the mean gas velocity with position will produce a similar result, the line shape will be affected.

All these effects will occur for any line that is emitted or absorbed but, for lines produced in stellar atmospheres or in fairly dense clouds, usually only the first two are of importance. In interstellar neutral hydrogen gas, however, the line emission is spread over such a wide volume that the large scale velocity gradient and their influence on the line shape are much greater than any local effects.

Let us therefore assume that neutral hydrogen gas has a bulk velocity $U(s)$ that is a function of the position s along the line-of-sight. For the sake of simplicity we will assume that the shape of the line does not depend on s. Then the line-of-sight element ds contributes $d\tau$ at the velocity V according to (11.7) and

$$d\tau(V) = -\, w \frac{N_{\mathrm{H}}(s)}{T_{\mathrm{K}}(s)} \varphi[V - U(s)]ds \ . \tag{11.25}$$

The coefficient w depends on the units used for s and V; if s is in cm and V in $\mathrm{km\,s^{-1}}$ we have $w = 5.4873(10) \cdot 10^{-19}$, while $w = 1.6932$ for s in pc and V in $\mathrm{km\,s^{-1}}$. For the total optical depth at the velocity V for gas between 0 and s we therefore obtain

$$\boxed{\tau(V,s) = w \int_0^s \frac{N_{\mathrm{H}}(x)}{T_{\mathrm{K}}(x)} \varphi[V - U(x)]dx} \tag{11.26}$$

or

$$\boxed{\tau(V,U) = w \int_{U(0)}^{U(s)} \frac{N_{\mathrm{H}}[s(U)]}{T_{\mathrm{K}}[s(U)]} \varphi(V - U) \frac{dU}{\left|\dfrac{dU}{ds}\right|}} \tag{11.27}$$

if

$$U = U(s) \ . \tag{11.28}$$

Two simple examples may illustrate the use of these expressions:

1) Gaussian line shape locally emitted in a homogeneous medium with a *linear velocity field*:

$$N_{\mathrm{H}} = \mathrm{const} \ , \qquad T_{\mathrm{K}} = \mathrm{const} \ , \qquad \varphi(V) = \frac{1}{\sigma\sqrt{2\pi}} \exp\left\{-\frac{V^2}{2\sigma^2}\right\}$$

$$U(s) = a\,s \ ;$$

$$\tau(V,s) = \frac{w\,N_{\mathrm{H}}}{\sigma\,T_{\mathrm{K}}\sqrt{2\pi}} \int_0^s \exp\left\{-\frac{(V-a\,x)^2}{2\,\sigma^2}\right\} dx$$

$$= \frac{w}{\sqrt{\pi}} \frac{N_{\mathrm{H}}}{T_{\mathrm{K}}} \frac{1}{a} \int_{\frac{1}{\sigma\sqrt{2}}(V-a\,s)}^{\frac{1}{\sigma\sqrt{2}}V} \mathrm{e}^{-t^2}\,dt \ .$$

$$\tau(V,s) = \frac{w}{2} \frac{N_{\mathrm{H}}}{T_{\mathrm{K}}} \frac{1}{a}\left[\mathrm{erf}\left(\frac{1}{\sigma\sqrt{2}}V\right) - \mathrm{erf}\left(\frac{1}{\sigma\sqrt{2}}(V-a\,s)\right)\right] \tag{11.29}$$

where we have introduced the error function

$$\mathrm{erf}\,x = \frac{2}{\sqrt{\pi}} \int_0^x \mathrm{e}^{-t^2}\,dt$$

[see Abramowitz and Stegun (1964), Eq. 7.1.1]. Since $-1 < \mathrm{erf}\,x < 1$ the optical depth of a line emitted by gas obeying such a linear velocity field will always remain finite

$$\tau \leqq \frac{w\,N_{\mathrm{H}}}{T_{\mathrm{K}}\,a} \tag{11.30}$$

even if $s \to \infty$. The reason is that, at any velocity V, only gas within a finite range of s will contribute to radiation of this velocity. The linewidth, on the other hand, increases linearly with s.

2) Gaussian line shape locally emitted in a homogeneous medium with a *quadratic velocity field*:

$$U(s) = U_c + b\,(s - s_c)^2 \ ;$$

then for $N_{\mathrm{H}} = \mathrm{const}$, $T_{\mathrm{K}} = \mathrm{const}$ and $s \to \infty$ we have

$$\tau(V) = \frac{w\,N_{\mathrm{H}}}{T_{\mathrm{K}}}\left[\int_0^{s_c} \varphi\,[V - U(x)]\,dx + \int_{s_c}^{\infty} \varphi\,[V - U(x)]\,dx\right] = I + II \ .$$

For the range of the integral I we find

$$s = s_c - \frac{1}{\sqrt{|b|}}|U - U_c|^{1/2}$$

while for II we have

$$s = s_c + \frac{1}{\sqrt{|b|}}|U - U_c|^{1/2}$$

and therefore

$$I = -\frac{1}{2}\frac{w\,N_{\mathrm{H}}}{T_{\mathrm{K}}}\frac{1}{\sqrt{|b|}} \int_{U_c + b s_c^2}^{U_c} \varphi\,(V - U)\frac{dU}{\sqrt{|U - U_c|}}$$

and

$$II = \frac{1}{2}\frac{w\,N_{\mathrm{H}}}{T_{\mathrm{K}}}\frac{1}{\sqrt{|b|}}\int\limits_{U_c}^{\infty}\varphi(V-U)\frac{dU}{\sqrt{|U-U_c|}} \; .$$

Assuming

$$\varphi(V) = \frac{1}{\sigma\sqrt{2\pi}}\exp\left\{-\frac{V^2}{2\sigma^2}\right\}$$

and

$$b\,s_c^2 \gg \sigma$$

so that we can adopt effectively $U_c + b\,s_c^2 \to \infty$, I and II can be taken together and we obtain

$$\tau(V) = \frac{w}{\sqrt{2\pi}}\frac{N_{\mathrm{H}}}{\sigma\,T_{\mathrm{K}}}\frac{1}{\sqrt{|b|}}\int\limits_{U_c}^{\infty}\frac{1}{\sqrt{|U-U_c|}}\exp\left\{-\frac{(V-U)^2}{2\sigma^2}\right\}dU \; .$$

Substituting

$$\sigma\,x = U - U_c$$

we find

$$\tau(V) = \frac{w}{\sqrt{2\pi}\,\sigma}\frac{N_{\mathrm{H}}}{T_{\mathrm{K}}}\frac{1}{\sqrt{|b|}}P\left(\frac{V-U_c}{\sigma}\right) \qquad (11.31)$$

$$P(\xi) = \int\limits_{0}^{\infty}\frac{1}{\sqrt{x}}\exp\{-\tfrac{1}{2}(\xi - x)^2\}\,dx \qquad (11.32)$$

The shape of $P(\xi)$ is for $\xi < 1$ quite similar to a shifted gaussian (Fig. 11.6). The values are given in Table 11.2 since they will be used to describe the shape of the 21 cm-line emission near the galactic equator close to those radial velocities that are measured at the tangential or subcentral points in the longitude range $270° < l < 360°$ and $0° < l < 90°$.

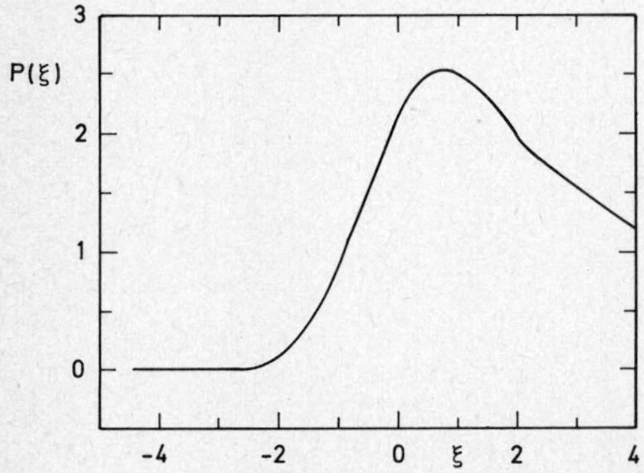

Fig. 11.6. The function $P(\xi)$

Table 11.2. The function $P(\xi) = \int\limits_0^\infty \frac{1}{\sqrt{x}} \exp[-\frac{1}{2}(\xi - x)^2]\,dx$

ξ	P	ξ	P	ξ	P
− 5.0	0.0000	− 1.0	0.9015	1.0	2.5263
− 4.0	0.0003	− 0.8	1.1467	1.2	2.4532
− 2.8	0.0110	− 0.6	1.4096	1.4	2.3516
− 2.6	0.0202	− 0.4	1.6762	1.6	2.2333
− 2.4	0.0358	− 0.2	1.9304	1.8	2.1095
− 2.2	0.0610	0.0	2.1560	2.0	1.9893
− 2.0	0.1002	0.2	2.3388	2.2	1.8793
− 1.8	0.2414	0.4	2.4690	2.4	1.7835
− 1.6	0.3545	0.6	2.5420	2.6	1.7026
− 1.4	0.5014	0.8	2.5591	2.8	1.6349
− 1.2	0.6844	1.0	2.5263	3.0	1.5758
				4.0	1.2285

11.5 The Galactic Velocity Field in the Interstellar Gas

One of the chief topics investigated using the line emission of neutral hydrogen has been the galactic kinematics of the gas. Neutral hydrogen gas is one of the main constituents of the interstellar medium; it is distributed all over the Galaxy and there are very few regions completely devoid of gas. Due to the low excitation energy of the 21 cm-line, H I is always observable if it exists, and therefore the 21 cm-line forms an almost ideal tool for galactic research. Its only shortcomings are the problems often encountered when the distance of certain observable features have to be determined. Therefore, the astronomer must often resort to the construction of models which can provide only the possible structure but can give no conclusive solution.

The large-scale kinematics of galactic interstellar gas is governed by galactic rotation, possibly supplemented by expanding (or contracting) motions close to the galactic center. In addition, the velocity field may be perturbed by density-wave induced streaming velocities and on a more local scale supernova explosions and the like may introduce irregularities in the velocity field. Here we will describe the large-scale field and how it influences the radial velocity and shape of the line radiation.

For the sake of simplicity we will assume that both the galactic rotation and the expansion (or contraction) velocity field are axially symmetric if seen from the galactic center. If $\Theta(r)$ is the (linear) rotational velocity at the galactic radius r, and $\Pi(r)$ the corresponding motion along r as defined by

$$\Theta(r) = r\,\Omega(r) \quad \text{and} \quad \Pi(r) = r\,H(r) \tag{11.33}$$

respectively, and if $\Omega(r)$ is the angular velocity and $H(r)$ is the expansion rate, then the radial velocity of the point P relative to the local standard of rest S is

$$V_r = \Theta(r)\sin(l+\vartheta) - \Theta(r_0)\sin l - \Pi(r)\cos(l+\vartheta) + \Pi(r_0)\cos l\ .$$

Now

$$r_0 \sin l = r \sin(l + \vartheta) \quad \text{and} \quad r \cos(l + \vartheta) = r_0 \cos l - s$$

so that

$$\boxed{V_r = r_0 [\Omega(r) - \Omega(r_0)] \sin l - r_0 [H(r) - H(r_0)] \cos l + H(r) s} \ . \tag{11.34}$$

This is the law of differential galactic rotation. Experience has shown $H(r_0)$ to be exceedingly small in the solar neighbourhood so that the velocity field there is well described by a pure rotation field.

Another result of observational experience with measurements in our own Galaxy and of other spiral galaxies is that $\Theta(r)$ varies very slowly with r if the immediate surroundings of the galactic center are avoided. Therefore the series expansion

$$\Theta(r) = \Theta(r_0) + \left.\frac{d\Theta}{dr}\right|_0 (r - r_0) \tag{11.35}$$

should give a good representation of $\Theta(r)$ in the solar neighbourhood. Now, from

$$r^2 = r_0^2 + s^2 - 2 r_0 s \cos l \ , \tag{11.36}$$

the approximations

$$r = r_0 \left(1 - \frac{s}{r_0} \cos l\right) \quad \text{and} \quad \frac{1}{r} = \frac{1}{r_0}\left(1 + \frac{s}{r_0} \cos l\right)$$

are obtained so that

$$\begin{aligned} V_r &= r_0 \left[\Theta_0 \left(\frac{1}{r} - \frac{1}{r_0}\right) - \left.\frac{d\Theta}{dr}\right|_0 \frac{s}{r_0} \cos l \right] \sin l \\ &= \left(\frac{\Theta_0}{r_0} - \left.\frac{d\Theta}{dr}\right|_0\right) s \cos l \sin l \ , \end{aligned}$$

and

$$\boxed{V_r = s A(r_0) \sin 2l} \tag{11.37}$$

when using

$$\boxed{A(r_0) = \frac{1}{2}\left(\frac{\Theta_0}{r_0} - \left.\frac{d\Theta}{dr}\right|_0\right) = -\frac{1}{2} r_0 \left.\frac{d\Omega}{dr}\right|_0} \ . \tag{11.38}$$

This is the famous Oort sin $2l$-relation describing the differential galactic rotation in the solar neighbourhood. For given l, V_r is proportional to s, but obviously this

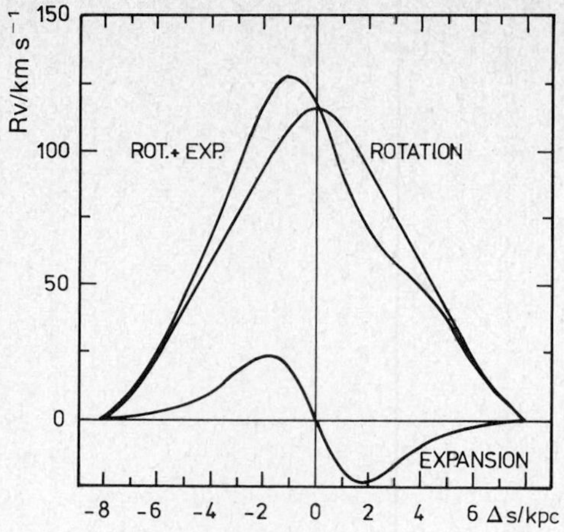

Fig. 11.7. Radial velocity in $\mathrm{km\,s^{-1}}$ caused by differential galactic rotation and expansion for $l = 20°$. The abscissa is the distance $\Delta s = s - s_c$ from the subcentral point defined by (11.40)

is valid only close to the sun. In the first galactic quadrant ($0° < l < 90°$) V_r, reaches a maximum (see Fig. 11.7) and in the fourth quadrant ($270° < l < 360°$) a minimum. This can be shown formally in the following way:

Along a given line of sight (l = const)

$$\frac{dV_r}{ds} = \frac{dV_r}{dr}\frac{dr}{ds}$$

and from (11.34) with $H \equiv 0$ using (11.38)

$$\frac{dV_r}{dr} = r_0 \left.\frac{d\Omega}{dr}\right|_0 \sin l = -\,2A(r_0)\sin l \qquad (11.39)$$

while (11.36) gives

$$\frac{dr}{ds} = \frac{s - r_0\cos l}{r} = \frac{s - s_c}{r} \quad \text{so that}$$

$$\frac{dV}{ds} = -\,2A(r_0)\sin l\,\frac{s - s_c}{r}\ .$$

Therefore $dV_r/ds = 0$ for

$$s = s_c = r_0\cos l \quad \text{and} \quad r_c = r_0\,|\sin l|\ . \qquad (11.40)$$

The measured radial velocity thus adopts an extreme value on the Thales circle and the angle SPC will be 90° there. For the radial velocity measured at this position we find from (11.34)

$$V_c = [\Theta(r_0\,|\sin l|) - \Theta_0\,|\sin l|]\,\frac{\sin l}{|\sin l|}\ . \qquad (11.41)$$

This relation can be used to construct the rotation curve point by point for $r < r_0$ from the extremes of the measured radial velocity for each longitude.

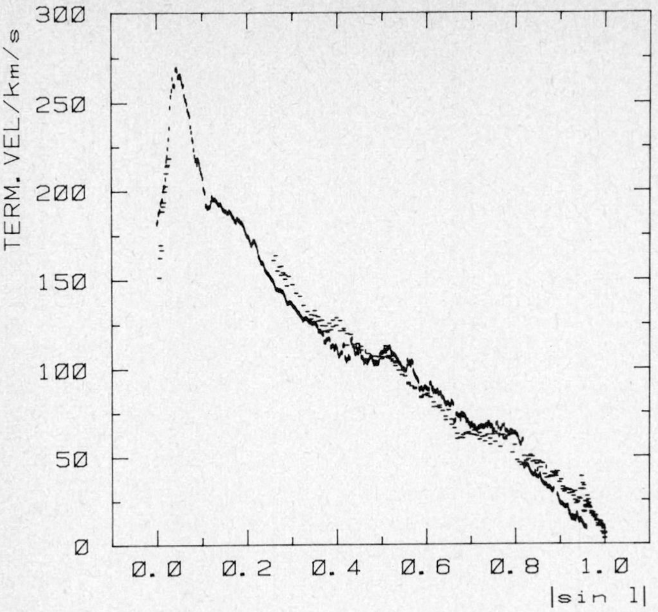

Fig. 11.8. Terminal velocity as function of $|\sin l|$. "|" for 1st galactic quadrant $0 < l < 90°$; "—" for 4th galactic quadrant $270° < l < 360°$

Equation (11.40) gives the galactic position of the gas that emitted this radiation, and

$$\Theta(r_0|\sin l|) = V_c \frac{\sin l}{|\sin l|} + \Theta_0 |\sin l| \ . \tag{11.42}$$

If the series expansion (11.35) is introduced into (11.41) we obtain using (11.38)

$$\boxed{V_c = 2A(r_0)\, r_0 (1 - |\sin l|) \frac{\sin l}{|\sin l|}} \ . \tag{11.43}$$

This relation is a good approximation for the conditions in our galaxy for $|\sin l| > 0.5$, that is, for $r > r_0/2$, and so a linear function in this range must be a good approximation for $\Theta(r)$, (see Fig. 11.8). In most investigations of the galactic velocity field a formula differing slightly from (11.43) is given, but it was shown by Gunn et al. (1979) that (11.43) is the correct expression for galaxies with almost straight rotation curves.

Close to the Thales point (called the subcentral point by some authors) the run of the radial velocity along the line-of-sight will not be represented well by a linear function; higher terms of the appropriate Taylor series have to be included.

Let

$$V_r(s) = V_c + \left.\frac{dV}{ds}\right|_c (s - s_c) + \frac{1}{2} \left.\frac{d^2 V}{ds^2}\right|_c (s - s_c)^2 + \dots \ ;$$

then from (11.39) we obtain changing the suffix 0 to c

$$\frac{d^2 V}{ds^2} = -\frac{2r_c}{r^2} A(r) - (s - s_c) \frac{d}{ds}\left(\frac{2r_c}{r^2} A(r)\right) .$$

Substituting this into the Taylor series we find for terms up to and including $(s - s_c)^2$:

$$\boxed{V_r(s) = V_c - \frac{A(r_c)}{r_c}(s - s_c)^2} \quad . \tag{11.44}$$

The large scale velocity field of galactic rotation thus appears as a linear velocity field with a $\sin 2l$-longitudinal dependence (11.37) in the solar neighbourhood, while it shows up as a quadratic velocity field (11.44) in the vicinity of the Thales point. According to the exposition in Sect. 11.4 such a velocity field will not only influence the observed frequency of the line, but also have an effect on the line shape.

Substituting (11.37) into (11.30) we find the peak optical depth for the local gas should have its lowest value in longitudes around $l \cong 45°$ with

$$\tau_{max} \leqq w \frac{N_H}{T_s} \frac{1}{A} \tag{11.45}$$

while $\tau_{max} \to \infty$ near $l \cong 90°$. Indeed the largest brightness temperatures in the local gas are observed near $l = 70°$ to $75°$ where values of $T_b \cong 125$ K are reached. If we therefore observe a local value of $T_b \cong 45$ K near $l = 45°$ the optical depth should be less than one in these regions resulting in

$$T_b = T_s \tau = w \frac{N_H}{A} \ . \tag{11.46}$$

With $A = 15\ \mathrm{km\,s^{-1}\,kpc^{-1}}$ this results in $N_H = 0.4\ \mathrm{cm^{-3}}$. That τ_{max} is indeed less than 1 in the local gas has been repeatedly stated by Burton (1974), p. 97. He concluded this from the intensity jump of the brightness temperature near $V = 0$ for longitudes $l < 90°$, $b = 0$. At these longitudes the radial velocity along the line-of-sight varies such that each radial velocity $V > 0$ is found at two positions, while for $V < 0$ this velocity is found only once. If the gas density remains reasonably constant, the resulting brightness temperature should be about a factor of 2 larger for $V > 0$ as for $V < 0$ if $\tau \ll 1$. Since such a ratio is just about observed, Burton concludes that $\tau \ll 1$.

Because we can assume such an optically thin situation, the results thus obtained remain true even if, in fact, the interstellar gas consists of several different phases, each with its own density and spin temperature. For the local gas the appropriate average values are $N_H = 0.4\ \mathrm{cm^{-3}}$, $T_s = 125$ K.

In the vicinity of the Thales points the velocity field is best described by the quadratic velocity field (11.44) so that the optical depth of the line radiation is described by substituting (11.44) into (11.31):

$$\tau(V) = \frac{w}{\sqrt{2\pi}\,\sigma}\,\frac{N_{\mathrm{H}}}{T_{\mathrm{s}}}\sqrt{\frac{r_c}{A_c}}\,P\left(\frac{V - V_c}{\sigma}\right) . \qquad (11.47)$$

If we may again assume that $\tau < 1$, then the high-velocity end of the line profiles may be described by

$$T_{\mathrm{b}}(V) = w\,\frac{N_{\mathrm{H}}}{\sqrt{2\pi}\,\sigma}\sqrt{\frac{r_c}{A_c}}\,P\left(\frac{V - V_c}{\sigma}\right) . \qquad (11.48)$$

In this expression V_c, σ, r_c and A_c all can be determined from the line measurements, therefore permitting the determination of $N_{\mathrm{H}}(r)$ from (11.48) for the subcentral points.

11.6 Modelling the Velocity and Density Distribution

Using (11.27) and simple velocity laws it is thus possible to obtain a general understanding of the influence of the large-scale velocity field on the shape and the radial velocity of the line radiation. If, however, we want to explain the detailed structure of the measured line profiles, we have to apply numerical procedures. We have to pay for this by the greater effort needed to find out which properties of the model are essential to the explanation and which are only subsidiary.

The inclusion of gas consisting of different phases with varying spin temperatures causes little added complication in the numerical models, but it is then necessary to leave the simple isothermal approach and to use the general solution of the equation of radiation transfer as given by (1.30). Usually no background radiation is present, so $T_{\mathrm{b}}(0) = 0$. The optical depth between the observer and the position s on the line of sight is given by (11.26); substituting this into (1.30) we obtain

$$T_{\mathrm{b}}(V) = w\int_0^\infty N_{\mathrm{H}}(s)\,\varphi[V - U(s)]\,\mathrm{e}^{-\tau(V,s)}\,ds \; . \qquad (11.49)$$

These two equations, (11.26) and (11.49), now permit the modelling of the line radiation $T_{\mathrm{b}}(V)$ to be measured if $N_{\mathrm{H}}(s)$, $T_{\mathrm{s}}(s)$ and $U(s)$ of the model are all specified. This has been used with considerable success by, amongst others, Burton (1971) by using for $U(s)$ the superposition of the general galactic rotation law and streaming motions as caused by spiral density waves with radial and tangential components

$$U_r = a_r \cos\chi(r,\theta) \quad \text{and}$$

$$U_\vartheta = a_\vartheta \sin\chi(r,\theta)$$

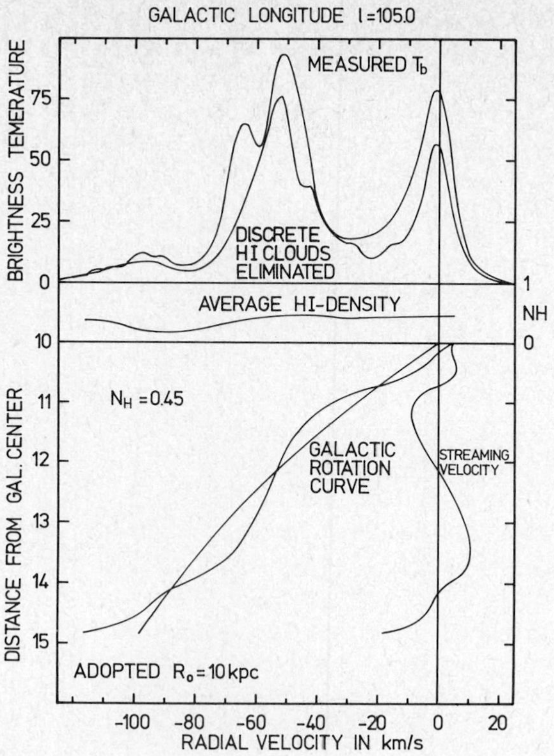

Fig. 11.9. The effect of differential galactic rotation and density wave streaming on the shape of line profiles

respectively, where $\chi(r, \theta)$ is the phase of the point (r, θ) in the spiral density wave and a_r and a_ϑ are streaming velocity amplitudes (see Fig. 11.9).

One of the results of these model computation was that the resulting line profiles differed very little, whether the gas density varied across the spiral arms or not; the kinematics of the gas was the governing influence. This can be understood if, in (11.26) and (11.49), the integration variable is changed into U to produce the expression (11.27) and

$$T_b(V) = w \int_{U(0)}^{U(\infty)} \frac{N_H(U)}{\left|\frac{dU}{ds}\right|} \varphi(V-U)\, e^{-\tau(V,U)} dU \quad . \tag{11.50}$$

We see now in both (11.27) and (11.50) that the observed $T_b(V)$ does not depend on N_H but that $N_H(U)/|dU/ds| = v(U)$ is significant. This is the gas density in velocity space, and small values of $|dU/ds|$ will produce what Burton calls "velocity crowding". These enhancements in the kinematic density are often much more important than true gas density variations. Burton (1972) made use of this by assuming $N_H = \text{const}$ along the line of sight and systematically adjusting $U(s)$ until a satisfactory fit of observed and computed line-profiles was obtained.

Modelling of the observed distribution of brightness temperature has become a tool that has been applied to various problems such as the explanation of the expanding 3 kpc arm by Simonson and Mader (1973), the large scale spiral structure of the Galaxy [Simonson (1976)] or the complicated kinematical state of the region near the galactic center [Peters (1975) and Burton and Liszt (1978)]. But as Rohlfs and Braunsfurth (1982) stressed, these methods need explicit statements of all model parameters. If some of the basic model features are still in dispute, it seems premature to use the model fitting approach.

12. Recombination Lines

12.1 Emission Nebulae

The interstellar medium pervades the whole galactic system, and interstellar gas is present practically everywhere as shown by the 21 cm-observations of neutral hydrogen. But the structure of this medium is rather irregular: on the one hand,

Fig. 12.1. The Orion nebula – a typical galactic H II region. ESO photograph obtained by unsharp masking technique to reveal the "inner" features from a plate obtained with the 2.2 m-telescope (courtesy ESO)

Fig. 12.2. The planetary nebula SP-1. ESO photograph with 3.6 m-telescope (courtesy ESO)

there are large regions with extremely low gas density and, on the other hand, we find large cloud complexes.

This medium is probably not in a state of equilibrium; violent internal motions are superposed on the general differential galactic rotation field. And the physical state of this medium varies just as violently from one region to the other because the gas temperature depends on the local energy input. There exist large, cool cloud complexes in which both dust grains and many different molecular species are abundant. The physics of the molecular line radiation will not be discussed until the next chapter, but we can state already now that the net effect of such radiation will be that the temperature and the gas pressure in some of these clouds become so low that the gas pressure can no longer support the cloud against the gravitational attraction of all the matter contained in it: the densest regions will collapse and new stars will form. These new stars are new sources of thermal energy and they will heat the gas surrounding them. If their surface temperature is sufficiently high, most of their energy will be emitted as photons with $\lambda < 912$ Å, that is, this radiation has sufficient energy to ionize hydrogen.

Occasionally such an ionized atom will recombine with one of the free electrons and, since the ionization rate is rather low, the time interval between two subsequent ionizations of one and the same atom will generally be much longer than the time it takes this excited atom to settle down to the ground state. In doing so this atom will emit the recombination lines. Young, luminous stars embedded in gas clouds therefore will be surrounded by emission regions (Fig. 12.1) in which the gas temperature and consequently the pressure will be high, much higher than in the cool nebulae. The emission nebulae therefore will expand, and this expansion is probably aided by strong stellar winds. Emission nebulae are thus transient phenomena which start off as dense, compact H II regions which expand more and more and end as fossil H II regions. The recombination time in fully evolved, low density H II regions is so long that these regions can survive as distinct objects for quite some time after the stellar energy sources have vanished from sight. H II regions are thus dynamical features with an evolutionary time scale comparable only to that of other extremely young objects.

There is, however, another class of H II regions that are related to objects of much greater age: the planetary nebulae (Fig. 12.2). Stars in a fairly advanced stage of evolution produce extended atmospheres which are only loosely bound to the parent stars and which have such large dimensions that they appear in visual observations with small telescopes as faintly luminous greenish disks resembling planets – hence their name.

In this chapter the physics of the line emission in these different objects will be discussed, and we will describe how the physical parameters of the nebulae can be derived from these observations.

12.2 Photoionization Structure of Gaseous Nebulae

Stellar photons with $\lambda < 912$ Å have sufficient energy to ionize neutral hydrogen if they are absorbed by the gas. After some time a stationary situation will form in the gas in which the recombinations will just balance the ionizations. The ionization rate by stellar radiation is fairly low in an average H II region resulting in a typical ionization time scale of 10^8 s; this is much longer than the time it takes an excited atom to settle down to the ground state by radiative transitions. If ionization by starlight is the dominant energy source of the gas, virtually all hydrogen atoms that are not ionized will be in the $1^2S_{1/2}$ ground level, and we need only to consider its photoionization cross section (Fig. 12.3). The detailed computations are complicated [see Rybicki and Lightman (1979), p. 282, or Spitzer (1978), p. 105, for further details]. Figure 12.3 shows the general frequency dependence of the ionization with $\sigma_\nu = 6.30 \cdot 10^{-22}\,\mathrm{m}^{-2}$ per atom for $\nu \cong \nu_1 = 3.3 \cdot 10^{15}$ Hz decreasing as ν^{-3} for $\nu > \nu_1$ and $\sigma_\nu \approx 0$ for $\nu < \nu_1$. The optical depth $\tau_\nu = \sigma_\nu N_{\mathrm{HI}} s$ therefore is large for $\nu \gtrsim \nu_1$, the mean free path (path s_0 for which $\tau = 1$) being

$$\left(\frac{N_{\mathrm{HI}}}{\mathrm{cm}^{-3}}\right)\left(\frac{s_0}{\mathrm{pc}}\right) = \frac{1}{20} \ . \tag{12.1}$$

Therefore, all neutral hydrogen atoms within s_0 will be ionized and N_{HI} will decrease; s_0 will subsequently increase until some equlibrium is reached. This is governed by the recombination rate of hydrogen.

When a proton recombines with a free electron the resulting hydrogen atom could be in any excited state. If α_i is the probability for recombination into the quantum state i, then the total recombination coefficient is

$$\alpha_t = \sum \alpha_i \ .$$

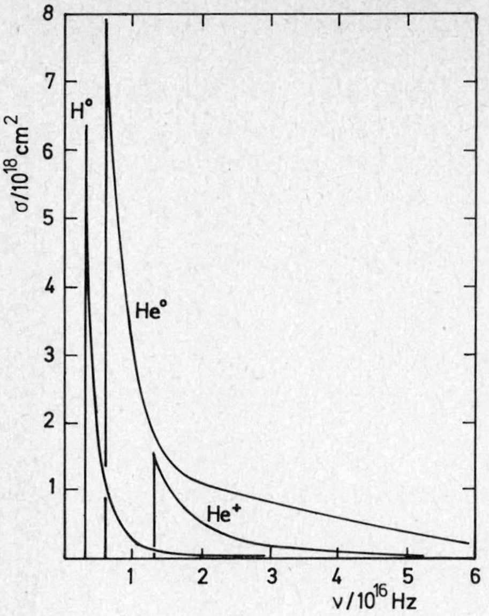

Fig. 12.3. The photoionization cross sections for H^0, He^0 and He^+ (Osterbrock 1974)

But since, on recombination, the excess energy of the free electron is radiated away as a photon, this photon energy is usually less than the ionization energy of hydrogen in the ground level; only if the atom recombines at this ground level, $h\nu_c > 13.6$ eV will always be true. Due to the large absorption coefficient this radiation will be quickly re-absorbed. It is therefore scattered throughout the nebula until the recombination occurs at $i > 1$. Therefore the effective recombination probability is given by

$$\alpha_t = \sum_{i=2}^{\infty} \alpha_i \ .$$

Surrounding a young, high temperature star we will therefore find an ionized region. For the details we would need to consider the detailed balance of the ionization equation; for this see the books by Spitzer (1978) or Osterbrock (1974).

Let the ionizing star emit N_{L_c} Lyman-continuum quanta per unit surface area, and let its radius be R_*, then the effective radius of the ionized region will be such that the total ionization rate caused by the star should just be equal to the total recombination rate of the surrounding H II region; that is,

$$4\pi R_*^2 N_{L_c} = \tfrac{4}{3}\pi N_e N_p \alpha_t s_0^3 \ . \tag{12.2}$$

If the nebula consists mainly of hydrogen, then $N_e \approx N_p$ and

$$\boxed{s_0 N_e^{2/3} = \left[\frac{3 R_*^2 N_{L_c}}{\alpha_t}\right]^{1/3} = U(SpT)} \ . \tag{12.3}$$

Except for some atomic properties the quantity in square brackets depends only on properties of the exciting stars; it does *not* depend on properties of the nebula.

Table 12.1. The flux of Lyman continuum photons N_{L_c} and the excitation parameter U for stars of spectral type O4-B1 (after Churchwell and Walmsley)

Sp T	T_{eff}/K	$\log(L_c/\text{photons s}^{-1})$	$U/\text{pc}\cdot\text{cm}^{2/3}$
O 4	52000	50.01	148
O 5	50200	49.76	122
O 6	48000	49.37	90
O 7	45200	48.99	68
O 8	41600	48.69	54
O 9	37200	48.35	41
O 9.5	34800	48.10	34
B 0	32200	47.62	24
B 0.5	28600	46.65	11
B 1	22600	45.18	3.5

N_{L_c} depends mainly on the surface temperature of the star, and this again is measured by its spectral type, so that the value of the right-hand side of (12.3) – the excitation parameter U will be a function of the spectral type only (see Table 12.1).

The average degree of ionization in the H II region can be estimated in the following way. Defining

$$x = \frac{N_p}{N_{HI} + N_p} \tag{12.4}$$

where N_{HI} is the neutral hydrogen gas density and N_p the number density of the protons, then using (12.1)

$$\frac{x}{1-x} = \frac{N_e}{N_{HI}} \cong 20\left(\frac{s_0}{\text{pc}}\right)\left(\frac{N_e}{\text{cm}^{-3}}\right)$$

or, with (12.3),

$$\boxed{\frac{x}{1-x} = 20\left(\frac{N_e}{\text{cm}^{-3}}\right)^{1/3} U\,(Sp\;T)} \quad . \tag{12.5}$$

Therefore $x/(1-x)$ is of the order of 10^3 in almost the whole of the H II region. The value of x changes abruptly only at the boundary between the H II and the H I region.

We can estimate the thickness of this transition layer by assuming a fractional ionization $x = 1/2$ in it, i.e. by putting $N_p \approx N_{HI}$. Then the thickness of this transition zone will be

$$\frac{\Delta s}{\text{pc}} = \frac{1}{20}\left(\frac{N_{HI}}{\text{cm}^{-3}}\right)^{-1} \approx 0.1\left(\frac{N_H}{\text{cm}^{-3}}\right)^{-1} , \tag{12.6}$$

a value that is usually quite small compared to that of s_0 in (12.3). The gas is therefore divided into two sharply separated regions: the almost completely ionized H II region surrounded by the almost completely nonionized H I region. The run of x with s can be determined explicitly using the ionization equation as first done by Strömgren (1939). But, for the more qualitative discussion here, these remarks should suffice.

So far we have not accounted for the fact that about 10% of the atoms in the interstellar gas are helium, not hydrogen. These will be ionized too, but, due to their ionization potential of 24.6 eV, the ionizing radiation has to be of shorter wavelength than needed for hydrogen. But each photon that can ionize He is able to ionize H too, while the reverse may not be true. For a precise discription of the situaion we would need a system of coupled differential equations, but we can obtain quite good approximations by doing some qualitative estimates.

The number of UV photons that can ionize H I is given by

$$\int_{\nu_1}^{\infty} \frac{L_\nu}{h\nu}\, d\nu = Q(\mathrm{H\,I})$$

and in a stationary situation, this number should equal the total number of H-recombinations. Therefore

$$Q(\mathrm{H\,I}) = \tfrac{4}{3}\pi s_0^3 N_\mathrm{e} N_\mathrm{p} \alpha_\mathrm{t}(\mathrm{H\,I}) \tag{12.7}$$

where s_0 is the radius of the H II zone and $\alpha_\mathrm{t}(\mathrm{H\,I})$ the total effective recombination coefficient for hydrogen. For He we obtain a quite similar expression

$$Q(\mathrm{He}) = \tfrac{4}{3}\pi s_1^3 N_\mathrm{e} N_\mathrm{He} \alpha_t(\mathrm{He}) \tag{12.8}$$

and, due to the different ionization potentials of H and He, $s_1 < s_0$. In the He II region the electrons can come from both H and He so, that $N_\mathrm{e}(\mathrm{He\,II}) = N_\mathrm{H} + N_\mathrm{He}$, while in the H I region outside s_1, only H contributes, so that $N_\mathrm{e}(\mathrm{H\,I}) = N_\mathrm{H}$. Dividing (12.7) and (12.8) we thus obtain

$$\boxed{\left(\frac{s_0}{s_1}\right)^3 = \frac{Q(\mathrm{H\,I})}{Q(\mathrm{He\,I})}\frac{N_\mathrm{He}}{N_\mathrm{H}}\left(1 + \frac{N_\mathrm{He}}{N_\mathrm{H}}\right)\frac{\alpha_t(\mathrm{He\,I})}{\alpha_t(\mathrm{H\,I})}}\ . \tag{12.9}$$

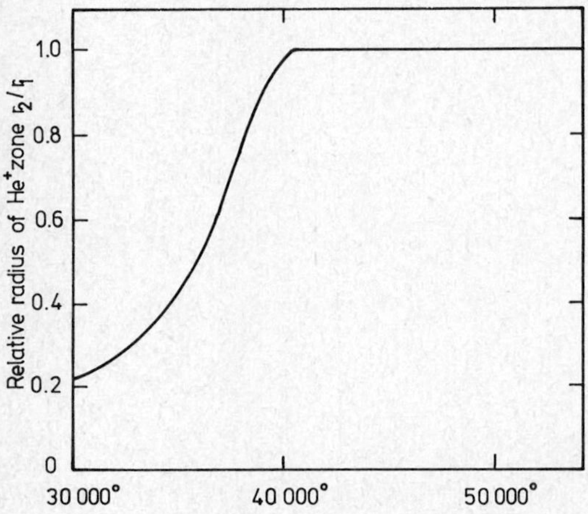

Fig. 12.4. The radius of the He II zone in units of the radius of the H II zone as function of the effective temperature of the exciting star (Osterbrock 1974)

In this, only the Q factors depend on the exciting star at the center of the nebula; the remaining factors depend only on the properties of the interstellar medium. In Fig. 12.4, Eq. (12.9) is plotted for different effective stellar temperatures, and we see that for $T_* > 37\,000$ K, that is, for stars earlier than O 7 the H II regions and the He II regions have the same size. This is confirmed by detailed solutions of the appropriate differential equations for x_{H} and x_{He}. But if the central star has a lower temperature, the size of the H II and the He II regions will differ greatly. The reason for this is found in the dependence of $Q\,(\mathrm{H\,I})/Q\,(\mathrm{He\,I})$ on the stellar temperature.

12.3 Physical Line Parameters

When ionized hydrogen recombines at some level with the principal quantum number $n > 1$, the atom will emit recombination line emission on cascading down to the ground state. Since the atomic diameter increases considerably with the quantum number, the radius of the nth Bohr orbit is

$$a_n = \frac{\varepsilon_0}{\pi}\frac{h^2}{m\,e^2}\,n^2\;, \tag{12.10}$$

and the energy levels with high principal quantum numbers are quite closely spaced. Pressure effects at large n caused by atomic collision may become important, eventually causing the different lines to merge. The Inglis-Teller formula gives a semi-empirical relation between the maximum number of resolvable lines $n_{\max}$ and the electron density

$$\log\left(\frac{N_{\mathrm{e}}}{\mathrm{cm}^{-3}}\right) = 23.26 - 7.5\log n_{\max}\;. \tag{12.11}$$

For $N_{\mathrm{e}} < 10^6\,\mathrm{cm}^{-3}$ this gives $n_{\max} > 200$ so that lines with very large quantum numbers should be observable. The frequency of the atomic lines of hydrogen-like atoms are given by the Rydberg formula

$$\nu_{\mathrm{ki}} = Z^2 R_M\left(\frac{1}{i^2} - \frac{1}{k^2}\right)\,, \qquad i < k \qquad \text{where} \tag{12.12}$$

$$R_M = \frac{R_\infty}{1 + \dfrac{m}{M}} \tag{12.13}$$

if m is the mass of the electron, M that of the nucleus and Z is the effective charge of the nucleus in units of the proton charge. For $n > 100$ we always have $Z \approx 1$ and the spectra of all atoms are quite hydrogen-like, the only difference being a slightly changed value of the Rydberg constant (Table 12.2).

Since so many different recombination lines are possible, but only certain kinds of transitions produce reasonably strong lines, a precise but concise desig-

Table 12.2. The Rydberg constant for the most abundant atoms

Atom	atomic mass (a.m.u.)	$R_A/(\mathrm{Hz})$	$\Delta V/(\mathrm{km\,s^{-1}})$
H^1	1.007825	$3.288\,051\,29(25) \cdot 10^{15}$	0.000
He^4	4.002603	3.289 391 18	122.166
C^{12}	12.000000	3.289 691 63	149.560
N^{14}	14.003074	3.289 713 14	151.521
O^{16}	15.994915	3.289 729 19	152.985
	∞	3.289 842 02	163.272

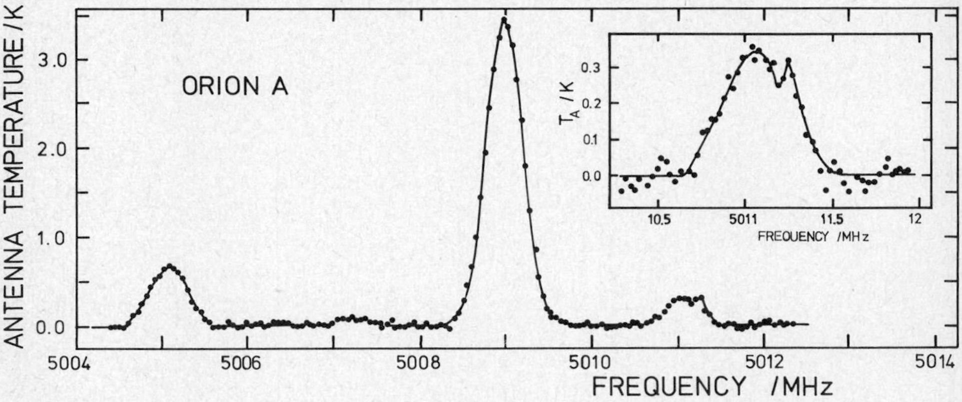

Fig. 12.5. Recombination lines in the Orion nebula near 5 GHz. The lines (from left to right) H 137 β, He 137 β, H 109 α, He 109 α, C 109 α are clearly visible. In the inset the lines He 109 α and C 109 α are shown in more detail

nation is needed. Lines corresponding to the transitions $n + 1 \rightarrow n$ are strongest and they are called α lines; those for transitions $n + 2 \rightarrow n$ are β lines; $n + 3 \rightarrow n$ transitions yield γ lines; etc. In addition both the element and the principle quantum number of the lower state are given: so H 109 α is the line corresponding to the transition $110 \rightarrow 109$ of H while He 137 β corresponds to $139 \rightarrow 137$ of He.

α-transitions with $n > 60$ produce lines with $\lambda > 1$ cm in the radio wavelength range. Kardashev (1959) first showed that such lines might be observable; they were first positively detected by Höglund and Mezger (1965). All radio recombination lines are fairly weak. Even in such bright H II regions as the Orion nebula M 42, peak line brightness temperatures of only 5 K are observed (Fig. 12.5); other H II regions give much weaker signals.

The linewidth of interstellar radio recombination lines is governed by external effects; neither the intrinsic linewidth nor the fine structure of the atomic levels has observable consequences. Broadening of the lines by the Stark effect due to inelastic collisions should be expected at the observed gas densities, but the situation is as yet unclear; the observed line width is fully explainable by Doppler broadening. One part of this is explainable by thermal Doppler broadening. The electrons will have a velocity distribution that is described very closely by a Maxwellian velocity distribution, and any deviation of this will decay with a relaxation time that is exceedingly short. This distribution is characterized by an electron temperature T_e, and, due to the electrostatic forces, the protons should have a similar distribution with the same temperature. The observed spectral line should therefore be of Gaussian shape with a full Doppler width at half intensity

$$\Delta\nu = \frac{2\,\nu_{ik}}{c}\left(\frac{2\,k\,T_e}{M}\ln 2\right)^{1/2} . \tag{12.14}$$

But, comparing (12.14) with the observed linewidths, we find that thermal Doppler motions corresponding to $T_e \cong 10^4$ K will produce a linewidth of only 21.4 km s^{-1} while about 30 km s^{-1} is observed. The shape of the line, however, is as closely Gaussian as can be expected. Therefore it is likely that nonthermal motions in the gas contribute to the broadening. These motions are usually called "micro-turbulence" and, with its halfwidth v_t (12.14) is generalized to

$$\Delta\nu = \frac{2\,\nu_{ik}}{c}\left[2\left(\frac{k\,T_e}{M} + v_t^2\right)\ln 2\right]^{1/2} . \tag{12.15}$$

12.4 Line Intensities Under Conditions of Local Thermodynamic Equilibrium

To compute the absorption coefficient as given by (10.17) for a given recombination line of hydrogen in local thermodynamic equilibrium several parameters for his transition must be further specified. For the statistical weight g_n of the level with the principal quantum number n, quantum theory gives

$$g_n = 2\,n^2 . \tag{12.16}$$

The transition probability A_{ki} for hydrogen has been determined by many different authors; a convenient review is given by Lang (1974), p. 118. But due to the correspondence principle, the data for high quantum numbers can be computed by using classical methods, and therefore we will use for A_{ki} the expression (10.24) for the electric dipole. For the change of the dipole moment in the transition $n+1 \rightarrow n$ we put

$$\mu_{n+1,n} = \frac{e\,a_n}{2} = \frac{\varepsilon_0}{2\pi}\frac{h^2}{m\,e}\,n^2$$

where $a_n = a_0\,n^2$ is the Bohr radius of hydrogen, and put correspondingly

$$\nu_{n+1,n} = \frac{m\,e^4}{4\,\varepsilon_0^2\,h^3\,n^3} .$$

Substituting this into (10.24) we obtain for the limit of large n

$$A_{n+1,n} = \frac{\pi\,m\,e^{10}}{48\,\varepsilon_0^5\,h^6\,c^3}\frac{1}{n^5} = \frac{5.36\cdot 10^9}{n^5}\,\mathrm{s}^{-1} . \tag{12.17}$$

For the line shape $\varphi(\nu)$ a Gaussian will be adopted. Introducing the full linewidth $\Delta\nu$ at half intensity points we obtain for the value of φ at the line center

$$\varphi(0) = \left(\frac{\ln 2}{\pi}\right)^{1/2} \frac{2}{\Delta\nu} . \tag{12.18}$$

Another factor in (10.17) is N_n, the density of atoms in the state n. This is given by the Saha-Boltzmann equation [cf. Osterbrock (1974), p. 61, Spitzer (1978), Sect. 2.4, or Rybicki and Lightman (1979), Eq. (9.45)]

$$N_n = n^2 \left(\frac{h^2}{2\pi m k T_e}\right)^{3/2} e^{X_n/k T_e} N_p N_e \tag{12.19}$$

where

$$X_n = h\nu_0 - \chi_n = \frac{h\nu_0}{n^2} \tag{12.20}$$

is the ionization potential of the level n.

Substituting Eqs. (12.16) to (12.19) into (10.17) and remembering that $X_n \ll k T_e$ for lines in the radio range so that $\exp(X_n/k T_e) \cong 1$ and $1 - \exp(-h\nu_{n+1,n}/k T_e) \cong h\nu_{n+1,n}/k T_e$, we obtain for the optical depth in the center of a line emitted in a region with the emission measure

$$\mathrm{EM} = \int N_e(s) N_p(s)\, ds = \int \left(\frac{N_e(s)}{\mathrm{cm}^{-3}}\right)^2 d\left(\frac{s}{\mathrm{pc}}\right) , \tag{12.21}$$

$$\tau_L = 1.92 \cdot 10^3 \left(\frac{T_e}{\mathrm{K}}\right)^{-5/2} \left(\frac{\mathrm{EM}}{\mathrm{cm}^{-6}\,\mathrm{pc}}\right) \left(\frac{\Delta\nu}{\mathrm{kHz}}\right)^{-1} . \tag{12.22}$$

Here we have put $N_p(s) \approx N_e(s)$ which should be true with good precision due to the large abundance of H. We will practically always find that $\tau_L \ll 1$, and therefore that $T_L = T_e \tau_L$, or

$$T_L = 1.92 \cdot 10^3 \left(\frac{T_e}{\mathrm{K}}\right)^{-3/2} \left(\frac{\mathrm{EM}}{\mathrm{cm}^{-6}\,\mathrm{pc}}\right) \left(\frac{\Delta\nu}{\mathrm{kHz}}\right)^{-1} . \tag{12.23}$$

Table 12.3. Dependence of T_L/T_e on the frequency

$\frac{\nu}{\mathrm{GHz}}$	$\frac{\lambda}{\mathrm{cm}}$	T_L/T_e
300	0.1	1.95
100	0.3	0.65
30	1	0.20
10	3	0.06
3	10	0.02
1.5	20	0.01
0.6	50	0.004

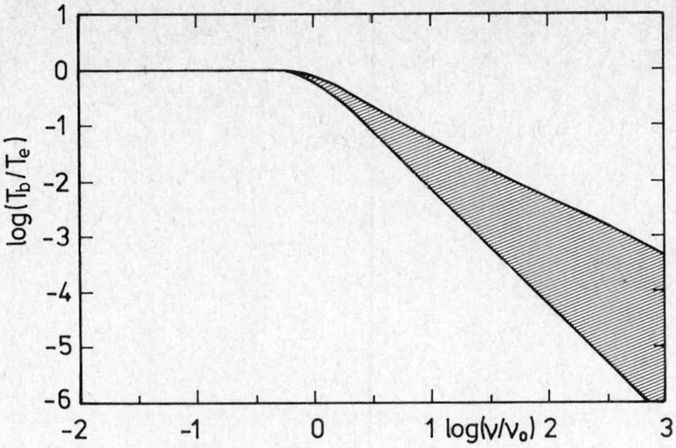

Fig. 12.6. Thermal continuum and recombination lines. Recombination lines are visible only in the hatched region

For frequencies above 1 GHz usually $\tau_c < 1$ for continuous radiation too, so that we obtain on dividing (12.22) by (8.28)

$$\frac{T_L}{T_c}\left(\frac{\Delta\nu}{\text{kHz}}\right) = \frac{2.330 \cdot 10^4}{a(\nu, T_e)}\left[\frac{\nu}{\text{GHz}}\right]^{2.1}\left[\frac{T_e}{\text{K}}\right]^{-1.15} \frac{1}{1 + N(\text{He})/N(\text{H})} \quad . \tag{12.24}$$

(Typical values are given in Table 12.3). The last factor is due to the fact that both N_H and N_{He} contribute to the electron density $N_e = N_H + N_{He}$ while N_p in (12.19) is due to N_H only. Equation (12.24) is valid only if both the line radiation and the continuous radiation are optically thin, but the effect of a finite optical depth is easily estimated.

Let both the line and the continuous radiation be emitted by the same cloud with an electron temperature T_e. At the line center, the brightness temperature

$$T_{bL} = T_e(1 - e^{-(\tau_L + \tau_c)})$$

is observed and, at frequencies adjacent to the line, we obtain

$$T_{bc} = T_e(1 - e^{-\tau_c})$$

so that, for the brightness temperature of the line alone, we find

$$T_L = T_{bL} - T_{bc} = T_e\, e^{-\tau_c}(1 - e^{-\tau_L}) \; . \tag{12.25}$$

Therefore, if $\tau_c \gg 1$, no recombination lines are visible (Fig. 12.6); they are observed only if τ_c is small! This is simply another version of the general principle that optically thick thermal radiation approaches black-body radiation and, in black-body radiation, there are no lines!

Recombination lines occur over an extremely wide frequency range, from the decimeter range down to the ultraviolet just to the long wavelength side of the Lyman limit at $\lambda = 912$ Å. It is therefore of interest to investigate at which frequency to observe for a given research subject. From the preceding considerations it is obvious that low frequencies, at which the continuous radiation becomes optically thick, must be avoided. But are there also limits for the highest frequency that can be usefully employed? In Sect. 12.3 we showed that the line-width $\Delta\nu$ is predominantly determined by the turbulent Doppler velocities in the gas. But then $\Delta\nu/\nu = v/c$, and thus (12.23) results in $T_L \propto \nu^{-1}$. Thus the amplitude of the recombination line decreases with the frequency, but from (12.24) we similarly derive that the contrast of the line to the continuous background increases with $T_L/T_c \propto \nu^{1.1}$. The data in Table 12.3 illustrating this are taken from Kardashev (1959). Therefore recombination line radiation of extended diffuse objects is best observed at low frequencies in order to maximise T_L/T_c, staying however well above the limiting frequency (8.29) where the source becomes optically thick. Compact H II regions with diameters less than the angular resolution of the telescope are in contrast best observed at the shortest possible wavelength permitted by both telescope, radiometer, and atmosphere in order to maximise T_{AL} leaving T_{AL}/T_{AC} unchanged.

12.5 Line Intensities when Conditions of Local Thermodynamic Equilibrium Do Not Apply

The diameter of hydrogen atoms is strongly dependent on the principle quantum number [cf. Eq. (12.10); numerical values are given in Table 12.4]. Atoms in highly excited states are large, very tenous objects and it is therefore to be expected that collisions could affect them more easily. A test for the presence of such influence of mutual collisions which is independent of instrumental influences therefore should be useful. Let us consider the ratio of the optical depth for two recombination lines which have approximately the same frequency but correspond to different upper (k, k') and lower (i, i') quantum levels. From (10.17) we have

$$\frac{\Delta\nu\, T_L}{\Delta\nu'\, T_L'} = \frac{g_i' g_k N_i A_{ki}}{g_k' g_i N_i' A_{ki}'}\,, \tag{12.26}$$

but at local thermodynamic equilibrium, the density of the different states is governed by the Boltzmann distribution (10.16)

$$\frac{N_i}{N_i'} = \frac{g_i}{g_i'} \exp\left\{-\frac{X_i - X_i'}{k\, T_e}\right\}\,.$$

Table 12.4. Diameter and average density of the hydrogen atom as a function of the principle quantum number

n	Diameter (m)	ϱ/ϱ_1
1	$1.06 \cdot 10^{-10}$	1
10	$1.06 \cdot 10^{-8}$	10^{-6}
100	$1.06 \cdot 10^{-6}$	10^{-12}
200	$4.24 \cdot 10^{-6}$	$1.6 \cdot 10^{-14}$
300	$9.52 \cdot 10^{-6}$	$1.4 \cdot 10^{-15}$

Since

$$X_i - X_i' \ll k\,T_e \ ,$$

we arrive at

$$\frac{\Delta\nu\, T_L}{\Delta\nu'\, T_L'} = \frac{g_k}{g_k'}\frac{A_{ki}}{A_{ki}'} \ . \tag{12.27}$$

The right-hand side of this equation contains only atomic quantities and can thus be computed theoretically, while the left-hand side contains observable quantities. There are many possible candidates for this test; amongst others the lines H 110 α, H 138 β, H 158 γ, H 173 δ and H 186 ε are all at frequencies close to 4.9 GHz, but numerous other combinations can be given.

The simplification from (12.26) to (12.27) works only, if the Boltzmann distribution applies; that is, it only holds under LTE conditions. If the observations give a result that differs from the theoretically expected one, we can be certain that LTE does not apply. Unfortunately the conclusion is not quite as unique the other way round. If the observations give a result that agrees with (12.27) we still cannot be certain that LTE applies since the different NLTE effects sometimes seem to be "conspiring" to force the fulfilment of (12.27). This has been discussed by Seaton (1980). But let us now investigate, how such deviations from LTE will affect the recombination line radiation.

By reversing the arguments leading to (10.16), it is clear that NLTE is equivalent to deviations from the Boltzmann distribution (10.16). We have already met this situation in low temperature gas when the brightness temperature T_b of the ambient radiation field and the kinetic temperature of the gas T_K differ. We can then describe the relative population of the two lowest states by introducing an excitation temperature T_{ex} that is some appropriate mean of T_b and T_K according to (10.39). This was the solution for the 21 cm-line of neutral hydrogen (11.13), and it would be quite possible to do this too for the recombination lines. But then we would find a different excitation temperature for each line. Therefore the procedure adopted by Menzel (1937) has been universally accepted. He introduced so-called "*departure coefficients* b_n" such that the actual level population N_n is given by

$$N_n = b_n N_n^* \tag{12.28}$$

where N_n^* is the level population for local thermodynamic equilibrium. Therefore $b_n \to 1$ for LTE, and

$$\frac{N_k}{N_i} = \frac{b_k}{b_i}\frac{g_k}{g_i}\exp\left\{-\frac{X_k - X_i}{k\,T_e}\right\} = \frac{b_k}{b_i}\frac{g_k}{g_i}\,e^{-h\nu_{ki}/kT_e} \ . \tag{12.29}$$

Similarly to (10.17) we obtain

$$\kappa_\nu = \frac{c^2}{8\pi}\frac{1}{\nu_{ki}^2}\frac{g_k}{g_i}N_i A_{ki}\left(1 - \frac{b_k}{b_i}e^{-h\nu_{ki}/kT_e}\right)\varphi(\nu) \tag{12.30}$$

so that we can write [Goldberg (1968)]

$$\boxed{\kappa_\nu = \kappa_\nu^* \, b_i \, \beta_{ik}} \quad \text{where} \tag{12.31}$$

$$\boxed{\beta_{ik} = \frac{1 - \dfrac{b_k}{b_i} \exp\left\{-\dfrac{h\,\nu_{ki}}{k\,T_e}\right\}}{1 - \exp\left\{-\dfrac{h\,\nu_{ki}}{k\,T_e}\right\}}} \; , \tag{12.32}$$

b_i is the departure coefficient of the lower level, and κ_ν^* is the absorption coefficient as given by (10.17) for local thermodynamic equilibrium. Again since $h\,\nu_{ki} \ll k\,T_e$ (12.32) can be simplified to

$$\beta_{ik} = \left[1 - \frac{b_k}{b_i}\left(1 - \frac{h\,\nu_{ki}}{k\,T_e}\right)\right] \frac{k\,T_e}{h\,\nu_{ki}} \quad \text{or}$$

$$\boxed{\beta_{ik} = \frac{b_k}{b_i}\left[1 - \frac{k\,T_e}{h\,\nu_{ki}} \frac{b_k - b_i}{b_k}\right]} \tag{12.33}$$

which for $|b_k - b_i| \ll b_k$ is equal to

$$\boxed{\beta_{ik} = \frac{b_k}{b_i}\left[1 - \frac{k\,T_e}{h\,\nu_{ki}} \left.\frac{d \ln b_n}{dn}\right|_i (k - i)\right]} \; . \tag{12.34}$$

Substituting numerical values for the physical constants we obtain for $\beta = (b_i/b_k)\,\beta_{ik}$

$$\boxed{\beta = 1 - 20.836 \left(\frac{T_e}{\mathrm{K}}\right)\left(\frac{\nu}{\mathrm{GHz}}\right)^{-1} \frac{d \ln b_n}{dn}\,\Delta n} \tag{12.35}$$

and

$$\boxed{\kappa_\nu = \kappa_\nu^* \, b_k \, \beta} \tag{12.36}$$

where the subscript k is the principal quantum number of the upper level. For all departure coefficients we find $b_n < 1$ but $b_n \to 1$ for $n \to \infty$; β however, can differ considerably from 1 and frequently attains negative values. This means that $\kappa_\nu < 0$; that is, we have maser amplification. In order to obtain an indication of how the line intensities are affected by this, the equation of transfer has to be solved. Inspecting the definition (10.15) for the emissivity ε_ν, we obviously must put

$$\varepsilon_L = \varepsilon_L^* \, b_k \tag{12.37}$$

where $\varepsilon_{\mathrm{L}}^{*}$ again is the appropriate value for LTE so that, according to Kirchhoff's law,

$$\frac{\varepsilon_{\mathrm{L}}^{*}}{\kappa_{\mathrm{L}}^{*}} = B_{\nu}(T) \ . \tag{12.38}$$

The equation of transfer (1.9) then becomes

$$-\frac{dI_{\nu}}{d\tau_{\nu}} = S_{\nu} - I_{\nu}$$

with the source function

$$S_{\nu} = \frac{\varepsilon_{\nu}}{\kappa_{\nu}} = \frac{\varepsilon_{\mathrm{L}}^{*}\, b_{k} + \varepsilon_{c}}{\kappa_{\mathrm{L}}^{*}\, b_{i}\, \beta_{ik} + \kappa_{c}} \ . \tag{12.39}$$

Using (12.38) this can be written as

$$S_{\nu} = \eta_{\nu}\, B_{\nu}(T) \qquad \text{where} \tag{12.40}$$

$$\eta_{\nu} = \frac{\kappa_{\mathrm{L}}^{*}\, b_{k} + \kappa_{c}}{\kappa_{\mathrm{L}}^{*}\, b_{i}\, \beta_{ik} + \kappa_{c}} \ . \tag{12.41}$$

Kirchhoff's law is therefore not valid at NLTE. For an isothermal slab of material with constant density, we then obtain for the brightness temperature at the line center

$$T_{\mathrm{L}} + T_{c} = \eta_{\nu}\, T_{\mathrm{e}}(1 - \mathrm{e}^{-\tau_{\mathrm{L}} - \tau_{c}}) \quad \text{or}$$

$$r = \frac{T_{\mathrm{L}}}{T_{c}} = \eta_{\nu}\, \frac{1 - \mathrm{e}^{-\tau_{\mathrm{L}} - \tau_{c}}}{1 - \mathrm{e}^{-\tau_{c}}} - 1 \ . \tag{12.42}$$

Under conditions of LTE $b_{i} = 1$ and $\beta_{ik} = 1$ so that $\eta_{\nu} = 1$ and

$$r^{*} = \frac{T_{\mathrm{L}}^{*}}{T_{c}} = \frac{1 - \mathrm{e}^{-\tau_{\mathrm{L}}^{*} - \tau_{c}}}{1 - \mathrm{e}^{-\tau_{c}}} - 1 \ . \tag{12.43}$$

Dividing (12.42) and (12.43) we obviously have

$$\frac{T_{\mathrm{L}}}{T_{\mathrm{L}}^{*}} = \frac{r}{r^{*}}$$

so that this ratio describes the influence of NLTE effects on the line intensity. Expanding the exponentials in (12.42) and retaining the quadratic terms in τ_{L} and τ_{c} we obtain

$$r = \eta_{\nu}\, \frac{\tau_{\mathrm{L}}(1 - \frac{1}{2}\tau_{\mathrm{L}})}{\tau_{c}(1 - \frac{1}{2}\tau_{c})} - 1 \ . \tag{12.44}$$

Substituting (12.41) for η_ν and (12.30) and (12.31) for κ_ν with $\tau_L = -\kappa_L s$ and $\tau_c = -\kappa_c s$ we find that

$$\frac{r}{r^*} = b_k - \frac{1}{2}\tau_c b_k \left[1 + \frac{b_i}{b_k} \beta_{ik} (1 + r^* b_k) \right] . \tag{12.45}$$

In most cases of interest $|\beta_{ik}| \gg 1$ and $r^* b_k \ll 1$; hence

$$\boxed{\frac{r}{r^*} = \frac{T_L}{T_L^*} = b_k (1 - \tfrac{1}{2}\tau_c \beta)} . \tag{12.46}$$

The first term of (12.46) accounts for NLTE line *formation* effects while the second describes NLTE *transfer* effects, that is, Maser amplification of the line radiation.

In order to be able to apply these concepts to real measurements the departure coefficients b_n for a given H II region must be known. But this requires that all important processes affecting the level population have to be investigated. These are

1) Radiative capture and cascade down to the ground level;
2) Collisional excitation and de-excitation by electrons and protons;
3) Collisional ionization and three-body recombination;
4) Redistribution of angular momentum by collison.

The rate equation (10.30) for level s may be written

$$\begin{aligned} &\sum_{r \neq s} (N_r C_{rs} + N_r B_{rs} U_{rs}) + \sum_{r > s} N_r A_{rs} + N_e N_i (\alpha_{is} + C_{is}) \\ &\quad = N_s \sum_{l < s} A_{sl} + N_s \sum_{l \neq s} (C_{sl} + B_{sl} U_{sl}) + N_s C_{si} \end{aligned} \tag{12.47}$$

where N_e is the electron density, N_i the ion density, A_{rs} the spontaneous Einstein coefficient for the transition $r \to s$ and B_{rs} and B_{sr} the corresponding coefficients for stimulated emission and absorption. $U_{rs} = 4\pi I_\nu / c$ is the radiation density at the frequency ν_{rs} corresponding to the transition $r \to s$, and α_{is} is the radiative recombination coefficient for transitions to the level s. Finally, C_{si} represents the collisional ionization rate for transitions for the level s, and C_{is} the corresponding three-body recombination rate.

Using a procedure described by Dupree (1969) this can be rewritten to arrive at the equation

$$\sum_{r = s - s_0}^{r = s + s_0} R_{rs} b_s = S_s \quad \text{where} \tag{12.48}$$

$$R_{rs} = \begin{cases} -\dfrac{g_r}{g_s} e^{\chi_r - \chi_s} (C_{rs} + B_{rs} U_{rs} + A_{rs}) ; & r < s \\ \displaystyle\sum_{l < s} A_{sl} + \sum_{\substack{l \neq s \\ s - s_0}}^{s + s_0} (C_{sl} + B_{sl} U_{sl}) + C_{si} ; & r = s \\ -\dfrac{g_r}{g_s} e^{\chi_r - \chi_s} (C_{rs} + B_{rs} U_{rs}) ; & r > s \end{cases} \tag{12.49}$$

and

$$S_n = \sum_{r=n+n_0+1}^{\infty} b_r \frac{g_r}{g_n} e^{X_r - X_n} A_{rn} + C_{ni} + \frac{(2\pi m k T_e)^{3/2}}{h^3} \frac{2 g_i}{g_n} e^{-X_n} \alpha_{in} . \qquad (12.50)$$

Here g_n and g_i are the statistical weights of an electron at the bound state n and in the continuum,

$$X_n = 1.58 \cdot 10^5 \left(\frac{T_e}{\mathrm{K}}\right)^{-1} \frac{1}{n^5}$$

and n_0 is the maximum value of $\Delta r = |r - n|$ for which collisions and stimulated radiative transitions are considered.

The rate equations (12.48) to (12.50) are written here in a form as given by Ungerechts and Walmsley (1978); the solution of this system is mainly a numerical problem once the transition rates A_{rs}, B_{rs}, I_ν and the collision rates are specified. Approximate numerical solutions were first given by Menzel and Pekeris (1935); in the meantime larger and larger systems have been solved, the most recent and complete being those of Sejnovskī and Hjellming (1969), Brocklehurst (1970), and for low temperature regions those of Ungerechts and Walmsley (1978).

Usually two major cases are considered:

Case A: The nebula is optically thin in all lines. Each photon leaves the nebula.

Case B: The Lyman photons are scattered many times until they are broken up into two photons belonging to higher transitions.

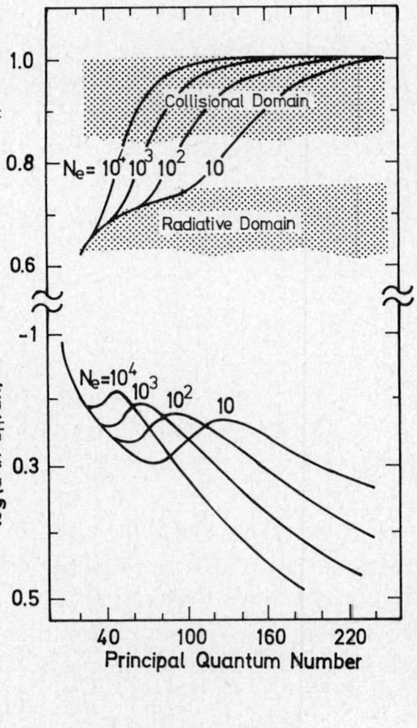

Fig. 12.7. Population departure coefficients b_n and its differential variation $d \ln b_n/dn$ plotted against the principal quantum number for $T_e = 10^4$ K and different N_e (Chaisson 1976)

In most nebulae case B most likely applies. The collision rates C_{rs} entering in (12.48) to (12.50) have to be computed by quantum mechanical methods, but according to (10.36) they obviously will depend on the ambient particle density. In low density regions the b_n factors therefore will be determined mainly by the radiation field. The relative importance of collisions and the radiation field are, however, strongly dependent on the principal quantum number of the level considered. In a diagram of b_n-values as functions of the quantum number n (Fig. 12.7) we can therefore distinguish two main regions: a *radiative domain* for low n independent of the ambient density and a *collisional domain* where the b_n values depend strongly on the electron density N_e.

12.6 Interpretation of Radio Recombination Line Observations

The intensity of the recombination lines as it results from the arguments presented in the preceding sections seems to be remarkably insensitive to gas density. It is true that τ_L or, if we use the terminology of the preceding section, τ_L^* as given by (12.22) depends on $\mathrm{EM} = \int N_e^2\, ds$ but, since τ_c also depends on EM, $r^* = T_L^*/T_c$ for an isothermal gas cloud is independent of the gas density according to (12.24). This is modified only slightly if the density dependence of the departure coefficients b_n and β is taken into account.

Nevertheless density variations are of great importance for a realistic interpretation of recombination line measurements. Rydberg atoms, that is atoms in highly excited states, are large diameter objects of low density that are easily perturbed by collisions. Their effect on the population of the energy levels is described by the b_n factors; the perturbation of the electron in the excited state will result in pressure-broadening of the emitted lines. This is not so much caused by the quasi-static electric fields of the colliding ions (Stark effect) as by random phase perturbation of the emitted line (impact effects) which have been investigated by Griem (1967). This results in a Lorentz dispersion profile but, since this will be convolved by the Doppler profile caused by the random velocities of the atoms, the line profile eventually will be a Voigt function [see Rybicki and Lightman (1979), Eq. (10.76)].

Brocklehurst and Seaton (1972) find

$$\frac{\Delta\nu_I}{\Delta\nu_D} = 0.142\left(\frac{n}{100}\right)^{7.4}\left(\frac{N_e}{10^4\,\mathrm{cm}^{-3}}\right)\left(\frac{T_e}{10^4\,\mathrm{K}}\right)^{-0.1}\left(\frac{T_D}{2\cdot 10^4\,\mathrm{K}}\right)^{-1/2} \tag{12.51}$$

for the ratio of the dispersion profile for the H recombination lines and the half-power linewidth of the Doppler broadening in a medium with $T_e = 1\cdot 10^4$ K and an equivalent Doppler temperature $T_D = 2\cdot 10^4$ K.

The impact linewidth thus depends very strongly on the principal quantum number of the line. For $N_e = 10^4\,\mathrm{cm}^{-3}$ we find $\Delta\nu_I/\Delta\nu_D = 0.14$ if $n = 100$, while for $n = 150$ $\Delta\nu_I/\Delta\nu_D = 20$. Lines with such large impact widths are not detectable with presently used techniques in which rather flexible baselines are fitted to the measured profiles in order to remove instrumental baseline ripple and this

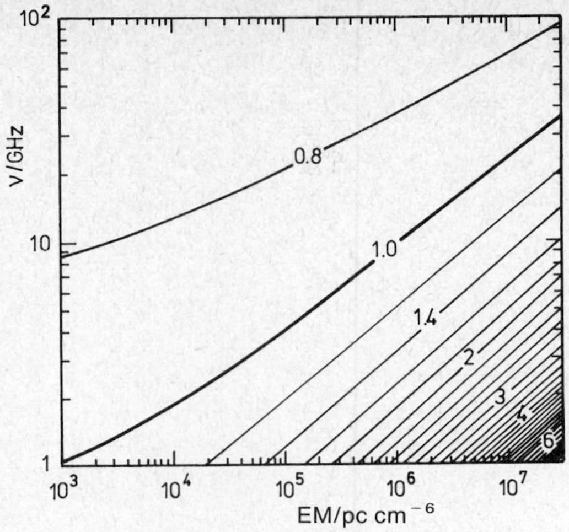

Fig. 12.8. The ratio T_e/T_e^* of the apparent electron temperature T_e^* according to an LTE analysis if emission measures EM and observing frequencies ν are given. An excitation parameter $U = 100$ pc cm$^{-2/3}$ and a true electron temperature $T_e = 10^4$ K have been adopted (Shaver 1980)

procedure automatically removes possibly existing wide shallow wings of the lines. Therefore there exists a rather sharply defined maximum principal quantum number n_{max} for which the recombination lines of gas of given density can be detected. Thus the Inglis-Teller formula (12.11) can be derived from (12.51) to give it a theoretical background.

If we therefore observe the recombination line radiation of a strongly clumped cloud, the line radiation of the high-density parts will not be measured if $n > n_{max}$ for this density. Lines of a given n can therefore only be detected for a gas with a density below a critical density N_n. High n lines are indicative of a gas of low gas density while high-density gas can only be detected by low n lines.

The interpretation of recombination line measurements is therefore not a straightforward thing, even if we disregard all NLTE-effects for the time being. This has been stressed particularly by Lockman and Brown (1978), and they show that we can only obtain insight into the complicated interrelation of the different factors that influence the intensity of the line by doing explicite model calculations. Summaries of these ideas are presented in the review article by Brown, Lockman and Knapp (1978) and in the proceedings of the recombination line workshop at Ottawa [Shaver (1980)] in the articles by Walmsley and by Brown.

One of the reasons why recombination line measurements meet with such great interest is that they seem to provide efficient means of determining the electron temperature T_e of gaseous nebulae (Fig. 12.8). However, as shown in this and the preceding section, T_e depends in a rather complicated way on T_L/T_e, $\langle N_e \rangle$ and $\langle N_e^2 \rangle$, that is, on the clumping of the gas. The only way to treat this complicated dependence properly is by explicite numerical model computations, but better insight into the physics is gained by the following approximate method. Let us consider a nebular model in which

1) the nebular structure is plane-parallel, homogeneous and isothermal;
2) all optical depths are small: $|\tau_L + \tau_c| \ll 1$ and $\tau_c \ll 1$;
3) the lines can be treated as if they were formed and transferred in LTE: $b_n = 1$, $\beta = 1$.

For this particular model, (12.24) is valid and can be solved for T_e which we will call T_e^* here:

$$\frac{T_e^*}{\mathrm{K}} = \left[\frac{2.330 \cdot 10^4}{a(\nu, T)} \left(\frac{\nu}{\mathrm{GHz}}\right)^{2.1} \frac{1}{1 + \dfrac{N(\mathrm{He})}{N(\mathrm{H})}} \left(\frac{\Delta\nu}{\mathrm{kHz}}\right)^{-1} \left(\frac{T_c}{T_\mathrm{L}}\right)\right]^{0.87} . \tag{12.52}$$

This is an approximation to the true electron temperature T_e, but one should not overestimate its precision. Even if assumptions 2 and 3 are valid, a different geometry can cause significant deviations of T_e from T_e^* [see e.g. Brown (1980)]. Non-LTE effects can be taken into account by using (12.46) resulting in

$$T_e = T_e^* [b_k(1 - \tfrac{1}{2}\beta\tau_c)]^{0.87} . \tag{12.53}$$

Clumping of the nebular gas will cause added difficulties, and what the relation between T_e^* and T_e will be if the nebula is not isothermal is still another problem.

Although there cannot be any doubt about the presence of NLTE effects in H II regions, their importance was overestimated by the early workers on this problem. In the first years after the recombination lines had been detected, T_e^* values between 5000 and 7000 K were found, values that differed significantly from the typical $T_e \cong 10\,000$ K that had been obtained from optical data. NLTE effects were then used to explain the difference, but modern, well calibrated measurements with large telescopes tend to give larger values for T_e^* that agree better with the optical values.

12.7 Recombination Lines from Other Elements

As all atoms with a single electron in a highly excited state are hydrogen-like, the radiative properties of these Rydberg atoms differ only by their different central mass. The Einstein coefficients A_{ik}, the statistical weights g_i and the departure coefficients b_i are identical for all Rydberg atoms, at least as long as the electrons in the inner atomic shells are not affected. Only the frequencies of the recombination lines are slightly affected, and if this frequency difference is expressed in terms of radial velocities, this Doppler velocity for a given element is independent of the quantum number (see Table 12.2).

The first recombination line measurements by Höglund and Mezger (1965) clearly showed a line that could be identified with a recombination line of He II. Both the radial velocity and the linewidth agree with those of the H II lines, and the line intensity is appropriate for the average cosmic abundance ratio of $N_\mathrm{He} : N_\mathrm{H} = 1 : 10$ by numbers. This identification is quite clear and beyond any doubt. This is an exciting result because it seems to offer the opportunity to determine the He abundance relative to hydrogen in interstellar space in a way that is reasonably free from many of the problems that are met when the He abundance is determined from optical data. Due to the large difference in the excitation energies needed for H and He lines in the optical range, it is very difficult to separate small abundance differences from differences in the physical state of the gas (temperature and pressure). For Rydberg atoms all atomic line parameters are identical, and therefore it should be possible to measure the abundance. This is of importance since the theory of cosmological element formation makes quite definite statements about the minimum value of $N_{\mathrm{He}}/N_{\mathrm{H}}$.

Making the usual assumption of small optical depth for the line we obtain from (12.30)

$$\frac{T_{\mathrm{L}}(\mathrm{H\,II})}{T_{\mathrm{L}}(\mathrm{He\,II})}=\frac{\int_{s_1}^{s_2} N_{\mathrm{i}}(\mathrm{H\,II})\,ds}{\int_{s'_1}^{s'_2} N_{\mathrm{i}}(\mathrm{He\,II})\,ds}=\frac{\int_{s_1}^{s_2} N_{\mathrm{e}}^2(s)\,ds}{\int_{s'_1}^{s'_2} N_{\mathrm{e}}^2(s)\,ds}\,. \tag{12.54}$$

Therefore we obtain

$$\frac{T_{\mathrm{L}}(\mathrm{H\,II})}{T_{\mathrm{L}}(\mathrm{He\,II})}=\left\langle\frac{N(\mathrm{H})}{N(\mathrm{He})}\right\rangle \tag{12.55}$$

provided that:

1) The He II regions have the same extent about the exciting stars. This will be the case if the number of UV photons emitted by the exciting stars which can ionize He is similar to those that can do this for H. As shown in Sect. 12.2 this is so if $T_* > 37\,000$ K.
2) $N(\mathrm{He}^{++})$ is negligible compared to $N(\mathrm{He}^{+})$. Since the ionization potential for He^{++} is $\chi_2 = 54.4$ eV compared to $\chi_1 = 24.6$ eV this can safely be adopted.
3) $N(\mathrm{H\,I}) \cong 0 \cong N(\mathrm{He\,I})$ in H II regions.

All these assumptions seem generally to be fulfilled approximately, thus $T_{\mathrm{L}}(\mathrm{He})/T_{\mathrm{L}}(\mathrm{H})$ gives a useful measure of $N(\mathrm{He})/N(\mathrm{H})$. But since there are systematic variations of $T_{\mathrm{L}}(\mathrm{He})/T_{\mathrm{L}}(\mathrm{H})$ with the distance from the sun, Brown (1980) in his discussion of the effect that clumping and other inhomogeneities of the gas have on $T_{\mathrm{L}}(\mathrm{He})/T_{\mathrm{L}}(\mathrm{H})$ casts doubts on the reality of the measured underabundance of $N(\mathrm{He})/N(\mathrm{H})$ near the galactic center. The He abundance derived from the line ratios is much closer to the average value if lines with small i are considered (Mezger 1980). Therefore some care is needed in interpreting the measured line ratios.

Recombination lines of other elements besides H and He have been observed too, but their unambiguous identification is more difficult because the radial velocity differences due to the atomic weight of the cores become less with increasing M_{A} and converge towards $\Delta V = 163.3\ \mathrm{km\,s^{-1}}$ for $M_{\mathrm{A}} \to \infty$. The line seen blended with the He line in the spectrum of Ori A (Fig. 12.5) could be due to carbon but for a definite identification, some additional evidence should be given. If this line originates in the same volume as the H and He lines, the line intensity is a factor of about 60 stronger than expected from the average abundance of carbon. To explain this, Dupree and Goldberg (1970) proposed that the b_n factors of C in the H II regions should be strongly perturbed due to dielectric recombination of C II atoms. The required large values for T_{e} and N_{e} in the H II regions do, however, contradict other measurements, and therefore this explanation had to be abandoned.

At this moment the following explanation is generally accepted: the line does not originate in the H II region but in the surrounding gas which is only partly ionized and has a much lower electron temperature ($T_e \cong 200$ K) than the H II region. Therefore, the lower the observing frequency, the stronger the C II lines become relative to the H II lines. This explains too why the radial velocity of the C II lines may differ from that of the H II lines by a few km s^{-1}, and usually agrees much better with the velocity of adjacent molecular clouds.

To explain why the line should belong to carbon and not to, say, oxygen which after all is more abundant than carbon, one has to remember that it is recombination line radiation that we observe: for a line to be emitted the atoms must first be ionized. Oxygen has an ionization potential of 13.6 eV while carbon needs only 11.3 eV. Therefore recombination line radiation of oxygen could only be emitted in H II regions, while carbon can quite well be ionized outside H II regions. All other elements that might be considered are less abundant than carbon by a large factor and their lines would hardly be detectable. Therefore the identification of this line as C$n\alpha$ is generally accepted but a detailed theory is still missing. Such a theory would need information on the distribution of gas density and temperature inside the C II regions. This information is not available, and therefore only model computations are possible. But it is not known how typical the parameters used for these are.

13. Interstellar Molecules and Their Line Radiation

13.1 Molecules in Interstellar Space

Two forms of material that emit or absorb line radiation exist in the universe: atoms and molecules. While in optical astronomy line radiation caused by atoms is by far the most important, interstellar molecular line radiation has become a very important research topic in radio astronomy.

In optical astronomy, molecular line radiation has also been detected in objects like late type stars with low surface temperature and in comets. Then in 1941 McKellar detected interstellar absorption lines of CN in the spectrum of ζ Oph. Later on lines belonging to CH^+ and CH were identified too, showing that in interstellar gas at least simple molecules exist if the physical conditions are right.

We expect molecular hydrogen (H_2) to be the most abundant molecule in the universe. Unfortunately there are no H_2 transitions in the visual or radio range that are excited under conditions which prevail in interstellar clouds. Therefore this molecule is not easily detected. Indeed, it was first observed in 1970 when G. Carruthers succeeded in detecting its ultraviolet absorption lines using a rocket to carry a spectrograph above the earth's atmosphere.

Molecular line radio spectroscopy began in 1963 when Weinreb et al. detected the close group of four OH lines in absorption in front of the radio source Cas A.

Shortly after, OH lines in emission were observed in the vicinity of some H II regions. But this emission had such high intensity, such strange line ratios (different from that expected for thermal equilibrium conditions), and such a high degree of polarization, that this radiation could hardly have originated in a thermal process: the corresponding temperature would have to be of the order of 10^{12} K. Therefore the proposal of Perkin, Gold and Salpeter that these emission lines originated by Maser action was generally accepted.

Until 1968 all interstellar molecules detected consisted of only two atoms, one of these being hydrogen, and this was believed to be a quite natural outcome of both the relative abundance of the various atomic species and the fact that the interstellar medium is in a gaseous state. But then in 1968 Townes' group detected the line radiation of ammonia (NH_3) and water vapour (H_2O). These were molecules consisting of 4 and 3 atoms respectively, and some of these sources, especially those emitting the $\lambda = 1.35$ cm water vapour line, showed exceedingly strong Maser action.

In the next year (1969) Snyder et al. discovered an absorption line of formaldehyde (H_2CO) a molecule consisting of 4 atoms. The absorption line also showed evidence of Maser action, but here it is not the line emission which is augmented by stimulated emission but the level population is changed such that line absorption is strengthened compared to thermal equilibrium.

The fact that formaldehyde (H_2CO) is generally considered to belong to organic chemistry caused quite some excitement and wild speculations in the press, and this became even stronger when additional organic molecules like methanol (CH_3OH) and even ethanol, common alcohol, (CH_3CH_2OH) were detected. Since then the number of different molecules detected in the interstellar medium has steadily increased and the total count is well above 50 by now. A definite identification of a given molecule in interstellar space is possible only if the relevant part of the energy level diagram is known so that the frequencies of the observed transitions can be determined. If in addition the excitation mechanism can be specified and the various excitation and de-excitation rates can be computed, the physics of a molecular line is understood and then this line, perhaps in comparison with certain other well understood lines, can be used as a probe for the physical state of the emitting or absorbing molecular cloud. Frequently the energy levels cannot be computed with sufficient precision and then extensive laboratory work is necessary. This is often rather difficult because many of the molecular species in interstellar space are exceedingly short lived under laboratory conditions, so that quite sophisticated laboratory techniques have to be used. Therefore this part of the identification is most often done by physical chemists. Of the 50 or so molecular species identified so far, 13 belong to what is commonly called inorganic chemistry. These are substances like H_2, OH, SO, CO, HNO, NH_3 etc., while 39 are usually included in organic chemistry. These are amongst others H_2CO, HCCCN, NH_2CHO and CH_3CH_2OH. It is remarkable that all these organic molecules are linear chains; not a single ring-shaped molecule has been detected yet, although extensive searches have been conducted. But it is not clear yet whether this means a true deficiency of such molecules or whether it only means that the excitation mechanisms for the lines emitted or absorbed by them are unfavourable.

The chemistry of the interstellar medium is still almost completely unknown, and it will not be discussed here at all. And because the physics of molecular line radiation is so much more complicated than that for atoms, even if compared with atoms of high atomic numbers, it will therefore be only touched on here; for a thorough treatment the fundamental books of Herzberg or Townes and Schawlow listed in the references should be consulted. Here we will only sketch out some of the basic physical principles without going into any depth. It will suffice to accept the explanations given by others for the form and characteristics of the term diagram of a selected molecule; if such a diagram is to be established ab initio, a much closer familiarity with the physics of molecular line radiation is needed.

13.2 Some Basic Concepts

Compared to atoms, molecules have a complicated structure and the Schrödinger equation of the system will be correspondingly complex involving positions and moments of all constituents, both the nuclei and the electrons. All particles, however, are confined to a volume with a diameter of the typical molecule diameter a, and therefore each particle will possess an average momentum $\hbar/a$ due to the uncertainty principle $\Delta p\,\Delta q \leqq \hbar$. The kinetic energy will then have states with a typical spacing $\Delta E \cong \Delta p^2/2m \cong \hbar^2/2ma^2$. For electrons these steps are of a few eV; for nuclei they are in the meV range.

In the Schrödinger equation of a molecule, therefore, those parts of the Hamiltonian operator that describe the kinetic energy of the nuclei can be neglected compared to the kinetic energy of the electrons. The nuclei enter into the Coulomb potential only through their position and therefore these positions enter as parameters into the solution. Because the motion of the nuclei is so slow, the electrons have sufficient time to adjust adiabatically to the new nuclear position. This separation of the nuclear and electronic motion in molecular quantum mechanics is called the Born-Oppenheimer approximation.

Transitions in a molecule can therefore be classified into three different sets:

a) electronic transitions with typical energies of a few eV – that is lines in the visual regions of the spectrum;
b) vibrational transitions caused by oscillations of the relative positions of nuclei with respect to their equilibrium positions. Typical energies are 0.1–0.01 eV, corresponding to lines in the infrared region of the spectrum;
c) rotational transitions caused by the rotation of the nuclei with typical energies of $\cong 10^{-3}$ eV corresponding to lines in the cm- and mm-wavelength range.

Molecular line radiation is greatly complicated by the fact that transitions are possible from one state described by one set of quantum numbers specifying (a), (b) and (c) to another state described by a different set. But if transitions of the electronic state a are involved, the corresponding line will lie in the optical range. If we confine ourselves to the radio range, only transitions between different rotational levels and sometimes different vibrational levels will be involved, resulting in much simpler considerations. Only occasionally will differences in the geometrical arrangement of the nuclei be involved resulting in doubling of the lines or other complications.

The position of the nuclei of a molecule in equilibrium can usually be approximated by some average distance r between the nuclei and the potential energy is then specified as $P(r)$ (Fig. 13.1). If r_e is the equilibrium value,

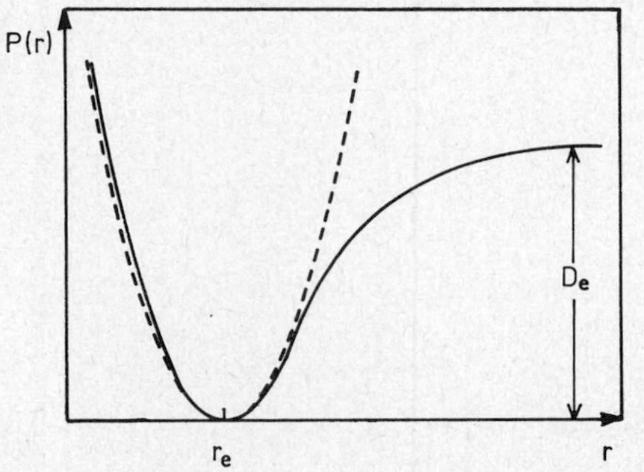

Fig. 13.1. The potential curve $P(r)$ of a binary molecule and the harmonic approximation

then

$$D_e = P(\infty) - P(r_e) \tag{13.1}$$

is the dissociation energy of the molecule. This is a unique value if we neglect for the time being that, in a molecule consisting of more than 2 nuclei, D_e might depend on which of the nuclei would increase its distance r.

For many molecules the potential curve $P(r)$ is well represented by an expression proposed by Morse

$$P(r) = D_e\{1 - \exp[-a(r - r_e)]\}^2 . \tag{13.2}$$

Often the even simpler harmonic approximation

$$P(r) = \frac{k}{2}(r - r_e)^2 = a^2 D_e (r - r_e)^2 \tag{13.3}$$

is sufficient.

13.3 Pure Rotation Spectra

Because the effective radius of even a simple molecule is about 10^5 times the radius of the nucleus of an atom, the moment of intertia Θ_e of such a molecule is at least 10^{10} times that of an atom of the same mass. The kinetic energy of rotation

$$H_{rot} = \tfrac{1}{2}\Theta_e \omega^2 = \boldsymbol{J}^2/2\,\Theta_e \tag{13.4}$$

where $\boldsymbol{J}$ is the angular momentum is then a quantity that cannot be neglected compared with the other internal energy states of the molecule, especially if the observations are made in the radio range.

For a simple molecule consisting of two nuclei A and B, the moment of inertia is

$$\Theta_e = m_A r_A^2 + m_B r_B^2 = m r_e^2 \quad \text{where} \tag{13.5}$$

$$\boldsymbol{r}_e = \boldsymbol{r}_A - \boldsymbol{r}_B \quad \text{and} \tag{13.6}$$

$$m = \frac{m_A m_B}{m_A + m_B}, \quad \text{and where} \tag{13.7}$$

$$\boldsymbol{J} = \Theta_e \boldsymbol{\omega} \tag{13.8}$$

is the angular momentum perpendicular to the line connecting the two nuclei.

For molecules consisting of three or more nuclei, similar though more complicated expressions can be given. Θ_e will depend on the relative orientation of the nuclei and will in general be a (three-axial) ellipsoid. In (13.8) then values of Θ_e appropriate for the direction of $\boldsymbol{\omega}$ will have to be used.

The solution of the Schrödinger equation then results in the *eigenvalues* for the rotational energy

$$E_{\text{rot}} = W(J) = \frac{\hbar^2}{2\,\Theta_e} J(J+1) \tag{13.9}$$

where J is the quantum number of angular momentum, which has integer values

$J = 0, 1, 2, \ldots$.

Equation (13.9) is true for a molecule that is completely rigid; for a slightly elastic molecule, r_e will increase with the rotational energy resulting in

$$E_{\text{rot}} = W(J) = \frac{\hbar^2}{2\,\Theta_e} J(J+1) - hD[J(J+1)]^2 \,. \tag{13.10}$$

Introducing the rotational constant

$$B_e = \frac{\hbar}{4\pi\,\Theta_e} \tag{13.11}$$

and the constant for centrifugal stretching D, the pure rotation spectrum for electric dipole transitions $\Delta J = -1$ (emission) or $\Delta J = +1$ (absorption) is (Figs. 13.2, 13.3)

$$\nu(J) = \frac{1}{h}[W(J+1) - W(J)] = 2B_e(J+1) - 4D(J+1)^3 \,. \tag{13.12}$$

But note that these are allowed electric dipole transitions: if the molecule does not possess a permanent electric dipole moment these lines can be neither emitted nor absorbed. Homonuclear diatomic molecules like H_2, O_2 or N_2 do not possess

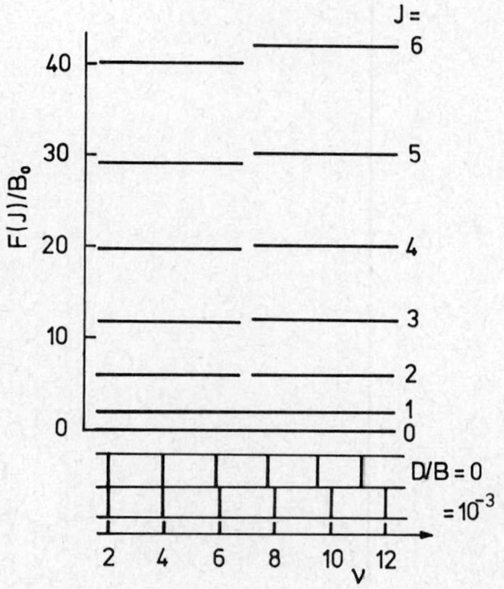

Fig. 13.2. Rotational energy levels for a rigid rotator (*left part*) and one deformed by centrifugal stretching with $D/B_e = 10^{-3}$. Realistic molecules have $D/B_e < 10^{-4}$. The resulting spectrum is indicated in the lower part

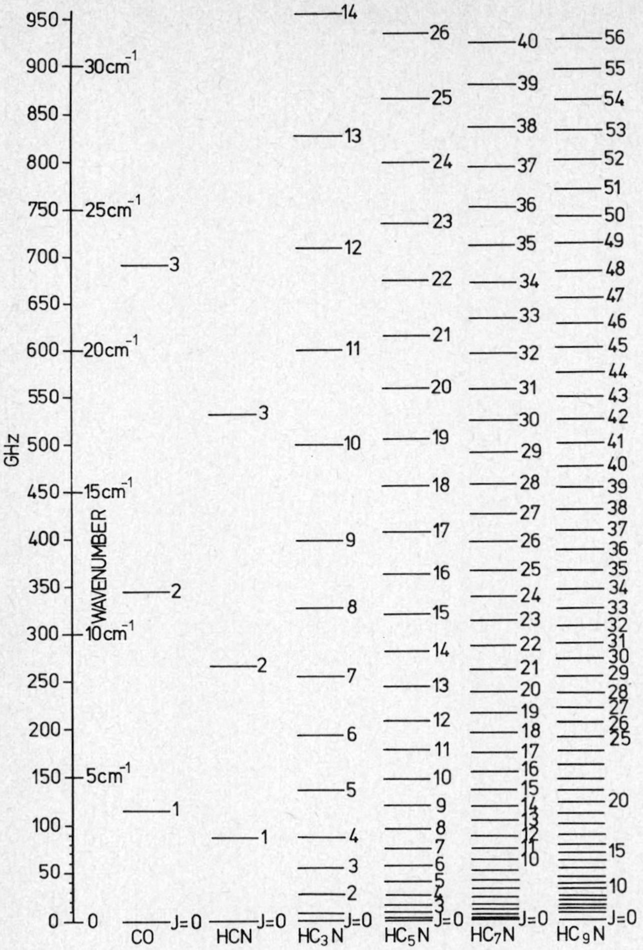

Fig. 13.3. Rotational energy levels of the vibrational ground states of some linear chain molecules detected in the interstellar medium [adapted from Avery (1980)]

such a permanent electric dipole moment, and they will therefore not produce such lines, either in emission or in absorption. This is the reason why it was so difficult to detect interstellar H_2.

The rotation of an arbitrary (rigid) molecule can be considered to be the superposition of three free rotations about the three principal axes of the intertial ellipsoid. Depending on the symmetry of the molecule these axes can be all different – this is the asymmetric top model, two axes can be equal – the symmetric top model, and finally all three principal axes can be equal – the spherical top model. In order to compute the rotation terms, the Hamiltonian operator entering the Schrödinger equation has to be set up and its eigenvalues have to be determined. For rotations it is only the angular momentum which enters here. But in order to have sharp eigenvalues for the energy, the appropriate angular momentum components have to be time-invariable according to classical mechanics.

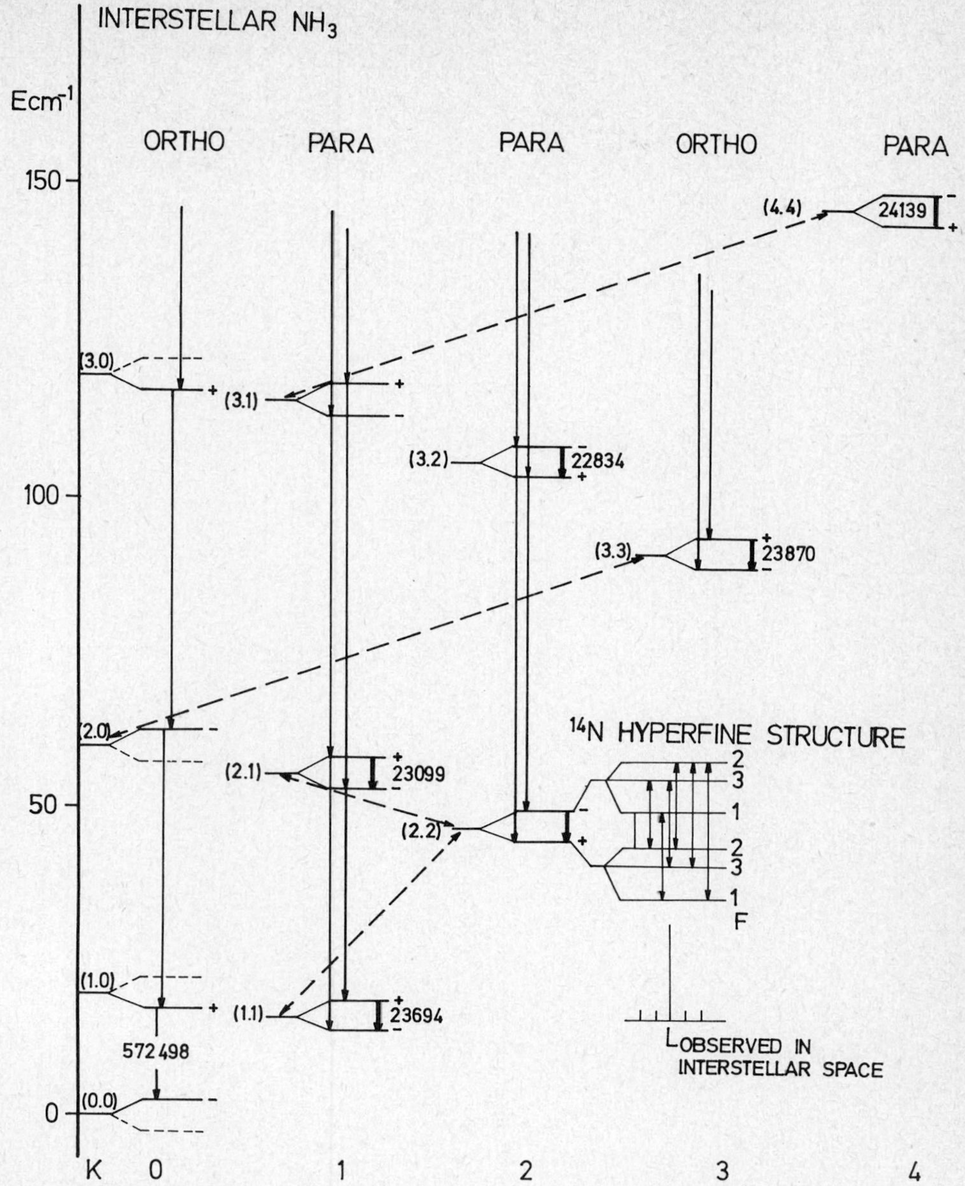

Fig. 13.4. The energy level diagrams of NH_3. This is a symmetric top molecule which shows inversion doubling. Interaction of the nuclear spin with that of the electronic shell causes hyperfine structure which is indicated too (Winnewisser et al. 1974)

For any top model, the total momentum $\boldsymbol{J}$ will remain constant with respect to both its absolute value and its direction. Known from atomic physics, this means that both $\boldsymbol{J}^2$ and the projection of $\boldsymbol{J}$ into an arbitrary but fixed direction, for example J_z, remain constant. If the molecule is in addition symmetric, the projection of $\boldsymbol{J}$ on the figure axis will be constant.

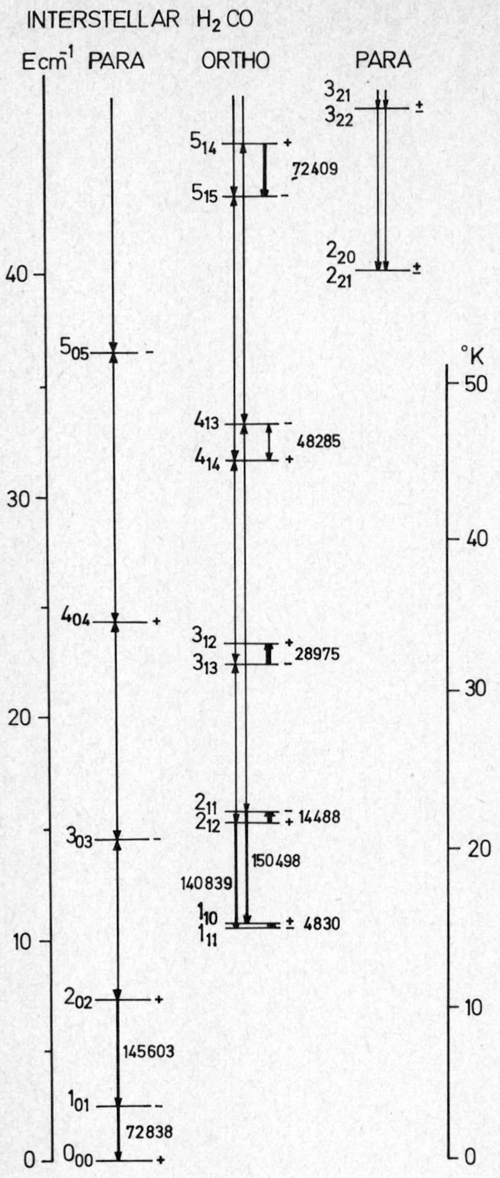

Fig. 13.5. Energy level diagram of formaldehyde, an example of an almost symmetric top molecule (Winnewisser et al. 1974)

Let us first consider the *symmetric top* molecule (Figs. 13.4, 13.5). Suppose $\boldsymbol{J}$ is inclined with respect to the axis of symmetry 3. Then the figure axis 3 will rotate around the direction $\boldsymbol{J}$ forming a constant angle with it, and the molecule will simultaneously rotate around the figure axis 3 with the constant angular momentum J_3. If $\Theta_1 = \Theta_2 = \Theta_\perp$ and $\Theta_3 = \Theta_\parallel$, the Hamiltonian operator will be

$$H = (J_1^2 + J_2^2)/2\,\Theta_\perp + J_3^2/2\,\Theta_\parallel = \boldsymbol{J}^2/2\,\Theta_\perp + J_3^2\left(\frac{1}{2\,\Theta_\parallel} - \frac{1}{2\,\Theta_\perp}\right) . \tag{13.13}$$

Its eigenvalues are

$$W(J,K) = J(J+1)\frac{\hbar^2}{2\Theta_\perp} + K^2\hbar^2\left(\frac{1}{2\Theta_\parallel} - \frac{1}{2\Theta_\perp}\right) \tag{13.14}$$

with the two quantum numbers

$$\mathrm{J} = 0, 1, 2, \ldots \qquad \mathrm{K} = 0, \pm 1, \pm 2, \ldots \pm J \,. \tag{13.15}$$

For linear molecules, and this applies for all molecules consisting of only two nuclei, $\Theta_\parallel \to 0$ so that $1/(2\Theta_\parallel) \to \infty$. Then finite energies in (13.14) are possible only if $K = 0$, so that for these cases the energies are given by Eq. (13.9). Each eigenvalue then has a multiplicity of $2J + 1$.

For an *asymmetric top molecule* no internal molecular axes with a time-invariable component of the angular momentum exist. So only the total angular momentum is conserved and we have only J and M as good quantum numbers. But neither the eigenstates nor the eigenvalues are easily expressed in explicit form.

13.4 Vibrational Transitions

If any of the nuclei of a molecule suffers a displacement from its equilibrium distance r_e, it will on release perform an oscillation about r_e. The Schrödinger equation for this is

$$\left(\frac{p^2}{2m} + P(r)\right)\psi^{\mathrm{vib}}(x) = W^{\mathrm{vib}}\psi^{\mathrm{vib}}(x) \,, \tag{13.16}$$

where $x = r - r_e$ and $P(r)$ is the potential function (13.2). If (13.3) can be used, we have the simple harmonic approximation (Fig. 13.6) with the classical oscillation frequency

$$\omega = 2\pi\nu = \sqrt{\frac{k}{m}} = a\sqrt{\frac{2D_e}{\mathrm{m}}} \,, \tag{13.17}$$

and (13.16) has the eigenvalues

$$W^{\mathrm{vib}} = W(v) = \hbar\omega(v + \tfrac{1}{2}) \quad \text{with} \tag{13.18}$$

$$v = 0, 1, 2, \ldots \,. \tag{13.19}$$

The solutions $\psi^{\mathrm{vib}}(x)$ can be expressed with the help of Hermite polynomials.

For large x the precision of the approximation (13.3) is no longer sufficient and the Morse potential (13.2), or even an empirical expression, will have to

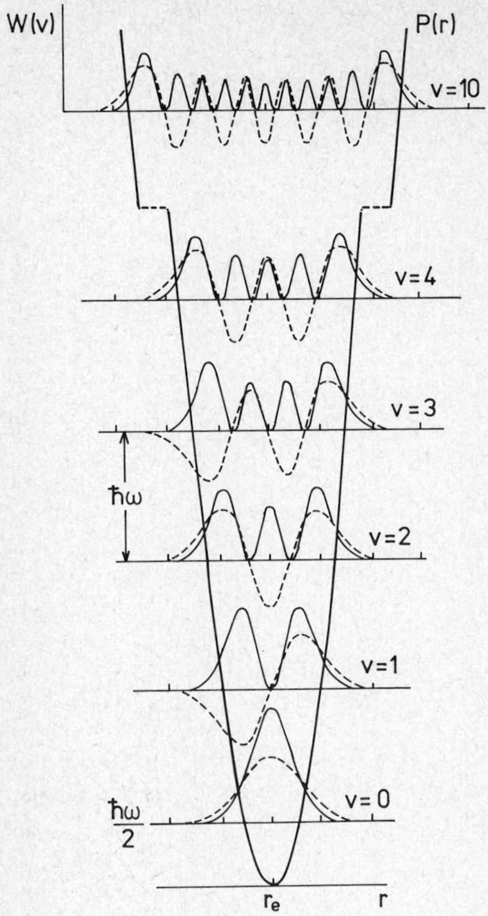

Fig. 13.6. Vibrational energy levels, eigenstates (----) and probability amplitudes (——) for a harmonic oscillator

be introduced into (13.16). The resulting differential equation can no longer be solved analytically, so that numerical methods will have to be used. Often it is sufficient to introduce *anharmonic* factors x_e, y_e into the solution corresponding to (13.18), viz.

$$W(v) = \hbar\,\omega\,(v + \tfrac{1}{2}) + x_e\,\hbar\,\omega\,(v + \tfrac{1}{2})^2 + y_e\,\hbar\,\omega\,(v + \tfrac{1}{2})^3 + \dots \,. \tag{13.20}$$

Usually x_e and y_e are small numbers which can be determined either empirically or by a fit to a numerical solution of (13.16). For the H_2 molecule in the ground state $^1\Sigma_g^+$ we have $x_e = -2.6 \cdot 10^{-2}$ and $y_e = 6.6 \cdot 10^{-5}$. The negative sign of x results in a decrease of the step size of the harmonic energy ladder.

A molecule consisting of only two nuclei can vibrate only in one direction; it has only one vibrational mode. This is different for molecules with three or more nuclei. Here a multitude of various vibrational modes may exist, each of which will result in its own ladder of vibrational states, some of which may be degenerate.

13.5 Line Radiation from Molecules

In the line spectrum of a molecule, transitions of all three kinds as classified by the Born-Oppenheimer approximation will be involved, that is electronic, vibrational and rotational:

$$W^{\rm tot} = W^{\rm el} + W^{\rm vib} + W^{\rm rot} \; . \tag{13.21}$$

$W^{\rm vib}$ and $W^{\rm rot}$ are the vibrational and rotational energies of the nuclei of the molecule and $W^{\rm el}$ is the energy of the electrons.

In many respects this electronic energy is similar to that of atomic fine structure as described by Russell-Saunders (LS) coupling. Each electronic state is designated by the symbol ${}^{2S+1}\Lambda_{\Omega}$, where $2S+1$ is the multiplicity of the state with S the electron spin and Λ is the projection of the electronic orbital angular momentum on the molecular axis in units of $\hbar$. The molecular state is described as Σ, Π, Δ etc. according to whether $\Lambda = 0, 1, 2, \ldots$.

Σ is the projection of the electron spin angular momentum on the molecular axis in units of $\hbar$ (not to be confused with the symbol Σ for $\Lambda = 0$). Finally, Ω is the total electronic angular momentum. For the Hund coupling case (a) $\Omega = |\Lambda + \Sigma|$ [see, e.g., Hellwege (1974)].

Since the frequencies emitted or absorbed by a molecule are in the optical range if electronic states change, many of the complications found in molecular spectra in the optical range are not met if lines in the cm- and mm-range are considered. But the electronic state of the electronic shell does affect the vibrational and rotational levels even in the radio range.

Of considerable practical importance is the so-called Λ-doubling of rotational levels (Fig. 13.7). The Π, Δ ... states of diatomic and linear polyatomic molecules are doubly degenerate if the molecule is not rotating. In the rotating molecule, the interaction of the rotation and the angular momentum of the electronic state causes a splitting of this degeneracy (Λ-doubling). This splitting can be quite important for Π states; for Δ and higher states it is usually negligible. The OH molecule is a prominent example for this effect.

Differences in the geometric orientation of the nuclei can be of importance too. In a non-planar symmetric top molecule a reflection of all particles at the origin (at the center of gravity) leads to a configuration which cannot be obtained by rotating the original molecule. Correspondingly, each rotational state of the molecule is doubly degenerate as long as the potential hill separating the two configurations in infinitely high. But in molecules like NH_3 for which the two configurations obtained by inversion are separated only by a rather small potential barrier, an appreciable doubling of the term levels will arise.

Finally the nuclear spin is sometimes important. Its interaction with the electronic shell is described as the hyperfine structure which is usually of importance when the detailed lineshape of a single line is of interest, but not for the linespacing. However, nuclei with an uneven number of nucleons are Fermi particles so that their spin function will be antisymmetric. The symmetry of the spin function of the molecule will depend on the relative orientation of the spin of its separate constituents. If it is symmetric, it is called the *ortho*-modification of the molecule;

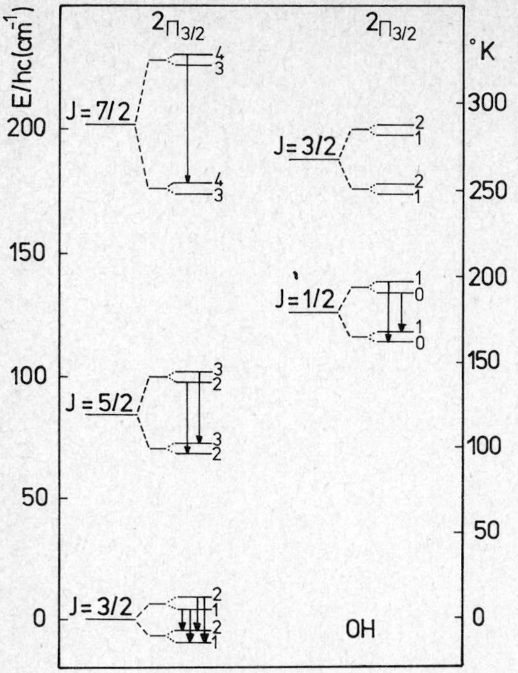

Fig. 13.7. The lowest energy levels of OH showing Λ-doubling (Moran 1976)

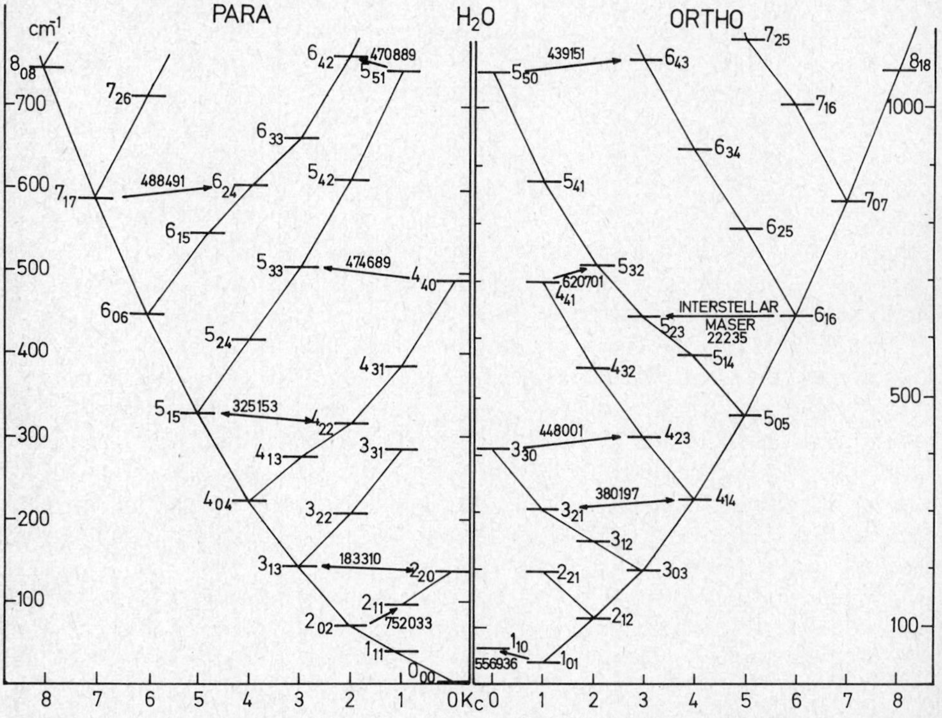

Fig. 13.8. Energy level diagrams for ortho- and para-H_2O. This is a symmetric top molecule, the energy level diagram is split into two parts that show almost no interaction under interstellar conditions (Winnewisser et al. 1974)

if antisymmetric it is called the *para*-modification. No change of the quantum numbers describing either electronic, vibrational or rotational states will affect this symmetry property of the molecule, and therefore the para- and ortho-modifications can be treated in many repects like different molecular species. In thermal equilibrium, collisions will change one modification into the other, and viceversa, and a statistical equilibrium can be computed. But at low temperatures the change of one into the other is rather slow indeed. Some of the most important molecular lines in interstellar space are emitted or absorbed by one modification only, like the 6 cm H_2CO line which is characteristic of the *ortho*-modification only, just as the interstellar H_2O maser line at $\lambda = 1.35$ cm is due to *ortho*-H_2O (see Fig. 13.8).

The computation of the molecular line intensities proceeds in complete analogy to the principles outlined in Chap. 10, if we neglect for the time being the effects of maser amplification. These will be discussed in the next section. What is needed, therefore, are expressions for the spontaneous transition probabilities.

As we observe only rotational transitions in the radio range, and as these are permitted dipole transitions, Eq. (10.24) is applicable and will be written, obviously simplified, as

$$A_J = \frac{16\,\pi^3}{3\,\varepsilon_0\,h\,c^3}\,\nu_r^3\,|\mu_J|^2 \tag{13.22}$$

where ν_r is the frequency of the transition $J \rightleftarrows J+1$ and the dipole moment μ_J is given by

$$|\mu_J|^2 = \mu^2\,\frac{J+1}{2J+1} \qquad \text{for } J \to J+1 \tag{13.23}$$

and

$$|\mu_J|^2 = \mu^2\,\frac{J+1}{2J+3} \qquad \text{for } J + J \to 1 \tag{13.24}$$

μ is the permanent electric dipole moment of the molecule. A table of measured values of μ for some molecules of astrophysical significance is given in Lang (1974) (Table 21, p. 162/163) together with a short discussion of the transition probabilities.

The statistical weight of the state J, which enters into the expression (10.17) for κ_ν, is given by

$$g_J = 2J+1\ . \tag{13.25}$$

The partition function used to determine the total molecular abundance from the abundance of molecules in a specific state is approximated by

$$U = \sum_{J=0}^{\infty} (2J+1)\exp\left[-\frac{h\,B_e\,J(J+1)}{k\,T}\right] \approx \frac{k\,T}{h\,B_e} \quad \text{for } h\,B_e \ll k\,T\ . \tag{13.26}$$

Here B_e is the rotation constant (13.11), and the molecule was assumed to be in thermal equilibrium so that the Boltzmann distribution could be applied. The situation may be more complicated when vibrations have to be taken into account, but the discussion would proceed along similar lines.

13.6 Cloud Densities and Temperatures

There are many reasons why molecular line observations are of such great importance in astrophysics. One is obviously that they bear indisputable witness to the existence of certain chemical compounds in interstellar space; and therefore they represent a challenge to astrophysicists as well as to chemists. Several groups in different parts of the world have responded to this challenge, and have proposed reaction chains that will lead to the observed molecular species as well as to a long list of unobserved and unobservable ones. But we are still far from a reasonable understanding of the reaction schemes; astrochemistry cannot yet be considered to be an established tool of astronomical research. But it will undoubtedly be so in the future, and then it definitely should be included in a discussion of the tools of radio astronomy.

Another reason why molecular line studies are of such great interest is that they can be used as probes to sample the physical state of those cloud regions where the molecules exist. Both the ambient gas density and the temperature are quantities that must be known for any successful modelling. And since molecular line radiation usually comes from fairly dense and cold cloud interiors where no atomic line radiation is emitted (or absorbed), there are no substitutes for molecular line radiation in this respect.

The methods and expressions used in this study have practically all been covered in Chap. 10, but since there seems to exist in the literature a widespread misunderstanding that the observation of a single line is sufficient to establish a minimum gas density, this problem will be outlined here following a discussion by Kegel (1976).

The emissivity of matter using the two-level approximation has been derived in Chap. 10 resulting in the expression (10.15):

$$\varepsilon_\nu = \frac{h\nu_0}{4\pi} N_2 A_{21} \varphi(\nu) \tag{13.27}$$

where N_2 is the population of the upper level 2 and $\varphi(\nu)$ the lineshape (cf. 10.1). This emissivity is proportional to A_{21}, and this seems to be the reason for the above-mentioned belief. A line with an exceedingly small A_{21} would be very weak.

But ε_ν is proportional to the population N_2 of the excited level, and in the limit of small A we will have to consider collisional excitation even in a low-density situation. The rate equation (10.30) then results in a stationary population as given by (10.35); that is

$$N_1(C_{12} + B_{12}\bar{U}) = N_2(A_{21} + B_{21}\bar{U} + C_{21}) \ . \tag{13.28}$$

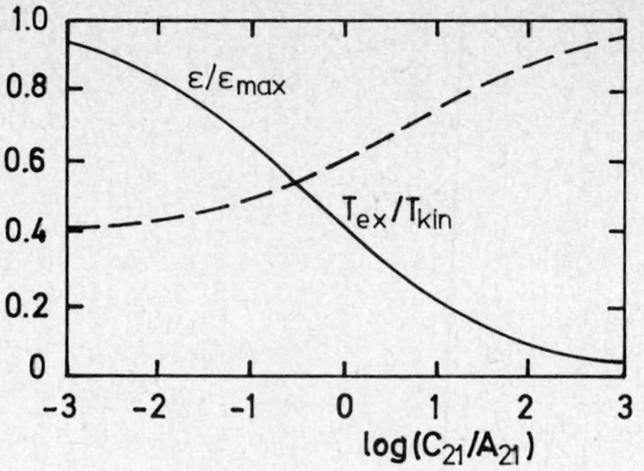

Fig. 13.9. Emissivity and excitation temperatures as function of C/A in the case of weak lines [after Kegel (1976)]

Substituting this into (13.27) we obtain

$$\varepsilon_\nu = \frac{h\nu_0}{4\pi}\frac{N_1(C_{12}+B_{12}\bar{U})}{A_{21}+B_{21}\bar{U}+C_{21}}\varphi(\nu) . \tag{13.29}$$

Now C_{12} and C_{21} will be related due to the principle of detailed balancing (10.37) by

$$\frac{C_{12}}{C_{21}} = \frac{g_2}{g_1}\mathrm{e}^{-h\nu_0/kT_{\mathrm{kin}}} \tag{13.30}$$

and the A and B coefficients are connected by the Einstein relations (10.8) and (10.9). Using these, and substituting for brevity

$$\tilde{S}_\nu = \frac{2h\nu_0^3}{c^2} , \tag{13.31}$$

(13.29) becomes (Fig. 13.9)

$$\varepsilon_\nu = \frac{h\nu_0}{4\pi}\frac{g_2N_1}{g_1}\frac{\bar{U}/\tilde{S}_\nu + \dfrac{C_{21}}{A_{21}}\exp(-h\nu_0/kT_{\mathrm{kin}})}{1+\bar{U}/\tilde{S}_\nu + C_{21}/A_{21}}\varphi(\nu) . \tag{13.32}$$

We then see that in the limit of small C_{21}/A_{21}, that is, if the level population is governed mainly by spontaneous transitions, ε_ν becomes independent of A_{21}. The fact that a certain molecular line excited by collision is observed does *not* mean that the ambient gas density has to be above the thermalization density. To obtain such information, several lines of the same molecule have to be observed. And since strong deviations from LTE are common in interstellar space, the observed lines have to be analyzed by NLTE calculations.

13.7 Galactic Masers

Masers have been discussed in Chap. 7 as low noise amplifiers used in microwave receiving systems. In radio astronomy they are of importance, too, in explaining a considerable part of the detected molecular line radiation.

Shortly after Weinreb et al. (1963) detected the OH line in absorption, Gardner et al. (1964) found such lines with intensity ratios that could not be explained at all by line radiation emitted under LTE conditions. This conclusion was strenghtened when some of this line radiation was found to be linearly and circularly polarized, and that the sources of the emission lines had exceedingly small angular diameters leading to very high brightness temperatures for the sources. All these observations could be explained by the cosmic maser theory of Litvak et al. (1966), and the general acceptance of these ideas was greatly strenghtened when cosmic H_2O masers were detected by Cheung et al. (1969). Since then many more masering transitions have been detected for many different molecules. Some of these result in strongly increased emission of line radiation, while others, like the 5 GHz H_2CO line, result in an increased strength of the absorption line, even at positions where no obvious background source is visible.

We will show in a slightly simplified version how such a galactic maser works, but it is again only the physical principle that we will consider. We will not describe a specific astrophysical maser model explaining how the energy is provided, how the pumping power is applied, etc. This discussion is based on the review by Moran (1976).

Let us consider a cloud of molecules with three energy levels 1, 2 and 3. For the sake of simplicity the statistical weights of the levels are taken to be the same. The radiation transfer equation (10.13) for radiation corresponding to the transition $1 \rightleftarrows 2$ with the frequency ν_0 is then

$$\frac{dI_\nu}{ds} = \frac{h\nu_0}{4\pi}\left[(N_2 - N_1)\,B\,\frac{4\pi}{c}\,I_\nu + N_2\,A\right]\varphi(\nu) \ , \tag{13.33}$$

where we write A for A_{21} and B for B_{12} and B_{21}, the last two being equal now because we assumed $g_1 = g_2$. The populations N_1 and N_2 of the levels can be changed by spontaneous emissions described by the transition rate A, by stimulated emissions M given by

$$M = \frac{4\pi}{c}\,B\,\bar{I} = \frac{4\pi}{c}\,B\,I\,\frac{\Omega_m}{4\pi} \tag{13.34}$$

where Ω_m is the beam solid angle of the radiation, or by collisions described by the rate C and finally by the pumping rates P_{12} and P_{21}. If the system is stationary then obviously, as in (10.35),

$$N_1(C_{12} + M_{12} + P_{12}) = N_2(C_{21} + M_{21} + P_{21} + A_{21}) \ . \tag{13.35}$$

In microwave masers we can assume as a good approximation that A is negligible compared to C and M, and that $C_{12} \approx C_{21} \approx C$ and $M_{12} \approx M_{21} \approx$ so that

$$\frac{N_2}{N_1} = \frac{P_{12} + M + C}{P_{21} + M + C} \ . \tag{13.36}$$

The population inversion which the pump would establish if $M = C = 0$, is then

$$\Delta N_0 = (N_2 - N_1)|_{M=C=0} = N\,\frac{P_{12} - P_{21}}{P_{12} + P_{21}} \tag{13.37}$$

with

$$N = N_1 + N_2 \ . \tag{13.38}$$

For $C \neq 0 \neq M$ we have from (13.36) with $P = P_{12} + P_{21}$:

$$\Delta N = \frac{\Delta N_0}{1 + \dfrac{2(C+M)}{P}} \ . \tag{13.39}$$

Substituting this into (13.33) results in

$$\frac{dI_\nu}{ds} = \frac{\alpha I_\nu}{1 + I_\nu/I_s} + \varepsilon \quad \text{where} \tag{13.40}$$

$$\alpha = \frac{h\nu_0}{\mathrm{c}} B \frac{\Delta N_0}{1 + \dfrac{2C}{P}} \varphi(\nu) \ , \tag{13.41}$$

$$I_s = \frac{cP}{2B\Omega_m}\left(1 + \frac{2C}{P}\right) , \quad \text{and} \tag{13.42}$$

$$\varepsilon = \frac{h\nu_0}{4\pi} N_2 A \varphi(\nu) \ . \tag{13.43}$$

In most astrophysical applications the term ε can be assumed to be a constant and quite often it can be dropped altogether.

For $I_\nu \ll I_s$ we have the solution for the *unsaturated maser* for $\nu = \nu_0$

$$I_{\nu_0} = I_0 \, \mathrm{e}^{\alpha_0 l} + \frac{\varepsilon}{\alpha_0} (\mathrm{e}^{\alpha_0 l} - 1) \tag{13.44}$$

where l is the length along the line-of sight within the maser region. Converting this into temperatures this becomes

$$T_\mathrm{b} = T_c \, \mathrm{e}^{\alpha_0 l} + |T_x| (\mathrm{e}^{\alpha_0 l} - 1) \ . \tag{13.45}$$

This is equal to the isothermal solution (1.31) with $\tau = -\alpha_0 l$.

For $I_\nu \gg I_s$ the right-hand side of (13.40) is a constant, and so the solution for the *saturated maser* becomes

$$I_{\nu_0} = I_0 + (\alpha_0 I_s + \varepsilon) \, l \ . \tag{13.46}$$

In this case the intensity increases linearly with l compared to the exponential growth with l in the unsaturated maser. If a maser is unsaturated the linewidth of the line will steadily decrease with increasing maser gain (Fig. 13.10). A wideband background source with a brightness temperature T_c will, according to (13.45),

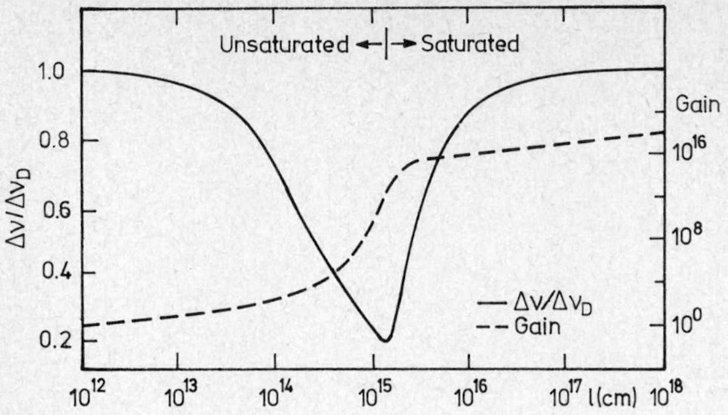

Fig. 13.10. Gain and linewidth in a linear maser (Moran 1976)

produce the output signal

$$T_b(\nu) = T_c\, e^{\alpha(\nu) l} . \tag{13.47}$$

Assuming a gaussian lineshape φ in (13.41), $\alpha(\nu)$ is given by

$$\alpha(\nu) = \alpha_0 \exp\left[-(\nu - \nu_0)^2/2\sigma_0^2\right] .$$

As long as we concentrate our attention on the center of the line this can be approximated by

$$\alpha(\nu) = \alpha_0 \left[1 - \frac{1}{2}\left(\frac{\nu - \nu_0}{\sigma_0}\right)^2\right] ,$$

so that (13.47) then becomes

$$T_b(\nu) = T_c\, e^{\alpha_0 l} \exp\left\{-\frac{\alpha_0 l (\nu - \nu_0)^2}{2\sigma_0^2}\right\} .$$

But this is a gaussian with the dispersion

$$\sigma(\alpha_0 l) = \frac{\sigma_0}{\sqrt{\alpha_0 l}} . \tag{13.48}$$

Hence the linewidth will decrease until the center of the line begins to saturate. The linewidth then broadens as the wings of the line continue to experience exponential growth until the line again has the original linewidth σ_0. In a saturated maser the linewidth remains constant.

What we have given here is only a much simplified phenomenological description of the maser effect. If the astrophysics of any given maser source is to be fully understood this is only the very first step. A detailed specification of the pump mechanism, the detailed solution of the rate equations and an account of the energy sources are needed just as a geometrical model. And finally it has to be specified into which evolutionary framework this model fits. We are still quite far from this goal.

13.8 The CO Line at $\lambda = 2.6$ mm

A considerable part of the gaseous material in the interstellar medium (ISM) of our galaxy exists in molecular form. Molecular hydrogen is believed to be fairly abundant. Unfortunately, H_2 is difficult to observe directly for several reasons: its rotational levels have energies much larger than the average kinetic temperature of the cold clouds; i.e. the level spacing of the $J = 2 \rightarrow 0$ rotational transition of ortho-H_2 corresponds to an energy of 509 K and the transition probabilities are very small ($A_{20} = 2.95 \cdot 10^{-11}\ \mathrm{s}^{-1}$) since H_2 has no permanent dipole moment. H_2 therefore has been detected only in "hot spot regions" of the ISM where the gas is locally heated by shocks so that the excitation of rotational and vibrational levels of H_2 is possible, resulting in the emission of lines in the infrared. Other exceptions are regions of low extinction close to a hot star where UV absorption lines can be detected.

Carbon monoxide (CO), on the other hand, is another simple molecule that should be found copiously in the ISM if the standard element abundance is considered. And this molecule is, in contrast to H_2, rather easily detectable by its microwave radiation because CO has a permanent dipole moment of $\mu = 0.112$ Debye $= 3.74 \cdot 10^{-14}$ C m. It is a binary molecule with a simple ladder of rotational levels spaced such that the lowest transitions are in the millimeter-wave region. Several isotopes exist for both C and O and therefore several isotopic species of CO have been observed. Table 13.1 gives some parameters for the most important ones.

A first approximation of the abundance of the CO molecules can be obtained by a standard LTE analysis of the CO line radiation. Line intensities in mm-wave radio astronomy are usually measured in terms of brightness temperature as defined in (1.28), viz.

$$T_\mathrm{b} = \frac{c^2}{2k} \frac{1}{\nu^2} I_\nu \ . \tag{13.49}$$

For the distribution of the CO gas, let us adopt the simplest configuration possible, that of an isothermal slab, so that the solution (1.31) may be used. Remembering that a baseline is usually subtracted from the measured line profile, and that the 2.7 K microwave background radiation is found everywhere, the

Table 13.1. Line parameters of the most abundant isotopic species of CO

Molecule	$J+1 \rightarrow J$	$\frac{\nu}{\mathrm{GHz}}$	$\frac{T_0}{\mathrm{K}}$ [a]	$\frac{T_A}{\mathrm{K}}$ [b]
$^{12}C^{18}O$	$1 \rightarrow 0$	109.782 182	5.269	1
$^{13}C^{16}O$	$1 \rightarrow 0$	110.201 370	5.289	7
$^{12}C^{16}O$	$1 \rightarrow 0$	115.271 203	5.532	75
$^{13}C^{16}O$	$2 \rightarrow 1$	220.398 714	10.578	16
$^{12}C^{16}O$	$2 \rightarrow 1$	230.538 001	11.064	76

[a] $T_0 = \frac{h\nu}{k}$; [b] T_A: observed antenna temperature for Ori A

appropriate formula is

$$T_b(\nu) = T_0\left(\frac{1}{\exp(-T_0/T_{ex}) - 1} - \frac{1}{\exp(-T_0/2.7) - 1}\right)(1 - e^{-\tau_\nu}) \; . \tag{13.50}$$

While $T_b(\nu)$ can be considered to be observable if all calibration problems as outlined in Chap. 10 have been solved, on the right-hand side of (13.50) we find the two unknown quantities: the excitation temperature of the line T_{ex} and the optical depth τ_ν. If τ_ν is known it is possible to solve for the column density $\mathcal{N}_{CO}$ as in the case of the line $\lambda = 21$ cm of H I. But in the case of CO we meet the difficulty that the strongest line $^{12}C^{16}O$ always seems to be optically thick with τ of the order of 100 or so. It is therefore almost impossible to derive information on the CO column density from this line; we have to use the weaker lines of the other isotopes. This procedure as outlined by Dickman (1978) is applicable if the following assumptions are valid:

1) All molecules along the line-of-sight possess a uniform excitation temperature in the $J = 1 \rightarrow 0$ transition.
2) The different isotopic species have the same excitation temperatures which are equal to the kinetic temperature of the gas.
3) The optical depth in the $^{12}C^{16}O$ line is large.

Then the excitation temperature can be determined from the measured T_L^{12} and the peak exitation temperature of the $^{12}C^{16}O$ line at 115.271 GHz is

$$\boxed{\frac{T_{ex}}{K} = 5.532 \ln\left(1 + \frac{5.532}{T_L^{12} - 0.8182}\right)} \; . \tag{13.51}$$

The optical depth of the $^{13}C^{16}O$ line at 110.201 GHz is then obtained by solving (13.50) for

$$\boxed{\tau_0^{13} = -\ln\left[1 - \frac{(T_L^{13}/K)}{5.289}\left\{\left[\exp\left(-\frac{5.289}{(T_{ex}/K)}\right) - 1\right]^{-1} - 0.1642\right\}^{-1}\right]} \; . \tag{13.52}$$

Using quite similar arguments to those in Sect. 11.1, we finally find that the column density $\mathcal{N}^{13}$ is related to τ_0^{13} by

$$\int_{\text{line}} \tau_\nu \, d\nu = \tau_0^{13} \, \Delta V_{13} \, \nu_{13}/c = \frac{c^2}{8\pi\nu_{13}^2}\frac{1}{g_1}\frac{g_2}{} A_J[1 - \exp(-T_0^{13}/T_{ex})]\frac{\mathcal{N}_{CO}^{13}}{Q} \; . \tag{13.53}$$

Here

$$A_J = 7.4 \cdot 10^{-8} \, s^{-1} \tag{13.54}$$

is the Einstein transition probability of the $J = 1 \rightarrow 0$ CO line and Q is the partition function of this molecule. Explicitly we have for rotational states of a linear molecule

$$Q = 1 + \sum_{L=1}^{\infty} (2L+1) \prod_{J=0}^{L} \exp[-h\nu(J)/k\,T_{ex}(J)] \tag{13.55}$$

where $\nu(J)$ and $T_{ex}(J)$ are the frequency and excitation temperature of the transition $J \to J-1$. For simplicity one usually assumes that $T_{ex}(J) = T_{ex}$ and evaluates the sum in (13.55) as if it were an integral, leading to

$$\boxed{Q = 2\,T_{ex}^{13}/T_0^{13}} \ . \tag{13.56}$$

Substituting this into (13.53) we finally obtain

$$\boxed{\frac{\mathcal{N}_{CO}^{13}}{\mathrm{cm}^{-2}} = 2.42 \cdot 10^{14} \frac{(T_{ex}/\mathrm{K}) \int \tau^{13}(V)\,dV}{1 - \exp[-5.289/(T_{ex}/\mathrm{K})]}} \tag{13.57}$$

where we have used

$$\tau_0^{13}\,\Delta V = \frac{c}{\nu_{13}} \int_0^{\infty} \tau_\nu\,d\nu = \int_{-\infty}^{\infty} \tau(V)\,dV \ . \tag{13.58}$$

In the limit of optically thin radiation this integral is equal to the integrated line intensity $\int T_A(V)\,dV$. If this is only approximately true, optical depth effects can be eliminated to some extent by using the approximation

$$\int_{-\infty}^{\infty} \tau(V)\,dV \cong \frac{\tau_0}{1 - e^{-\tau_0}} \int_{\infty}^{\infty} T_A(V)\,dV \ . \tag{13.59}$$

This formula is good to within 15% for $\tau_0 < 2$, and it always overestimates $\mathcal{N}$ when $\tau_0 > 1$. The formulae (13.51), (13.52), (13.57), and (13.59) permit an evaluation of the column density $\mathcal{N}_{CO}^{13}$ under the assumption of LTE.

The uncertainties and errors of the ^{13}CO column densities derived in this way come from several sources which can be collected together under the general heading NLTE effects. Perhaps more important is the uncertainty in the excitation temperature as derived from the optically thick ^{13}CO peak antenna temperature. While the ^{12}CO emission might be thermalized even at low densities, the less abundant isotopes may be characterized by a lower excitation temperature if they are subthermally excited or if their emission arises primarily from the colder cloud interiors. The optically thick ^{12}CO emission would not reflect these interior conditions. Frerking et al. (1982) estimate this effect to be at most $T_{ex}^{12}/2$ for densities larger than $10^3\ \mathrm{cm}^{-3}$ and $T_K = 10$ K.

Another effect is that, although T_{ex} describes the population of the $J = 0$ and $J = 1$ states well, it fails to do so for $J > 1$. The higher rotational levels might not be thermalized because their larger Einstein A coefficients depopulate them faster than they are populated collisionally, and this lack of knowledge of the population of the upper states leads to an uncertainty in the partition function. But a strict lower limit can be determined by assuming a two-level molecular model with the corresponding partition function.

Another kind of NLTE effect becomes important if the molecular gas is not assumed to be static, but shows systematic velocity gradients as in collapsing cloud models. Scoville et al. (1974) described such numerical models using the concept of photon escape probability. Leung and Liszt (1976) and Dickman (1978) gave additional models of this kind. The ^{12}CO and ^{13}CO radiation leaving such collapsing clouds was computed by using cloud models consisting of many concentric shells and by considering a large number of rotational levels. The effect of the NLTE effects can then be estimated if this radiation is analyzed by the standard LTE method and the resulting column densities are subsequently compared with the true column densities input into the model. Over most of the extent of the cloud models, LTE gives underestimates of the true ^{13}CO column densities by a factor varying from 1 to 4 depending on the properties of the model and of the position in the cloud. Failures of LTE by factors in excess of two occur only for visual extinctions along the line-of-sight below $A_V = 1^{m}.5$.

A problem of an entirely different kind is met if an attempt is made to derive ^{12}CO column densities from the ^{12}CO lines. Since they are optically thick any amount of ^{12}CO could be hidden in such clouds, but the close correspondence of the line shapes of ^{12}CO and the ^{13}CO lines seems to indicate that there is a relation between the two. Solomon and Sanders (1980) showed that the galactic "emissivity" for the longitude range $270° < l < 90°$ given by

$$J_0(r) = \left\langle \frac{T_L}{2} \frac{dV}{dr} \right\rangle \tag{13.60}$$

where $\langle\rangle$ stands for the average along the line-of-sight is remarkably similar for both ^{12}CO and ^{13}CO line radiation with an average ratio of 5.5. This can be explained if the total emission depends primarily on the number of clouds. If this is true, ^{12}CO line measurements can be used to obtain estimates of $\mathcal{N}_{CO}^{12}$.

13.9 Correlation of CO Line Intensities with Other Constituents of the Interstellar Medium

CO is only one of the constituents of the ISM, and it is obviously of interest to investigate whether there is a relation between these different components. If there is, then CO could be used as a tracer. Actually this is a question which belongs to the astrophysics of the ISM, but since since the CO line measurements are used as the standard research tool for the investigation of the ISM, we may just as well include this as a radio astronomical tool.

That there should indeed be a relation between $\mathcal{N}_{CO}$ and the visual extinction A_V can be rather easily understood. Outside dust clouds, the interstellar radiation field contains enough UV quanta able to photodestruct most molecules so that the molecular fractional abundance will be kept low. But with increasing depth into the clouds, the radiation will be attenuated resulting in a corresponding increase of CO fractional abundance. This is further increased by the self-shielding of the CO molecules.

From this qualitative sketch it should be expected that CO will be observed only if A_V is above a certain threshold value. Investigations of this kind have been made by various authors; to quote just one fairly recent investigation, Frerking et al. (1982) find the relations

$$\begin{aligned} \mathcal{N}_{CO}^{18} &= 0.7 \cdot 10^{14} (A_V - 1.9)\ \mathrm{cm}^{-2} \quad \text{for } 2 < A_V < 4 \ , \\ &= 2.4 \cdot 10^{14} (A_V - 2.9)\ \mathrm{cm}^{-2} \quad \text{for } 5 < A_V < 11 \ , \end{aligned} \tag{13.61}$$

and

$$\begin{aligned} \mathcal{N}_{CO}^{13} &= 1.4 \cdot 10^{15} (A_V - 1.0)\ \mathrm{cm}^{-2} \quad \text{for } 1 < A_V < 5 \ , \\ &= 2.7 \cdot 10^{15} (A_V - 1.6)\ \mathrm{cm}^{-2} \quad \text{for } 4 < A_V < 15 \ . \end{aligned} \tag{13.62}$$

There is some scatter around these relations, and they tend to vary somewhat from one place on the sky to another.

Of even greater importance is the observation that CO can be used as a tracer of molecular hydrogen in interstellar clouds where $A_V > 1^{\mathrm{mag}}$. The mass of the molecular cloud should be determined from observations of an isotope of CO that is readily observable, optically thin and not affected by enhancements of the isotope by some process of fractionization. Comparing CO abundances with H_2 column densities obtained in the infrared and the UV, Frerking et al. (1982) give for dense cores of clouds

$$\begin{aligned} \mathcal{N}(H_2) &= \left[\frac{\mathcal{N}(C^{18}O)}{1.7 \cdot 10^{14}} + 1.3\right] \cdot 10^{21}\ \mathrm{cm}^{-2} \quad \text{for } \mathcal{N}(C^{18}O) > 3 \cdot 10^{14}\ \mathrm{cm}^{-2} \\ &= \left[\frac{\mathcal{N}(C^{18}O)}{0.7 \cdot 10^{14}} + 1.9\right] \cdot 10^{21}\ \mathrm{cm}^{-2} \quad \text{for } \mathcal{N}(C^{18}O) < 3 \cdot 10^{14}\ \mathrm{cm}^{-2} \ . \end{aligned} \tag{13.63}$$

Liszt (1982) finds that even ^{12}CO can be used to estimate the column density of H_2 with remarkable success, and he derives for

$$\mathcal{N}(H) = \mathcal{N}(\mathrm{H\,I}) + 2\mathcal{N}(H_2) \tag{13.64}$$

the ratio

$$\frac{\mathcal{N}(H)}{\mathrm{cm}^{-2}} = (1.0 \pm 0.5) \cdot 10^{21} \int_{-\infty}^{\infty} \frac{T_L^{12}(V)}{\mathrm{K}}\, d\left(\frac{V}{\mathrm{km\,s}^{-1}}\right) . \tag{13.65}$$

The ratio of these two quantities varies remarkably little for different regions of the sky and from one author to the other. (13.65) therefore should be a useful approximation.

Appendix A. Some Useful Vector Relations [3]

Let $\boldsymbol{A}$, $\boldsymbol{B}$, $\boldsymbol{C}$, $\boldsymbol{D}$ be arbitrary vector fields assumed to be continuous and differentiable everywhere except at a finite number of points, and let ϕ and ψ be arbitrary scalar fields for which the same assumptions are adopted. If $\boldsymbol{A}\cdot\mathbf{B}$ is the scalar product and $\boldsymbol{A}\times\boldsymbol{B}$ the vector product the following algebraic relations are true:

$$\begin{aligned}\boldsymbol{A}\cdot(\boldsymbol{B}\times\boldsymbol{C}) &= (\boldsymbol{A}\times\boldsymbol{B})\cdot\boldsymbol{C} = (\boldsymbol{A},\boldsymbol{B},\boldsymbol{C}) = (\boldsymbol{B},\boldsymbol{C},\boldsymbol{A}) = (\boldsymbol{C},\boldsymbol{A},\boldsymbol{B})\\ &= -(\boldsymbol{A},\boldsymbol{C},\boldsymbol{B}) = -(\boldsymbol{C},\boldsymbol{B},\boldsymbol{A}) = -(\boldsymbol{B},\boldsymbol{A},\boldsymbol{C})\end{aligned} \tag{A.1}$$

$$\boldsymbol{A}\times(\boldsymbol{B}\times\boldsymbol{C}) = (\boldsymbol{A}\cdot\boldsymbol{C})\boldsymbol{B} - (\boldsymbol{A}\cdot\boldsymbol{B})\boldsymbol{C} \tag{A.2}$$

$$\boldsymbol{A}\times(\boldsymbol{B}\times\boldsymbol{C}) + \boldsymbol{B}\times(\boldsymbol{C}\times\boldsymbol{A}) + \boldsymbol{C}\times(\boldsymbol{A}\times\boldsymbol{B}) = 0 \tag{A.3}$$

$$\begin{aligned}(\boldsymbol{A}\times\boldsymbol{B})\cdot(\boldsymbol{C}\times\boldsymbol{B}) &= \boldsymbol{A}\cdot[\boldsymbol{B}\times(\boldsymbol{C}\times\boldsymbol{D})]\\ &= (\boldsymbol{A}\cdot\boldsymbol{C})(\boldsymbol{B}\cdot\boldsymbol{D}) - (\boldsymbol{A}\cdot\boldsymbol{D})(\boldsymbol{B}\cdot\boldsymbol{C})\end{aligned} \tag{A.4}$$

$$(\boldsymbol{A}\times\boldsymbol{B})\times(\boldsymbol{C}\times\boldsymbol{D}) = [(\boldsymbol{A}\times\boldsymbol{B})\cdot\boldsymbol{D}]\boldsymbol{C} - [(\boldsymbol{A}\times\boldsymbol{B})\cdot\boldsymbol{C}]\boldsymbol{D}\ . \tag{A.5}$$

Introducing the gradient of a scalar as $\nabla\phi$, ∇ considered as a differential operator obeys the following identities

$$\operatorname{grad}(\phi\psi) = \nabla(\phi\psi) = \phi\nabla\psi + \psi\nabla\phi \tag{A.6}$$

$$\operatorname{div}(\phi\boldsymbol{A}) = \nabla\cdot(\phi\boldsymbol{A}) = \boldsymbol{A}\cdot\nabla\phi + \phi\nabla\cdot\boldsymbol{A} \tag{A.7}$$

$$\operatorname{curl}(\phi\boldsymbol{A}) = \operatorname{rot}(\phi\boldsymbol{A}) = \nabla\times(\phi\boldsymbol{A}) = \phi\nabla\times\boldsymbol{A} - \boldsymbol{A}\times\nabla\phi \tag{A.8}$$

$$\operatorname{div}(\boldsymbol{A}\times\boldsymbol{B}) = \nabla\cdot(\boldsymbol{A}\times\boldsymbol{B}) = \boldsymbol{B}\cdot(\nabla\times\boldsymbol{A}) - \boldsymbol{A}\cdot(\nabla\times\boldsymbol{B}) \tag{A.9}$$

$$\begin{aligned}\operatorname{curl}(\boldsymbol{A}\times\boldsymbol{B}) &= \operatorname{rot}(\boldsymbol{A}\times\boldsymbol{B}) = \nabla\times(\boldsymbol{A}\times\boldsymbol{B})\\ &= \boldsymbol{A}(\nabla\cdot\boldsymbol{B}) - \boldsymbol{B}(\nabla\cdot\boldsymbol{A}) + (\boldsymbol{B}\cdot\nabla)\boldsymbol{A} - (\boldsymbol{A}\cdot\nabla)\boldsymbol{B}\end{aligned} \tag{A.10}$$

$$\begin{aligned}\operatorname{grad}(\boldsymbol{A}\cdot\boldsymbol{B}) &= \nabla(\boldsymbol{A}\cdot\boldsymbol{B})\\ &= \boldsymbol{A}\times(\nabla\times\boldsymbol{B}) + \boldsymbol{B}\times(\nabla\times\boldsymbol{A}) + (\boldsymbol{B}\cdot\nabla)\boldsymbol{A} + (\boldsymbol{A}\cdot\nabla)\boldsymbol{B}\ .\end{aligned} \tag{A.11}$$

---

[3] Mainly adopted from Panofsky, W., Phillips, M. (1962): *Classical Electricity and Magnetism* (Addison-Wesley, Reading MA).

For ∇ some second order formulae are useful

$$\nabla^2\phi = \nabla \cdot \nabla\phi = \Delta\phi \tag{A.12}$$

$$\nabla^2 \boldsymbol{A} = \nabla(\nabla \cdot \boldsymbol{A}) - \nabla \times (\nabla \times \boldsymbol{A}) \tag{A.13}$$

$$\nabla \times \nabla\phi = 0 \tag{A.14}$$

$$\nabla \cdot (\nabla \times \boldsymbol{A}) = 0 \ . \tag{A.15}$$

Relations for Special Functions: Let $\boldsymbol{r}$ be the radius vector from the origin to the point x, y, z. Then

$$\nabla \cdot \boldsymbol{r} = 3 \tag{A.16}$$

$$\nabla \times \boldsymbol{r} = 0 \tag{A.17}$$

$$\nabla r = \nabla|\boldsymbol{r}| = \boldsymbol{r}/|\boldsymbol{r}| \tag{A.18}$$

$$\nabla(1/r) = -\,\boldsymbol{r}/r^3 \tag{A.19}$$

$$\nabla \cdot (\boldsymbol{r}/r^3) = -\,\nabla^2(1/r) = 4\pi\,\delta(r) \ . \tag{A.20}$$

Integral Relations: Let a vector field $\boldsymbol{A}$ and its divergence $\nabla \cdot \boldsymbol{A}$ be continuous over a closed region V with the surface S, the surface element $d\boldsymbol{S}$ being counted positive in direction outward from the enclosed volume. Then Gauss' theorem states

$$\oint_S \boldsymbol{A} \cdot d\boldsymbol{S} = \int_V (\nabla \cdot \boldsymbol{A})\, dv \tag{A.21}$$

while Stokes' theorem postulates

$$\oint_S d\boldsymbol{S} \times \boldsymbol{A} = \int_V (\nabla \times \boldsymbol{A})\, dv \ . \tag{A.22}$$

Green's theorem is

$$\int_V (\phi\,\nabla \cdot \nabla\psi - \psi\,\nabla \cdot \nabla\phi)\, dv = \oint_S (\phi\,\nabla\psi - \psi\,\nabla\phi) \cdot d\boldsymbol{S} \ . \tag{A.23}$$

The vector components of ∇ in cylindrical and spherical polar coordinates are often of importance. They are:

	Cartesian coord.	Cylindrical coord.	Spherical polar coord.	
Coordinates	x, y, z	r, ϕ, z	r, θ, ϕ	
Orthogonal line element	dx, dy, dz	$dr, r\,d\phi, dz$	$dr, r\,d\theta, r \sin\theta\, d\phi$	
Gradient ∇	$(\nabla\psi)_x = d\psi/dx$	$(\nabla\psi)_r = d\psi/dr$	$(\nabla\psi)_r = d\psi/dr$	
	$(\nabla\psi)_y = d\psi/dy$	$(\nabla\psi)_\varphi = \frac{1}{r}\frac{d\psi}{d\varphi}$	$(\nabla\psi)_\theta = \frac{1}{r}\frac{d\psi}{d\theta}$	(A.24)
	$(\nabla\psi)_z = d\psi/dz$	$(\nabla\psi)_z = d\psi/dz$	$(\nabla\psi)_\varphi = \frac{1}{r\sin\theta}\frac{d\psi}{d\varphi}$	
Divergence $\nabla \cdot \boldsymbol{A}$	$\frac{\partial A_x}{\partial x} + \frac{\partial A_y}{\partial y} + \frac{\partial A_z}{\partial z}$	$\frac{1}{r}\left(\frac{\partial (r\,A_r)}{\partial r} + \frac{\partial A_\varphi}{\partial \varphi}\right) + \frac{\partial A_z}{\partial z}$	$\frac{1}{r^2}\frac{\partial (r^2 A_r)}{\partial r} + \frac{1}{r\sin\theta}\left(\frac{\partial (\sin\theta\, A_\theta)}{\partial\theta} + \frac{\partial A_\varphi}{\partial\varphi}\right)$	(A.25)
Components of curl $\boldsymbol{A} = \nabla \times \boldsymbol{A}$	$(\nabla\times\boldsymbol{A})_x = \frac{\partial A_z}{\partial y} - \frac{\partial A_y}{\partial z}$	$(\nabla\times\boldsymbol{A})_r = \frac{1}{r}\frac{\partial A_z}{\partial\varphi} - \frac{\partial A_\varphi}{\partial z}$	$(\nabla\times\boldsymbol{A})_r = \frac{1}{r\sin\theta}\left(\frac{\partial(\sin\theta\, A_\varphi)}{\partial\theta} - \frac{\partial A_\theta}{\partial\varphi}\right)$	
	$(\nabla\times\boldsymbol{A})_y = \frac{\partial A_x}{\partial z} - \frac{\partial A_z}{\partial x}$	$(\nabla\times\boldsymbol{A})_\varphi = \frac{\partial A_r}{\partial z} - \frac{\partial A_z}{\partial r}$	$(\nabla\times\boldsymbol{A})_\theta = \frac{1}{r\sin\theta}\frac{\partial A_r}{\partial\varphi} - \frac{1}{r}\frac{\partial(r\,A_\varphi)}{\partial r}$	(A.26)
	$(\nabla\times\boldsymbol{A})_z = \frac{\partial A_y}{\partial x} - \frac{\partial A_x}{\partial y}$	$(\nabla\times\boldsymbol{A})_z = \frac{1}{r}\left(\frac{\partial(r\,A_\varphi)}{\partial r} - \frac{\partial A_r}{\partial\varphi}\right)$	$(\nabla\times\boldsymbol{A})_\varphi = \frac{1}{r}\left(\frac{\partial(r\,A_\theta)}{\partial r} - \frac{\partial A_r}{\partial\theta}\right)$	
Laplacian of ψ div grad $\psi = \nabla^2\psi = \Delta\psi$	$\frac{\partial^2\psi}{\partial x^2} + \frac{\partial^2\psi}{\partial y^2} + \frac{\partial^2\psi}{\partial z^2}$	$\frac{1}{r}\frac{\partial}{\partial r}\left(r\frac{\partial\psi}{\partial r}\right) + \frac{1}{r^2}\frac{\partial^2\psi}{\partial\varphi^2} + \frac{\partial^2\psi}{\partial z^2}$	$\frac{1}{r^2}\frac{\partial}{\partial r}\left(r^2\frac{\partial\psi}{\partial r}\right) + \frac{1}{r^2\sin\theta}\left[\frac{\partial}{\partial\theta}\left(\sin\theta\frac{\partial\psi}{\partial\theta}\right) + \frac{1}{\sin\theta}\frac{\partial^2\psi}{\partial\varphi^2}\right]$	(A.27)

Appendix B. Fourier Transform[4]

Fourier transform

$$F(s) = \int_{-\infty}^{\infty} f(x)\, e^{-i 2\pi s x}\, dx \; . \tag{B.1}$$

Inverse Fourier transform

$$f(x) = \int_{-\infty}^{\infty} F(s)\, e^{i 2\pi x s}\, ds \; . \tag{B.2}$$

Theorems for the Fourier transform:

Theorem	$f(x)$	$F(s)$	
Similarity	$f(a x)$	$\frac{1}{\lvert a\rvert} F\left(\frac{s}{a}\right)$	(B.3)
Addition	$f(x) + g(x)$	$F(s) + G(s)$	(B.4)
Shift	$f(x - a)$	$e^{-i 2\pi a s} F(s)$	(B.5)
Modulation	$f(x) \cos x$	$\frac{1}{2} F\left(s - \frac{\omega}{2\pi}\right) + \frac{1}{2} F\left(s + \frac{\omega}{2\pi}\right)$	(B.6)
Convolution	$f(x) \otimes g(x)$	$F(s)\, G(s)$	(B.7)
Autocorrelation	$f(x) \otimes f^*(-x)$	$\lvert F(s)\rvert^2$	(B.8)
Derivative	$f'(x)$	$i 2\pi s F(s)$	(B.9)

Rayleigh theorem

$$\int_{-\infty}^{\infty} \lvert f(x)\rvert^2 dx = \int_{-\infty}^{\infty} \lvert F(s)\rvert^2 ds \; . \tag{B.10}$$

Power theorem

$$\int_{-\infty}^{\infty} f(x)\, g^*(x)\, dx = \int_{-\infty}^{\infty} F(s)\, G^*(s)\, ds \; . \tag{B.11}$$

---

[4] Adopted from Bracewell, R. (1965): *The Fourier Transform and its Applications* 1965 (McGraw Hill, New York).

A short list of Fourier transform pairs:

$f(x)$	$F(s)$
$\mathrm{e}^{-\pi x^2}$	$\mathrm{e}^{-\pi s^2}$
$x\,\mathrm{e}^{-\pi x^2}$	$-\mathrm{i}\,s\,\mathrm{e}^{-\pi s^2}$
1	$\delta(s)$
$\cos \pi x$	$\mathrm{II}(s)$
$\sin \pi x$	$\mathrm{i}\,{}^{I}{}_{I}(s)$
$\mathrm{III}(x)$	$\mathrm{III}(s)$
$\operatorname{sinc} x$	$\Pi(s)$
$\operatorname{sinc}^2 x$	$\Pi(s) \otimes \Pi(s)$
$\operatorname{sinc}^3 x$	$\Pi(s) \otimes \Pi(s) \otimes \Pi(s)$
$\mathrm{e}^{-\|x\|}$	$\dfrac{2}{1+(2\pi s)^2}$
$\mathrm{e}^{-\|x\|}\dfrac{\sin x}{x}$	$\arctan \dfrac{1}{2\pi^2 s^2}$

$f(x)$	$F(s)$
$\|x\|^{-1/2}$	$\|s\|^{-1/2}$
$\mathrm{e}^{-\|x\|}\cos \pi x$	$\dfrac{2}{1+(2\pi s)^2} \otimes \Pi(s)$
$\operatorname{sech} \pi x$	$\operatorname{sech} \pi s$
$\operatorname{sech}^2 \pi x$	$2\,s \operatorname{cosech} \pi s$
$H(x)$	$\dfrac{1}{2}\delta(s) - \dfrac{i}{2\pi s}$
$J_0(2\pi x)$	$\dfrac{\Pi(s/2)}{\pi(1-s^2)^{1/2}}$
$J_1(2\pi x)/2x$	$(1-s^2)^{1/2}\,\Pi\left(\dfrac{s}{2}\right)$

where

$$\operatorname{sinc} x = \frac{\sin \pi x}{\pi x}$$

$$\Pi(x) = \begin{cases} 1 & \text{for } |x| < \frac{1}{2} \\ 0 & \text{for } |x| > \frac{1}{2} \end{cases}$$

$$H(x) = \begin{cases} 1 & \text{for } x > 0 \\ 0 & \text{for } x < 0 \end{cases}$$

$$\text{Ш}(x) = \sum_{n=-\infty}^{\infty} \delta(x-n)$$

$$\mathrm{II}(x) = \tfrac{1}{2}\delta(x+\tfrac{1}{2}) + \tfrac{1}{2}\delta(x-\tfrac{1}{2})$$

$${}^{I}{}_{I}(x) = \tfrac{1}{2}\delta(x+\tfrac{1}{2}) - \tfrac{1}{2}\delta(x-\tfrac{1}{2})$$

Appendix C. Hankel Transform [5]

Hankel transform

$$F(q) = 2\pi \int_0^\infty f(r)\, J_0(2\pi q r)\, r\, dr \quad .$$

Inverse Hankel transform

$$f(r) = 2\pi \int_0^\infty F(q)\, J_0(2\pi q r)\, q\, dq \quad .$$

Theorems for the Hankel transforms:

Theorem	$f(r)$	$F(q)$
Similarity	$f(ar)$	$\frac{1}{a^2} F\left(\frac{q}{a}\right)$
Addition	$f(r) + g(r)$	$F(q) + G(q)$
Shift	shift of origin destroys circular symmetry	
Convolution	$\int_0^\infty \int_0^{2\pi} f(r')\, g(R)\, r'\, dr'\, d\theta$ $R^2 = r^2 + r'^2 - 2 r r' \cos\theta$	$F(q)\, G(q)$

Rayleigh theorem

$$\int_0^\infty |f(r)|^2\, r\, dr = \int_0^\infty |F(q)|^2\, q\, dq \quad .$$

Power theorem

$$\int_0^\infty f(x)\, g^*(r)\, r\, dr = \int_0^\infty F(q)\, G^*(q)\, q\, dq \quad .$$

[5] Adopted from Bracewell, R. (1965): *The Fourier Transform and its Applications* 1965 (McGraw Hill, New York).

Some Hankel transforms:

$f(r)$	$F(q)$
$\Pi\left(\frac{r}{2a}\right)$	$\frac{a}{q} J_1(2\pi a q)$
$\frac{1}{r}\sin(2\pi a r)$	$\frac{\Pi(q/2a)}{(a^2-q^2)^{1/2}}$
$\frac{1}{2}\delta(r-a)$	$\pi a J_0(2\pi a q)$
$e^{-\pi r^2}$	$e^{-\pi q^2}$
$(a^2+r^2)^{-1/2}$	$\frac{1}{q} e^{-2\pi a q}$
$(a^2+r^2)^{-3/2}$	$\frac{2\pi}{a} e^{-2\pi a q}$
$(a^2+r^2)^{-1}$	$2\pi K_0(2\pi a q)$
$2a^2(a^2+r^2)^{-2}$	$4\pi^2 a q K_1(2\pi a q)$
$(a^2-r^2)\Pi\left(\frac{r}{2a}\right)$	$\frac{a^2}{\pi q^2} J_2(2\pi a q)$
$\frac{1}{r}$	$\frac{1}{q}$
e^{-ar}	$2\pi a(4\pi^2 q^2 + a^2)^{-3/2}$
$\frac{1}{r} e^{-ar}$	$2\pi(4\pi^2 q^2 + a^2)^{-1/2}$
$\frac{1}{\pi r}\delta(r)$	1
$\frac{1}{2a^2}\left[\Pi\left(\frac{r}{2a}\right)\otimes\Pi\left(\frac{r}{2a}\right)\right]$	$\frac{1}{2a^2}\lvert J_1(2\pi a q)\rvert^2$
$r^2 e^{-\pi r^2}$	$\frac{1}{\pi}\left(\frac{1}{\pi}-q^2\right) e^{-\pi q^2}$

Appendix D. Electromagnetic Field Quantities

The international system of units SI that was introduced by the II. Conférence Générale des Poids et Mésures (CGPM) in 1960, it is the only legal system of units in most countries. Only in some astronomical applications, other units which are more suited for the situation and which do not belong to the SI-System will be employed in agreement with the general usage.

In SI the Ampere A is introduced as a basic unit of electric current which produces a specified force between two wires carrying such a current, while the unit of electromotive force (emf), the Volt, is a derived quantity obtained by considering the current needed to arrive at a given amount of power (measured in W): 1 Volt = 1 V = 1 W/1 A = 1 m^2 kg s^{-3} A^{-1}. The unit of charge, the Coulomb, is defined as 1 C = 1 A s. All other electromagnetic field quantities now can conveniently be expressed in these units.

The charge is obviously an additive scalar quantity, and therefore it makes sense to define a scalar *charge density* ϱ by forming the limit

$$\varrho = \lim_{V_n \to 0} \frac{Q_n}{V_n}$$

where Q_n is the charge in the volume V_n. ϱ obviously is measured in C m^{-3} = A s m^{-3}.

For the current I the appropriate *current density* $\boldsymbol{J}$ is the current per unit area resulting in

$$|\boldsymbol{J}| = \lim_{F_n \to 0} \frac{I_n}{F_n}$$

measured in A m^{-2}, where F_n is the cross-section of the current.

The electric field intensity $\boldsymbol{E}$ again is a vector quantity. It measures the change of the emf along a given direction

$$|\boldsymbol{E}| = \lim_{l_n \to 0} \frac{U_n}{l_n}$$

where U_n is the change of the voltage over the length l_n. $\boldsymbol{E}$ is measured in V m^{-1}.

Another property of the electric field is described by the *electric flux* $\boldsymbol{D}$. It measures the power of the field to induce charges on the surfaces of a small condensor. If Q_n is the charge induced on the condensor plate of surface F_n then

$$|\boldsymbol{D}| = \lim_{F_n \to 0} \frac{Q_n}{F_n} .$$

$\boldsymbol{D}$ is again a vector quantity since the induced charge can depend on the orientation of F_n, it is measured in $\mathrm{C\,m^{-2} = A\,s\,m^{-2}}$.

The magnetic field is also described by two vector fields, $\boldsymbol{H}$ and $\boldsymbol{B}$. By 1820, Ørsted had already shown that an electric current is always accompagnied by a magnetic field. The field intensity on the axis of a long homogeneous solenoid is given by

$$|\boldsymbol{H}| = \lim_{l_n \to 0} \frac{|\boldsymbol{I}|}{l_n}$$

where $\boldsymbol{I}$ is the toal current (if there are n windings, I is n times the current in a single wire), and l the length of the solenoid. Therefore the field intensity of an arbitrary field is measured in units of an equivalent field in a selenoid given in $\mathrm{A\,m^{-1}}$.

The magnetic flux density $\boldsymbol{B}$ is a measure of the ability of a given magnetic field to produce a voltage by induction. If a given magnetic field $\boldsymbol{B}$ is switched on or off, it induces a flux of Φ_n Weber or V s in a conducting loop of area F_n.

Therefore

$$|\boldsymbol{B}| = \lim_{F_n \to 0} \frac{\Phi_n}{F_n}$$

is measured in units of Tesla or Weber per square meter ($1\,\mathrm{T} = 1\,\mathrm{Wb\,m^{-2}} = \mathrm{V\,s\,m^{-2}}$) and is dependent on the orientation of the loop area F_n. $\boldsymbol{B}$ is also a vector quantity.

An electromagnetic field therefore will be described by four Vector fields:

the electric field intensity $\boldsymbol{E}$ measured in $\mathrm{V\,m^{-1}}$;
the electric flux density $\boldsymbol{D}$ measured in $\mathrm{C\,m^{-2} = A\,s\,m^{-2}}$;
the magnetic field intensity $\boldsymbol{H}$ measured in $\mathrm{A\,m^{-1}}$;
the magnetic flux density $\boldsymbol{B}$ measured in $T = \mathrm{Wb\,m^{-2} = V\,s\,m^{-2}}$.

In addition there are

the current density $\boldsymbol{J}$ measured in $\mathrm{A\,m^{-2}}$ and
the charge density ϱ measured in $\mathrm{C\,m^{-3} = A\,s\,m^{-3}}$.

Appendix E. A List of Calibration Radio Sources

The usual method to determine the flux of radio sources is to measure the ratio of the response of the antenna/receiver combination to the source and to that of a well-known calibrator. In order to avoid as much as possible the effects of varying telescope efficiencies it is desirable to have a set of calibrators reasonably well distributed over the sky, such that radio source and calibration source can be measured at nearly equal zenith angles. The list of calibration sources given here has been taken from Baars, J.W.M., Genzel, R., Pauliny-Toth, I.I.K., Witzel, A. (1977): Astron. Astrophys. **61,** 99 and it has been used as calibration standard in most radio observatories.

Source	RA (1950.0) [h m s]	Dec (1950.0) [° ′ ″]	b^{II} [°]	S_{1400} [Jy]	S_{1665} [Jy]	S_{2700} [Jy]	S_{5000} [Jy]
3C48	01 34 49.8	+32 54 20	−29	15.9	13.9	9.20	5.24
3C123	04 33 55.2	+29 34 14	−12	48.7	42.4	28.5	16.5
3C147	05 38 43.5	+49 49 42	+10	22.4	19.8	13.6	7.98
3C161	06 24 43.1	−05 51 14	−8	19.0	16.8	11.4	6.62
3C218	09 15 41.5	−11 53 06	+25	43.1	36.8	23.7	13.5
3C227	09 45 07.8	+07 39 09	+42	7.21	6.25	4.19	2.52
3C249.1	11 00 25.0	+77 15 11	+39	2.48	2.14	1.40	0.77
3C274	12 28 17.7	+12 39 55	+74	214	184	122	71.9
3C286	13 28 49.7	+30 45 58	+81	14.8	13.6	10.5	7.30
3C295	14 09 33.5	+52 26 13	+61	22.3	19.2	12.2	6.36
3C348	16 48 40.1	+05 04 28	+29	45.0	37.5	22.6	11.8
3C353	17 17 54.6	−00 55 55		57.3	50.5	35.0	21.2
Dr 21	20 37 14.2	+42 09 07	+1	–	–	–	–
NGC 7027[a]	21 05 09.4	+42 02 03	−3	1.35	1.65	3.5	5.7

[a] Data up to 5 GHz are the direct measurements, not calculated from fit

Source	S_{8000} [Jy]	$S_{10\,700}$ [Jy]	$S_{15\,000}$ [Jy]	$S_{22\,235}$ [Jy]	Spec.	Ident.	Polar. (at 5 GHz) %	Ang. size (at 1.4 GHz) ″
3C 48	3.31	2.46	1.72	1.11	C^-	QSS	5	< 1
3C 123	10.6	7.94	5.63	3.71	C^-	GAL	2	20
3C 147	5.10	3.80	2.65	1.71	C^-	QSS	< 1	< 1
3C 161	4.18	3.09	2.14	–	C^-	GAL	5	< 3
3C 218	8.81	6.77	–	–	*S*	GAL	1	core 25 halo 200
3C 227	1.71	1.34	1.02	0.73	*S*	GAL	7	180
3C 249.1	0.47	0.34	0.23	–	*S*	QSS	–	15
3C 274	48.1	37.5	28.1	20.0	*S*	GAL	1	halo 400 [a]
3C 286	5.38	4.40	3.44	2.55	C^-	QSS	11	< 5
3C 295	3.65	2.53	1.61	0.92	C^-	GAL	0.1	4
3C 348	7.19	5.30	–	–	*S*	GAL	8	115 [b]
3C 353	14.2	10.9	–	–	C^-	GAL	5	150
Dr 21	21.6	20.8	20.0	19.0	*Th*	H II	–	20 [c]
NGC 7027	–	6.43	6.16	5.86	*Th*	PN	< 1	10

[a] Halo has steep spectral index, so for $\lambda \leqq 6$ cm, more than 90% of the flux is in the core
[b] Angular distance between the two components
[c] Angular size at 2 cm, but consists of 5 smaller components

List of Symbols

The numbers given after the explanation of each symbol indicate the pages on which the symbol is defined

$\mathbf{A}$	Vector potential 50
A_e	Effective telescope aperture 66
A_g	Geometric aperture 66
$A(r)$	Oort constant at r 238
$A(t)$	Envelope of analytic signal 40
A_{ik}, B_{ik}	Einstein transition probabilities 208
A_{klmn}	Closure amplitude 121
$\mathbf{B}$	Magnetic flux 16
$\mathbf{B}$	Baseline vector of interferometer 108
B	Total brightness 10
B_e	Rotational constant of molecule 271
B_ν	Brightness, Planck law 4
$C(\omega)$	Power spectrum 175
C_{ik}	Collision rate 215
$\mathbf{D}$	Electric flux 16
D	Telescope diameter 75
D_e	Centrifugal stretching 271
$\mathscr{D}$	Directivity of antenna 65
DM	Dispersion measure 31
$\mathbf{E}$	Electric field strength 16
E	Cosmic ray energy 188
$E\{x\}$	Expected value 123
EM	Emission measure 178
G, g	Green's functions 53
$G(\nu), H(\nu), W(\nu)$	Power transfer functions 134
$G(\vartheta, \varphi)$	Gain of antenna 63
$\mathbf{H}$	Magnetic field strength 16
HM	Hydrogen measure 227
HPBW	Half power beam width 65
I, Q, U, V	Stokes parameter 37, 38, 44
I_ν	Intensity 5
$\mathbf{J}$	Current density 16
$\mathbf{J}$	Coherence matrix 41
Jy	Jansky (flux unit) 5
N_e	Electron density 194
N_i	Volume density of atoms in the state i 208
N_n^*	Level population for local thermodynamic equilibrium (LTE) 257
$\mathcal{N}_{HI}$	Column density of H I 226

$\mathcal{N}_{CO}$	Column density of CO	287
$P(r)$	Potential curve	270
$P(\xi)$	Normalized profile shape	236
$P(\vartheta, \varphi)$	Power pattern of antenna	63
$P_n(\vartheta, \varphi)$	Normalized power pattern	63, 73
R	Rydberg constant	251
$R(\tau)$	Correlation function	125
$R(\boldsymbol{B})$	Correlation response for baseline $\boldsymbol{B}$	108
R_S	Radiation impedance	60
S_i	Stokes parameter	27, 38, 44
$S(\nu)$	Power spectral density (PSD)	125
$\boldsymbol{S}$	Poynting vector	17
S_ν	Flux	5
T_A	Antenna temperature	68, 94, 221
T_b	Brightness temperature	12
T_{ex}	Excitation temperature	216
T_K	Kinetic temperature	228
T_N	Noise temperature	12
T_S	Spin temperature	225
$\bar{U}$	Average energy density	209
U_s	Expansion velocity of supernova remnants (SNR)	201
$U(SpT)$	Excitation parameter	248
$V(\boldsymbol{B})$	Visibility function	109, 111
$V(t)$	Analytic signal	40
X_i	Ionization potential of the level i	254
$X(z)$	Air mass	165
W	Energy	4
Z_0	Intrinsic impedance	24
$a(\nu, T)$	Correction factor for optical depth approximation	178
b_n	Departure coefficients	257
c	Velocity of light	21
e	Specific internal energy of gas	200
d	Correlation distance of phase errors	85
d	Electric dipole moment	172
$g(x, y)$	Illumination grading	71, 73
k	Wave number	22
n	Refraction index	22, 30
p	Gas pressure	200
p	Degree of polarization	45
p	Collision parameter	173
$p(x)$	Probability density	129
q	Signal to noise ratio	98
r_S	Radius of SNR	202
s_0	Normalized radius of Strömgren sphere	248
u	Arbitrary component of electromagnetic field	21, 25
u_ν	Energy density	6
v	Phase velocity	22, 29

v_g	Group velocity 26, 29
x	Degree of ionization 249
x_e, y_e	Anharmonic factors 276
α	Spectral index 188
β	Velocity in units of c 181
β	Correction for NLTE 258
β_{ik}	Correction of κ_ν for NLTE 258
Γ	Sensitivity of antenna 93, 221
$\Gamma(P_1, P_2, \tau)$	Mutual coherence function 101, 104
γ	Ratio of specific heats 200
γ	Lorentz factor 181
γ_{mn}	Collision probability 215
Δf	Noise bandwidth 135
$\Delta\psi$	Faraday rotation 49
δ	Phase difference 33
δ	Phase error in aperture 84
δ	Exponent in power law distribution of relativistic electrons 203
δ	Dirac's delta function 53
ε	Permittivity, dielectric constant 16
ε	Surface tolerance of a reflector 84
ε_ν	Emissivity 7
ε_p	Energy density of relativistic particles 190
ε_e	Energy density of relativistic electrons 190
η	$\eta = \varepsilon_p/\varepsilon_e$ 190
η_R	Radiation efficiency 68, 94
η_M	Main beam efficiency 65, 222
η_ν	Deviation of source function from Planck function 259
$\Theta(r)$	Linear galactic rotation velocity 237
Θ_e	Moment of inertia of molecule 270
κ_ν	Opacity, absorption coefficient 7
κ_ν^*	Absorption coefficient for LTE 258
Λ	Wave length of sound wave 161
Λ	Lambda function 85
λ	Wave length 26
μ	Permeability 16
μ_{mn}	Mean electric dipole moment 213
μ_{mn}^*	Mean magnetic dipole moment 213
ν_p, ω_p	Plasma frequency 4, 29
$\Pi(r)$	Galactic expansion velocity 237
ϱ	Electric charge density 16
ϱ	Density 199
ϱ	Ray parameter 195
σ	Electric conductivity 16, 29
τ_ν	Optical depth 8
$\Phi(t)$	Mean phase of analytic signal 40
Φ	Electric potential 50
φ	Phase of harmonic wave 22

$\varphi(\nu)$	Line profile function	208
Ψ_{ikl}	Closure phase	120
Ω	Solid angle	4
Ω_A	Beam solid angle	64
Ω_M	Main beam solid angle	64
$\Omega(r)$	Galactic angular rotational velocity	237
ω_c	Cyclotron frequency	183
ω_G	Gyro frequency	184
$\langle \cdots \rangle$	Time average	17
$\langle g_{ff} \rangle$	Gaunt factor	178
Ш	Sha function	115

References

Chapter 1

a) General

Kraus, J.D. (1966): *Radio Astronomy* (McGraw Hill, New York)
Rybicki, G.B., Lightman, A.P. (1979): *Radiative Processes in Astrophysics* (Wiley, New York)

b) Special

Berkhuijsen, E.M. (1975): Astron. Astrophys. **40,** 311
Evans, J.V., Hagford, T. (1968): *Radar Astronomy* (McGraw Hill, New York) Chap. 2
Reif, F. (1965): *Statistical and Thermal Physics* (McGraw Hill, New York) **A** 11

Chapter 2

Becker, R., Sauter, F. (1962): *Theorie der Elektrizität,* Vol. 1 (Teubner, Stuttgart)
Jackson, J.D. (1975): *Classical Electrodynamics* (Wiley, New York)
Kühn, R. (1964): *Mikrowellen-Antennen* (VEB Verlag Technik, Berlin)
Manchester, R.N., Taylor, J.H. (1977): *Pulsars* (Freeman, San Francisco)
Panofsky, W., Phillips, M. (1962): *Classical Electricity and Magnetism* (Addison-Wesley, Reading, MA)
Sommerfeld, A. (1964): *Optik* (Akademische Verlags-Gesellschaft, Leipzig)
Sommerfeld, A. (1959): *Elektrodynamik* (Akademische Verlags-Gesellschaft, Leipzig)

Chapter 3

Born, M., Wolf, E. (1965): *Principles of Optics* (Pergamon, Oxford)
Spitzer, Jr., L. (1968): *Diffuse Matter in Space* (Interscience, New York)
Wiener, N. (1949): *Extrapolation, Interpolation and Smoothing of Stationary Time Series* (MIT Cambridge, MA)

Wave propagation in a plasma is treated more or less extensively in most texts on plasma physics. Some wellknown textbooks are

Chen, F.F. (1974): *Introduction to Plasma Physics* (Plenum, New York)
Krall, N.A., Trivelpiece, A.W. (1973): *Principles of Plasma Physics* (McGraw Hill, New York)
Sommerfeld, A. (1964): *Optik* (Akademische Verlags-Gesellschaft, Leipzig)
Spitzer, Jr., L. (1962): *Physics of Fully Ionized Gas,* 2nd ed. (Wiley, New York)

Chapter 4

Abramowitz, M., Stegun, I.A. (1964): *Handbook of Mathematical Functions* (National Bureau of Standards, Washington, DC)

Kraus, J.D. (1950): *Antennas* (McGraw Hill, New York)

Kühn, R. (1964): *Mikrowellen-Antennen* (VEB Verlag Technik, Berlin)

Pawsey, J.L., Bracewell, R.N. (1954): *Radio Astronomy* (Oxford University Press, Oxford)

Silver, S. (ed). (1949): *Microwave Antenna Theory and Design* (McGraw Hill, New York)

Chapter 5

a) General

Christiansen, W.N., Högbom, J.A. (1985): *Radiotelescopes,* 2nd ed. (Cambridge University Press, Cambridge)

Heilmann, A (1970): *Antennen I–III* (Bibliographisches Institut, Mannheim)

Kraus, J.D. (1950): *Antennas* (McGraw Hill, New York)

Kraus, J.D. (1966): *Radio Astronomy* (McGraw Hill, New York)

Kühn, R. (1964): *Mikrowellen-Antennen* (VEB Verlag Technik, Berlin)

Meeks, M.L. (ed.) (1976): *Astrophysics Part B; Radiotelescopes, Methods of Experimental Physics,* Vol. 12 B (Academic, New York)

Meinke, H., Gundlach, F.N. (eds.) (1968): *Taschenbuch der Hochfrequenztechnik* (Springer, Berlin, Heidelberg)

Silver, S. (ed.) (1949): *Microwave Antenna Theory and Design* (McGraw Hill, New York)

Stutzman, W.L., Thiele, G.A. (1981): *Antenna Theory and Design* (Wiley, New York)

b) Special

Abramowitz, M., Stegun, J.A. (1964): *Handbook of Mathematical Functions* (National Bureau of Standards, Washington, DC)

Bao, V.T. (1969): Proc. IEEE **116,** 195

Bracewell, R.N. (1962): „Radio Astronomy Techniques", in *Handbuch der Physik,* Vol. 54 ed. by S. Flügge (Springer, Berlin, Heidelberg) p. 42

Bracewell, R.N. (1965): *The Fourier Transform and its Application*s (McGraw Hill, New York)

Burns, W.R. (1972): Astron. Astrophys. **19,** 41

Gradshteyn, I.S., Ryzhik, I.M. (1965): *Tables of Integrals, Series and Products* (Academic, New York)

Höglund, B. (1967): Pencil beam survey of radio sources between declinations + 18° and + 20° at 750 and 1410 MHz. Acta Polytech. Scand. PH **48**

Hoerner, von, S. (1961): Very large antennas for the cosmological problem, 1. Basic Considerations. Publ. NRAO **1**, 19

Keller, J.B. (1962): J. Opt. Soc. Am. **52,** 116

Mittra, R. (ed.) (1975): *Numerical and Asymptotic Techniques in Electromagnetics,* Topics Appl. Phys., Vol. 3 (Springer, Berlin, Heidelberg)

Nash, R.T. (1964): IEEE Trans. Antennas Propag. **12,** 918

Ricardi, L.J. (1977): Proc. IEEE **65,** 356

Rice, S.O. (1954): "Mathematical Analysis of Random Noise", in *Selected Papers on Noise and Stochastical Processes,* ed. by N. Wax (Dover, New York)

Rush, W.V.T., Potter, P.D. (1970): *Analysis of Reflector Antennas* (Academic, New York)

Rush, W.V.T., Sörensen, O. (1975): IEEE Trans. Antennas Propag. **23,** 414

Ruze, J. (1952): Nuovo Cimento Vol. IX, Ser. IX, Suppl., 364

Ruze, J (1966): Proc. IEEE **54,** 633

Chapter 6

a) General

Born, M., Wolf, E. (1965): *Principles of Optics* (Pergamon, Oxford) Chap. X
Christansen, W.N., Högbom, J.A. (1985): *Radiotelescopes,* 2nd ed. (Cambridge University Press, Cambridge) Chap. 5–7
Fomalont, E.B., Wright, M.C.H., (1974): "Interferometry and Aperture Synthesis" in *Galactic and Extragalactic Radio Astronomy,* ed. by G.L. Verschuur, K.I. Kellermann (Springer, New York, Heidelberg, Berlin) p. 256
Schoonefeld, van, C. (ed.) (1979): *Image Formation from Coherence Functions in Radio Astronomy* (Reidel, Dordrecht)
Steel, W.H. (1967): *Interferometry* (Cambridge University Press, Cambridge)

b) Special

Born, M., Wolf, E. (1965): *Principles of Optics* (Pergamon, Oxford)
Clark, B.G. (1979): "Digital Processing Methods for Aperture Synthesis Observations", in *Image Formation from Coherence Functions in Radio Astronomy,* ed. by C. van Schoonefeld (Reidel, Dordrecht) p. 113
Cohen, M.H. (1969): High resolution observations of radio sources. Annu. Rev. Astron. Astrophys. **7,** 619
Cohen, M.H. (1973): Proc. IEEE **61,** No. 9
Cohen, M.H. et al. (1975): Astrophys. J. **201,** 249
Cornwell, T.J. (1982): "Image Restoration (and the CLEAN Technique)", in *Synthesis Mapping,* ed. by A.R. Thompson, L.R. D'Addario (NRAO, Green Bank) Chap. 9
Högbom, J.A., Brouw, W.N. (1974): Astron. Astrophys. **33,** 289
Pearson, T.J., Readhead, A.C.S. (1984): Image Formation by Self-Calibration in Radio Astronomy. Annu. Rev Astron. Astrophys. **22,** 97
Readhead, A.C.S., Napier, P.J., Bignell, R.C. (1980): Astroph. J. **237,** L 55
Schwarz, U.J. (1977): "The Method CLEAN – Use, Misuse and Variations", in *Image Formation from Coherence Functions in Radio Astronomy,* ed. by C. van Schoonefeld (Reidel, Dordrecht) p. 261
Sramek, R.A. (1982): "Map Plane – UV Plane Relationships", in *Synthesis Mapping,* ed. by A.R. Thompson, L.R. D'Addario (NRAO, Green Bank) Chap. 2
Steel, W.H. (1967): *Interferometry* (Cambridge University Press, Cambridge) p. 40 ff.
Thompson, A.R. (1982): "Introduction and Basic Theory", in *Synthesis Mapping,* ed. by A.R. Thompson, L.R. D'Addario (NRAO, Green Bank) Chap. 1

Chapter 7

a) General

The subject of this chapter draws material from many heterogeneous sources and correspondingly there is no single source that covers all aspects. But the following books contain most of it.

Bracewell, R.N. (1962): "Radio Astronomy Techniques", in *Handbuch der Physik,* Vol. 54, ed. by S. Flügge (Springer, Berlin, Heidelberg) p. 42
Hachenberg, O., Vowinkel, B. (1982): *Technische Grundlagen der Radioastronomie* (Bibliographisches Institut, Mannheim) Chaps. 2, 3
Kingston, H., Blake, C. (1968): "Receivers", in *Radar Astronomy* ed. by J. Evans, T. Hagfors, (McGraw Hill, New York) pp. 465–496

Meeks, M.L. (ed.) (1976): *Astrophysics Part B: Radiotelescopes, Methods of Experimental Physics,* Vol. 12 B (Academic, New York) Chap. 3, p. 201
Tiuri, M.E. (1966): "Radio Telescope Receivers", in J.D. Kraus: *Radio Astronomy* (McGraw Hill, New York) Chap. 7

b) Special

Ables, J.G. et al. (1975): Rev. Sci. Instrum. **46,** 284
Abramowitz, M., Stegun, J.A. (1964): *Handbook of Mathematical Functions* (National Bureau of Standards, Washington DC)
Blachman, N.M. (1966): *Noise and its Effect on Communication* (McGraw Hill, New York)
Born, M., Wolf, E. (1965): *Principles of Optics* (Pergamon, Oxford)
Bracewell, R.N. (1965): *The Fourier Transform and its Application* (McGraw Hill, New York)
Chang, I.C. (1976): IEEE Trans. Sonics Ultrason. SU **23,** 2
Cole, T.W., Ables, J.G. (1974): Astron. Astrophys. **34,** 149
Colvin, R.S. (1961): "A study of radio astronomy receivers", Thesis, Stanford Electronics Lab. Reports 18
Cooper, B.F.C. (1976): "Autocorrelation Spectrometers", in *Astrophysics Part B: Radiotelescopes, Methods of Experimental Physics,* Vol. 12B, ed. by M.L. Meeks (Academic, New York) p. 280
Davenport, W.B., Root, W.L. (1958): *An Introduction to the Theory of Random Signals and Noise* (McGraw Hill, New York)
Dicke, R.H. (1946): Rev. Sci. Instrum. **17,** 268
Edrich, J. (1977): IEEE Transact. Microwave Theory and Techniques MIT **25,** No. 4
Emerson, D.T., Klein, U., Haslam, C.G.T. (1979): Astron. Astrophys. **16,** 92
Khinchin, A.I. (1949): *Mathematical Foundations of Statistical Mechanics* (Dover, New York)
Papoulis, A. (1965): *Probability, Random Variables and Stochastic Processes* (McGraw Hill, New York)
Schoenberg, E. (1929): *Theoretische Photometrie, Handb. d. Astrophysik* Bd. II/1 ed. by K.F. Bottlinger et al. (Springer, Berlin) p. 1
Steiner, K.-H., Pungs, L. (1965): *Parametrische Systeme* (Hirzel, Stuttgart)
Weinreb, S. (1963): A digital spectral analysis technique and its application to radio astronomy. MIT Tech. Rep. 412
Whittaker, E.T., Watson, G.N. (1963): *A Course of Modern Analysis* (Cambridge University Press, Cambridge)
Yngvesson, K.S. (1976): "Maser Amplifiers", in *Astrophysics Part B: Radiotelescopes, Methods of Experimental Physics* Vol. 12B, ed. by M.L. Meeks (Academic, New York) p. 246

Chapter 8

a) General

Bekefi, G. (1966): *Radiation Processes in Plasmas* (Wiley, New York)
Jackson, J.D. (1962): *Classical Electrodynamics* (Wiley, New York)
Landau, L.D., Lifschitz, E.M. (1967): *Lehrbuch der Theoretischen Physik,* Vol. 2 (Akademie, Berlin)
Lang, K. (1974): *Astrophysical Formulae* (Springer, New York, Heidelberg, Berlin) Chap. 1
Longair, M.S. (1981): *High Energy Astrophysics* (Cambridge University Press, Cambridge) Chap. 3, 18
Pacholczyk, A.G. (1970): *Radio Astrophysics* (Freeman, San Francisco)
Pacholczyk, A.G. (1977): *Radio Galaxies* (Pergamon, Oxford)
Panofsky, W.K., Phillips, M. (1962): *Classical Electricity and Magnetism* (Addison-Wesley, Reading, MA)
Rybicki, G.B., Lightman, A.P. (1979): *Radiative Processes in Astrophysics* (Wiley, New York)
Tucker, W.H. (1975): *Radiation Processes in Astrophysics* (MIT, Cambridge, MA)
Unsöld, A. (1955): *Physik der Sternatmosphären,* 2. Aufl. (Springer, Berlin, Heidelberg)

b) Special

Alfvén, H., Herlofson, N. (1950): Phys. Rev. **78,** 616 (Letter)
Blewett, J.P. (1946): Phys. Rev. **69,** 87
Elder, F. et al. (1947): Phys. Rev. **71,** 829
Elder, F., Langmuir, R.V., Pollack, H.C. (1948): Phys. Rev. **74,** 52
Hirschfield, J.L., Baldwin, D.E., Brown, S.C. (1961): Phys. Fluids **4,** 198
Ivanenko, D.D., Sokolov, A.A. (1948): Dokl. Akad. Nauk SSSR **59,** 1551
Kiepenheuer, K.O. (1950): Phys. Rev. **79,** 138 (Letter)
Kramers, H.A. (1923): Philos. Mag. **46,** 836
Meyer, P. (1969): Annu. Rev. Astron. Astrophys. **7,** 1
Mezger, P.G., Henderson, A.P. (1967): Astrophys. J. **147,** 471
Oster, L. (1959): Z. f. Astrophys. **47,** 169
Oster, L. (1961): Rev. Mod. Phys. **33,** 525
Pfleiderer, J., Priester, W., Köhnlein, W. (1974): in "Processes of Continuous Radio Emission", in *Lectures in Space Physics,* Vol. 2, ed. by A. Bruzek, H. Pilkuhn (Bertelsmann, Gütersloh) p. 153
Scheuer, P.A.G. (1967): "Radiation" in *Plasma Astrophysics,* ed. by P.A. Sturrock, Proc. Int. Sch. Phys. "Enrico Fermi", **39,** 289
Schott, G.A. (1907): Ann. Phys., 4. Folge, **24,** 635
Schott, G.A. (1912): *Electromagnetic Radiation* (Cambridge University Press, Cambridge)
Schwinger, J. (1949): Phys. Rev. **75,** 1912
Theimer, O. (1963): Ann. Phys. **22,** 102
Vladimirsky, V.V. (1948): Zh. Eksp. Teor. Fiz. **18,** 392

Chapter 9

Caswell, J.L., Lerche, I. (1979): Proc. Astron. Soc. Aust. **3,** 343
Caswell, J.L., Lerche, I. (1979a): Mon. Not. Astron. Soc. **187,** 201
Chevalier, R.A. (1974): Astrophys. J. **188,** 501
Chevalier, R.A. (1975): Astrophys. J. **198,** 355
Clark, D.H., Caswell, J.L. (1976): Mon. Not. R. Astron. Soc. **174,** 267
Hogg, D.E. (1974): "Supernova Remnants", in *Galactic and Extragalactic Radio Astronomy,* ed. by G. Verschuur, K.I. Kellermann (Springer, Berlin, Heidelberg)
Jaeger, J.C., Westfold, K.C. (1949): Aust. J. Sci. Res. **2 A,** 322
Krüger, A. (1979): *Introduction to Solar Radio Astronomy and Radio Physics* (Reidel, Dordrecht)
Kundu, M.R. (1965): *Solar Radio Astronomy* (Interscience, New York)
Landau, L.D., Lifschitz, E.M. (1967): *Lehrbuch der Theoretischen Physik VI, Hydrodynamik* (Akademie, Berlin)
Rees, M.J., Stoneham, R.J. (1982): *Supernovae: A Survey of Current Research* (Reidel, Dordrecht)
Shklovsky, J.S. (1960): Sov. Astron. AJ. **4,** 243
Smerd, S.F. (1950): Aust. J. Sci. Res. **3 A,** 34
Spitzer, L. (1978): *Physical Processes in the Interstellar Medium* (Wiley, New York)
Woltjer, L. (1970): "Supernovae and the Interstellar Medium", Proc. IAU Symposium No. 39, ed. by H.J. Habing (Reidel, Dordrecht) p. 229
Woltjer, L. (1972): Annu. Rev. Astron. Astrophys. **10,** 129
Zheleznyakov, V.V. (1970): *Radio Emission of the Sun and Planets* (Oxford University Press, Oxford)

Chapter 10

Crovisier, J. (1978): Astron. Astrophys. **70,** 43
Einstein, A. (1916): Verh. Dtsch. Phys. Ges. **18,** 318

Herrick, S. (1935): Lick Obs. Bull. **XVII,** 85
Kalberla, P.M.W., Mebold, U., Reich, W. (1980): Astron. Astrophys. **82,** 275
Kalberla, P.M.W., Mebold, U., Reif, K. (1982): Astron. Astrophys. **106,** 190
Lang, K. (1974): *Astrophysical Formulae* (Springer, New York, Heidelberg, Berlin) Chap. 2
MacRae, D.A., Westerhout, G. (1956): *Tables for the Reduction of Velocities to the Local Standard of Rest* (Lund Observatory)
Spitzer, L. (1978): *Physical Processes in the Interstellar Medium* (Wiley, New York)
Stumpff, P. (1979): Astron. Astrophys. **78,** 229
Stumpff, P. (1980): Astron. Astrophys. Suppl. Ser. **41,** 1

Chapter 11

a) General

Burton, W.B. (1974): "The Large-Scale Dimension of Neutral Hydrogen in the Galaxy", in *Galactic and Extra-Galactic Radio Astronomy,* ed. by G.L. Verschuur, K.I. Kellermann (Springer, New York, Heidelberg, Berlin) p. 82
Kerr, F.J. (1968): "Radio Line Emission and Absorption by the Interstellar Gas," in *Stars & Stellar System VII,* ed. by B.M. Middlehurst, L.H. Aller (University of Chicago Press, Chicago, London) p. 575
Verschuur, G.L. (1974): "Interstellar Neutral Hydrogen and its Small-Scale Structure", in *Galactic and Extra-Galactic Radio Astronomy,* ed. by G.L. Verschuur, K.I. Kellermann (Springer, New York, Heidelberg, Berlin) p. 27

b) Special

Abramowitz, M., Stegun, I.A. (1964): "*Handbook of Mathematical Functions*" (National Bureau of Standards, Washington, DC)
Burton, W.B. (1971): Astron. Astrophys. **10,** 76
Burton, W.B. (1972): Astron. Astrophys. **19,** 51
Burton, W.B. (1974): "The Large Scale Distribution of Neutral Hydrogen in the Galaxy", in *Galactic and Extragalactic Radio Astronomy,* ed. by G.L. Verschuur, K.I. Kellermann (Springer, New York, Heidelberg, Berlin) p. 82
Burton, W.B., Liszt, H.S. (1978): Astrophys. J. **225,** 815
Celnik, W., Rohlfs, K. Braunsfurth, E. (1979): Astron. Astrophys. **76,** 24
Elwert, G. (1959): Ergeb. Exakten Naturwiss. **32,** 1
Field, G.B. (1958): Proc. IRE **46,** 240
Gunn, J.E., Knapp, G.R., Tremaine, S.D. (1979): Astron. J. **84,** 1181
Kahn, F.D. (1955): "Gasdynamics of Cosmic Clouds", Proc. IAU Symposium Nr. 2 (North-Holland, Amsterdam)
Manchester, R.N., Taylor, J.H. (1977): *Pulsars* (Freeman, Reading, MA)
Peters, H.E. et al. (1965): Appl. Phys. Lett. **7,** 34
Peters, W.L.H. (1975): Astrophys. J. **195,** 617
Purcell, E.M., Field, G.B. (1956): Astrophys. J. **124,** 542
Radhakrishnan, V. et al. (1972): Astroph. J. Suppl. Ser. **24,** 1
Rohlfs, K.: (1971): Astron. Astrophys. **12,** 43
Rohlfs, K., Braunsfurth, E. (1982): Astron. Astrophys. **113,** 237
Simonson, S.C. (1976): Astron. Astrophys. **46,** 261
Simonson, S.C., Mader, G.L. (1973): Astron. Astrophys. **27,** 337

Chapter 12

a) General

Brown, R.L., Lockman, F.J., Knapp, G.R. (1978): Annu. Rev. Astron. Astrophys. **16,** 445
Chaisson, E.J. (1976): "Gaseous Nebulae and their Interstellar Environment", in *Frontiers of Astrophysics,* ed. by E.H. Avrett (Harvard University Press, Cambridge, MA)
Dupree, A.K. (1969): Astrophys. J. **158,** 491
Dupree, A.K., Goldberg, L. (1970): Annu. Rev. Astron. Astrophys. **8,** 231
Lang, K.L. (1974): *Astrophysical Formulae* (Springer, New York, Heidelberg, Berlin) Chap. 2
Osterbrock, D.E. (1974): *Astrophysics of Gaseous Nebulae* (Freeman, San Francisco)
Rybicki, G.B., Lightman, A. P. (1979): *Radiative Processes in Astrophysics* (Wiley, New York)
Shaver, P. (ed.) (1980): *Radio Recombination Lines* (Reidel, Dordrecht)
Spitzer, L. (1978): *Physical Processes in the Interstellar Medium* (Wiley, New York)

b) Special

Brocklehurst, M. (1970): Mon. Not. R. Astron. Soc. **148,** 417
Brocklehurst, M., Seaton, M.J. (1972): Mon. Not. R. Astron. Soc. **157,** 179
Brown, R.L. (1980): "The Importance of Non-LTE Effects to the Interpretation of Radio Recombination Lines", in *Radio Recombination Lines,* ed. by P.A. Shaver (Reidel, Dordrecht)
Goldberg, L. (1968): "Theoretical Intensities of Recombination Lines", in *Interstellar Ionized Hydrogen,* ed. by Y. Terzian (Benjamin, New York) p. 373
Griem, H.R. (1967): Astrophys. J. **148,** 547
Höglund, B., Mezger, P.G. (1965): Science **150,** 339
Kardashev, N.S. (1959): Astron. Zh. **36,** 838 [English transl.: Sov. Astron. A.J. **3,** 813 (1960)]
Lockman, F.J., Brown, R.L. (1978): Astrophys. J. **222,** 153
Menzel, D.H. (1937): Astrophys. J., **85,** 330
Menzel, D.H., Pekeris, Ch.L. (1935): Mon. Not. R. Astron. Soc. **96,** 77
Mezger, P.G. (1972): "Interstellar Matter: An Observer's View", in *Interstellar Matter*, ed. by N.C. Wickramasinghe et al. (Geneva Observatory, Sauverny) p. 1
Mezger, P.G. (1980): "Helium Recombination Lines", in *Radio Recombination Lines,* ed. by P.A. Shaver (Reidel, Dordrecht)
Pankonin, V. (1980): "The Partially Ionized Medium Adjacent to H II Regions", in *Radio Recombination Lines,* ed. by P.A. Shaver (Reidel, Dordrecht) p. 111
Seaton, M.J. (1980): "Theory of Recombination Lines", in *Radio Recombination Lines,* ed. by P.A. Shaver (Reidel, Dordrecht) p. 3
Sejnowskī, T.J., Hjellming, R.M. (1969): Astrophys. J. **156,** 915
Strömgren, B. (1939): Astrophys. J. **89,** 529
Ungerechts, H., Walmsley, C.M. (1978): *Tables of b_n-Factors for Radio Recombination Lines at Low Temperatures,* Tech. Ber. 45 (MPI für Radioastronomie, Bonn)

Chapter 13

Avery, L.W. (1980): Int. Astron. Union Symp. **87,** 47
Balian, R., Encranaz, P., Lequeux, J. (1975): *Atomic and Molecular Physics and the Interstellar Matter, Les Houches XXVI* (North Holland, Amsterdam)
Bingel, W.A. (1969): *Theory of Molecular Spectra* (Wiley, New York)
Cheung, A.C. et al. (1969): Nature **221,** 626
Cook, A.H. (1977): *Celestial Masers* (Cambridge University Press, Cambridge)
Dickman, R.L. (1978): Astrophys. J., Suppl. Ser. **37,** 407
Frerking, M.A., Langer, W.D., Wilson, R.W. (1982): Astrophys. J. **262,** 590

Gardner, F.F. et al. (1964): Phys. Rev. Lett. **13,** 3
Hellwege, K.H. (1974): *Einführung in die Physik der Molekeln,* Heidelberger Taschenbücher, Bd. 146 (Springer, Berlin, Heidelberg)
Herzberg, G. (1950): *Molecular Spectra and Molecular Structure,* Vol. 1 (Van Nostrand, New York)
Herzberg, L., Herzberg, G. (1960): "Molecular Spectra", in *Fundamental Formulas of Physics,* ed. by D.H. Menzel (Dover, New York) p. 465
Kegel, W.H. (1976): Astron. Astrophys. **50,** 293
Lang, K.R. (1974): *Astrophysical Formulae* (Springer, New York, Heidelberg, Berlin) Sect. 2.15
Leung, C.M. Liszt, H.S. (1976): Astrophys. J. **208,** 732
Liszt, H.S. (1982): Astrophys. J. **262,** 198
Liszt, H.S., Delin, X., Burton, W.B. (1981): Astrophys. J. **249,** 532
Litvak, M.M. et al. (1966): Phys. Rev. Lett. **17,** 821
Moran, J.M. (1976): "Radio Observations of Galactic Masers", in *Frontiers of Astrophysics,* ed. by E.H. Avrett (Harvard University Press, Cambridge, MA)
Sanders, D.M., Solomon, P.M., Scoville, N.Z. (1984): Astrophys. J. **276,** 182
Scoville, N.Z., Solomon, P.M. (1974): Astrophys. J. **187,** L 67
Solomon, P.M., Sanders, D.B. (1980): "Giant Molecular Clouds as a Dominant Component of the Interstellar Medium of the Galaxy", in *Giant Molecular Clouds in the Galaxy,* ed. by P.M. Solomon, M.G. Edmunds (Pergamon, Oxford) p. 41
Townes, C.H., Schawlow, A.L. (1955): *Microwave Spectroscopy* (McGraw Hill, New York)
Turner, B.E. (1974): "Interstellar Molecules" in *Galactic and Extragalactic Radio Astronomy,* ed. by G. Verschuur, K. Kellermann (Springer, New York, Heidelberg, Berlin) p. 199
Weinreb, S. et al. (1963): Nature **200,** 829
Winnewisser, G., Churchwell, E., Walmsley, C.M. (1979): "Astrophysics of Interstellar Molecules", in *Modern Aspects of Microwave Spectroscopy,* ed. by G.W. Chantry (Academic, London) pp. 313–503
Winnewisser, G., Mezger, P.G., Breuer, H.D. (1974): *Interstellar Molecules,* in Topics Curr. Chem., Vol. 44 (Springer, Berlin, Heidelberg)
Zuckerman, B., Palmer, P. (1974) Annu. Rev. Astron. Astrophys. **12,** 279

Subject Index

C. Hoffmeister, G. Richter, W. Wenzel

Variable Stars

Translated from the German by S. Dunlop

1985. 170 figures, 64 tables. XV, 328 pages.
ISBN 3-540-13403-4

Contents: General Introduction. – Pulsating Variables. – Eruptive Variables. – Eclipsing Stars. – Supplement to the Classification. – The Discovery of Variable Stars. – The Significance of Variable Stars for Research on the Structure of the Galaxy and Stellar Evolution. – Observational Methods and Organizations. – Literature. – Subject Index. – Star Index.

This is a book about research into variable stars. Particular emphasis has been laid on the brightness variations in the "optical" spectral range which can often be observed using fairly simple instrumentation. The topics covered span from the development and structure of stars and the galaxy to mathematical statistics, modern x-ray satellite astronomy and the role of amateur observations.
Variable Stars should be useful for scientists and students working in astronomy as well as for suitably-informed amateur observers with an interest in this field.

Springer-Verlag
Berlin Heidelberg
New York Tokyo

Springer

Astronomy in Lecture Notes in Physics

G. Serra (Ed.)

Nearby Molecular Clouds

Proceedings of a Specialized Colloquium of the Eighth IAU European Regional Astronomy Meeting
Toulouse, September 17–21, 1984

1985. IX, 242 pages. (Lecture Notes in Physics, Volume 237).
ISBN 3-540-15991-6

Contents: Nearby Molecular Clouds as Part of Larger Structure. – Compared Internal Structure of Nearby Molecular Clouds. – Star Formation in Nearby Molecular Clouds. – Miscellaneous Related Topics. – Stellar Content of Nearby Molecular Clouds.

Much of modern astrophysical research concentrates on the study of the interstellar medium and on star formation. This colloquium shows the impact of the study of nearby molecular clouds on these problems. Recent infrared and millimetric observations make it possible to compare among each other nearby objects in order to understand general trends in cloud evolution and the physics of the relation between the stars, the gas and the dust in the cloud. This volume presents a host of experimental material and some theoretical work which should be seen as the start of exciting future research in this field. The book will be of interest to specialists and graduates alike.

In preparation:

M. Zeilik, D. M. Gibson (Eds.)

Cool Stars, Stellar Systems, and the Sun

Proceedings of the Fourth Cambridge Workshop, Santa Fé, New Mexico, October 16–18, 1985
(Lecture Notes in Physics, Volume 254) ISBN 3-540-16763-3

D. Mihalas, K.-H. A. Winkler, (Eds.)

Radiation Hydrodynamics in Stars and Compact Objects

Proceedings of the IAU Colloquium No. 89, Copenhagen, Denmark, June 11–20, 1985
1986. VI, 454 pages. (Lecture Notes in Physics, Volume 255)
ISBN 3-540-16764-1

Springer-Verlag
Berlin Heidelberg
New York Tokyo

Springer

QB
475
.R63
1986